Some Definite Integrals

$$\int_0^\infty x^{n-1}e^{-x}\,dx = \Gamma(n) \qquad \Gamma(n) = (n-1)\Gamma(n-1) = (n-1)! \qquad \Gamma(1/2) = \pi^{1/2}$$

$$\int_0^\infty e^{-ax^2}dx = \frac{1}{2}\left(\frac{\pi}{a}\right)^{1/2} \qquad a > 0$$

$$\int_0^\infty x^{2n}e^{-ax^2}dx = \frac{1\cdot 3\cdot 5\cdots(2n-1)}{2^{n+1}a^n}\left(\frac{\pi}{a}\right)^{1/2} \qquad (n \text{ a positive integer and } a > 0)$$

$$\int_0^\infty x^{2n+1}e^{-ax^2}dx = \frac{n!}{2a^{n+1}} \qquad (n \text{ a positive integer and } a > 0)$$

$$\int_0^\infty e^{-ax}\cos bx\,dx = \frac{a}{a^2+b^2} \qquad a > 0$$

$$\int_0^\infty e^{-ax}\sin bx\,dx = \frac{b}{a^2+b^2} \qquad a > 0$$

$$\int_0^\infty e^{-ax^2}\cos bx\,dx = \left(\frac{\pi}{4a}\right)^{1/2}e^{-b^2/4a} \qquad a > 0$$

$$\int_0^1 x^{n-1}(1-x)^{m-1}dx = 2\int_0^{\pi/2}\sin^{2n-1}\theta\,\cos^{2m-1}\theta\,d\theta = B(n,m) = \frac{\Gamma(n)\Gamma(m)}{\Gamma(n+m)} \qquad n > 1,\ m > 1$$

$$\frac{2}{\sqrt{\pi}}\int_0^x e^{-u^2}du = \text{erf}(x) = 1 - \text{erfc}(x)$$

$$\int_0^a \sin\frac{n\pi x}{a}\sin\frac{m\pi x}{a}\,dx = \int_0^a \cos\frac{n\pi x}{a}\cos\frac{m\pi x}{a}\,dx = \frac{a}{2}\delta_{nm} \qquad (m \text{ and } n \text{ integers})$$

$$\int_0^a \cos\frac{n\pi x}{a}\sin\frac{m\pi x}{a}\,dx = 0 \quad (m \text{ and } n \text{ integers})$$

$$\int_0^\pi \cos^n\theta\,\sin\theta\,d\theta = \int_{-1}^1 x^n\,dx = \begin{cases} 0 & \text{if } n \text{ is an odd integer} \\ \dfrac{2}{n+1} & \text{if } n \text{ is an even integer} \end{cases}$$

$$\int_0^\pi \cos^n\theta\,\sin^3\theta\,d\theta = \int_{-1}^1 x^n(1-x^2)\,dx = \begin{cases} 0 & \text{if } n \text{ is an odd integer} \\ \dfrac{4}{(n+1)(n+3)} & \text{if } n \text{ is an even integer} \end{cases}$$

$$\int_{-\infty}^\infty \frac{\sin x}{x}\,dx = \int_{-\infty}^\infty \frac{\sin^2 x}{x^2}\,dx = \pi$$

Mathematics for Physical Chemistry

Mathematics for Physical Chemistry

DONALD A. McQUARRIE

Department of Chemistry
University of California, Davis

University Science Books
Mill Valley, California

University Science Books
www.uscibooks.com

Production Manager: Jennifer Uhlich at Wilsted & Taylor
Manuscript Editor: Jennifer McClain
Design: Yvonne Tsang at Wilsted & Taylor
Illustrator: Mervin Hanson
Compositor: ICC Macmillan Inc.
Printer & Binder: Victor Graphics, Inc.

This book is printed on acid-free paper.

Library of Congress Cataloging-in-Publication Data

McQuarrie, Donald A. (Donald Allan)
  Mathematics for physical chemistry : opening doors / Donald A. McQuarrie.
     p. cm.
  Includes index.
  ISBN 978-1-891389-56-6 (alk. paper)
1. Chemistry, Physical and theoretical—Mathematics. I. Title.
   QD455.3.M3M385   2008
   510.2′454–dc22

                                                          2008060850

Printed in the United States of America
10 9 8 7 6 5 4 3 2

For Rhona,
to whom I owe so much.

I advise my students to listen carefully the moment
they decide to take no more mathematics courses.
They might be able to hear the sound of closing doors.

James Caballero, *CAIP Quarterly* 2 (Fall, 1989)

CONTENTS

Contents

xi

PREFACE

From years of advising undergraduate students in the sciences, a favorite quotation of mine is from James Caballero and appears as the epigraph to this book: "I advise my students to listen carefully the moment they decide to take no more mathematics courses. They might be able to hear the sound of closing doors." This book is written for those students. It is the outgrowth of a collection of MathChapters from my *Physical Chemistry: A Molecular Approach*, which I wrote with John Simon several years ago, and from my *Quantum Chemistry*, which recently was published in its second edition. These MathChapters consist of concise reviews of mathematical topics, discussing only the minimum amount that you need to know to understand subsequent material. From years of publishing scientific texts, my publisher says that physical chemistry is difficult because of the mathematics, but it is impossibly difficult without it. The point of the MathChapters is that by reading these reviews before the mathematics is applied to physical chemistry topics, you will be able to spend less time worrying about the math and more time learning the physical chemistry. A number of people suggested that I expand these MathChapters into a single volume, and this book is the result.

One thing that makes mathematics courses difficult for many science students is that they are taught by professional mathematicians, whose primary interest is to develop mathematics for its own sake, often without regard to applications. Consequently, they appreciate and recognize the need for rigor by carefully specifying each and every condition for the validity of theorems and the use of certain techniques. Most science students simply want to apply mathematics to physical problems and bring a certain degree of physical intuition into their mathematics courses and feel that the rigor is excessive. Unfortunately, this intuition is not always correct. Since the development of calculus in the 17th and 18th centuries, mathematicians have discovered many counterintuitive examples of functions that are supposed to display certain behavior. For example, there is a function that is continuous at every point but has a derivative nowhere. In fact, there is an entire book on mathematical counterexamples, *Counterexamples in Analysis*, by B.R. Gelbaum and J.M.H. Olmsted, published by Dover. These counterexamples rarely arise in physical problems but are of central importance in mathematics. Thus, there is a natural and justifiable dichotomy between the people who teach mathematics and many of the students who take mathematics. I don't know the source of this saying, but it goes, "Applied mathematicians don't understand pure mathematicians and pure mathematicians don't trust applied mathematicians."

There are 23 chapters in the book. Each one is fairly short, with the longest one being 20 pages, and is meant to be read at a single sitting. The material is presented at a practical level with an emphasis on applications to physical problems, although a few theorems along with their conditions are given. Each chapter contains several Examples, which serve to illustrate the techniques that are discussed. No one can learn physical chemistry (nor anything else in the physical sciences for that matter) without doing lots of problems. For this reason, I have included about 30 problems at the end of each chapter. These problems range from filling in gaps to extending the material presented in the chapter, but most illustrate applications to physical problems. All told, there are over 600 problems, and I have provided answers to most of them at the end of the book.

Throughout the book, I encourage you to learn how to use one of a number of general mathematics programs such as *Mathematica*, *Maple*, or *MathCad*, which are collectively called computer algebra systems (CAS). For a comparison of these various CAS, see *http://en.wikipedia.org/wiki/Computer_algebra_system*. These CAS make it easy to do calculations routinely that were formerly a drudgery. Most chemistry departments have a license for one of these programs. These programs not only perform numerical calculations but also can perform algebraic manipulations. They are relatively easy to learn and to use, and every serious science student should know how to use one of them. They allow you to focus on the underlying physical ideas and free you from getting bogged down in algebra. They also allow you to explore the properties of equations by varying parameters and plotting the results. In fact, all the figures in the book were produced by Professor Mervin Hanson of Humboldt State University using *Mathematica*. There are a number of problems that require the use of one of these programs, and many others are made much easier by using one of them.

Another product of the computer age is the availability of websites for most any topic, and that is true for mathematics as well. At one time, most physical chemistry students owned some sort of math handbook such as the *CRC Standard Math Tables*, which contained tables of trigonometric formulas, a table of integrals, and many other tables, but these are all available online nowadays. I have suggested websites for many topics throughout the book. Websites have the distressing habit of disappearing now and then, and I have tried to include only those that I think might still exist when you go to them. If by any chance you can't open one of them, you'll probably find it and many others by going to Google. There is a complete list of the websites that I refer to in the References at the end of the book.

It always amazes me just how many people are involved in the production of a book. Foremost are the reviewers, who frequently save me from promulgating my misunderstandings and misconceptions. I wish to thank Scott Feller of Wabash College, Mervin Hanson of Humboldt State University, Helen Leung and Mark Marshall of Amherst College, and John Taylor of the University of Colorado for very helpful reviews. I also wish to thank Christine Taylor and her crew at Wilsted & Taylor Publishing Services and particularly Jennifer Uhlich, who could not have done a more conscientious or competent job of turning a manuscript into a beautiful-looking and inviting book, and Yvonne Tsang for designing a great-looking book; Jennifer McClain for superb copyediting; and Bill Clark and Gunjan Chandola at ICC Macmillan for one of the best jobs of composition that I have experienced. I also wish to thank Mervin Hanson for rendering and rerendering all the figures in *Mathematica* without one crusty word, Jane Ellis of University Science Books for overseeing many of the production details and for suggesting the cover that ties in so well with the epigraph by James Caballero, and Bruce Armbruster and his wife and associate, Kathy, for being the best publishers around and good friends in addition. Finally, I wish to thank my wife, Carole, for preparing the manuscript in TeX, for reading the entire manuscript, and for being my best critic in general (in all things).

There are bound to be both typographical and conceptual errors in the book, and I would appreciate your letting me know about them so that they can be corrected in subsequent printings. I would also welcome general comments, questions, and suggestions at *mquarrie@mcn.org*, or through the University Science Books website, *www.uscibooks.com*.

Mathematics for Physical Chemistry

CHAPTER 1

Functions of a Single Variable: Differentiation

In the first two chapters, we shall review some of the essential features of your calculus course. We shall review the idea of a function and then the processes of differentiation and integration. It's clearly not possible to review an entire course in a single chapter, but most problems in your physical chemistry courses actually use only a small part of what you covered in calculus. Although you spent some time determining the derivatives of a variety of functions from first principles, these results are well tabulated in a number of places, including numerous websites. Using these tables along with a few general rules such as the derivative of a product, you can differentiate almost anything. The inverse of differentiation, in other words, the determination of which function when differentiated yields a certain result, is called antidifferentiation, or integration. You undoubtedly spent a great deal of time in your calculus course learning to integrate various functions. Fortunately, these results are well tabulated, as they are for differentiation, and are available at a number of websites. One of the aims of these two chapters is to introduce you to these various websites and encourage you to use them with confidence. As we stated in the Preface, the mathematics at the level of most physical chemistry courses requires about 20% talent and 80% confidence, which any chemistry student can gain by doing problems.

1.1 Functions

Recall from calculus that a function is a rule that relates one number, x, to another, y. We express this relationship by writing $y = f(x)$, where f represents the function.

If only one value of y is produced from each value of x, then the function is said to be *single-valued*. If more than one value of y is produced from a value of x, then f is said to be *multiple-valued*. Some authors require that a function be single-valued, but we'll adopt the somewhat more liberal definition.

Let's look at some examples. Consider the relation $y = x^2$, or $y = f(x) = x^2$, for values of x given by $-2 \le x \le 2$. In this case, f is single-valued because each value of x leads to only one value of y. Now consider $y^2 = x$, where $0 \le x \le 1$. Solving for y, we obtain $y = \pm\sqrt{x}$, showing that there are two values of y for each value of x. We can view this relationship as corresponding to either a multiple-valued function $\pm\sqrt{x}$ or to two single-valued functions, $y = f_1(x) = \sqrt{x}$ and $y = f_2(x) = -\sqrt{x}$.

Strictly speaking, a function is denoted by f, and the value obtained when f is applied to x is denoted by $y = f(x)$. However, it is common practice to call $f(x)$ a function, and a "function of x," in particular. We even write $y = y(x)$ to indicate that the value y results when the rule for sending x into y is applied to x. This notation is common and very convenient. In any case, x is called the *independent variable* and y is called the *dependent variable*.

There are two broad classes of functions, *algebraic functions* and *transcendental functions*. Algebraic functions can be expressed in terms of a finite number of the algebraic operations of addition, subtraction, multiplication, division, and taking roots. For example, $y(x) = (x^2+2)/(x-1)$ and $y(x) = (x^3+x^2+3)/\sqrt{x^2-3}$ are algebraic functions. They are essentially polynomials or ratios of polynomials, or ratios of powers (even fractional powers) of polynomials. Functions that are not algebraic functions are called transcendental functions. Trigonometric functions, exponential functions, and logarithmic functions are examples of transcendental functions. They are called transcendental functions because they cannot be expressed in terms of a finite number of the algebraic operations of addition, subtraction, multiplication, division, and taking roots. In a sense, they transcend algebraic operations.

Although we first learn about the trigonometric functions in terms of ratios of sides of triangles, it is better to think of them in terms of a unit circle (a circle with radius equal to one). Figure 1.1 shows a unit circle with an inscribed angle θ. Note that θ can be measured in terms of the arclength that it subtends. Thus, $90°$ corresponds to an arclength of $2\pi/4 = \pi/2$; $180°$ corresponds to an arclength of $2\pi/2 = \pi$; and so forth. The sine function is just the length of the heavy vertical line shown in the figure, taking positive values if it lies above the horizontal axis and negative values if it lies below it. Figure 1.2 shows $\sin\theta$ plotted against θ. Note that $\sin\theta$ takes on the values 0 for $\theta = 0$ or π, $+1$ for $\theta = \pi/2$, and -1 for $\theta = 3\pi/2$ and then starts repeating at $\theta = 2\pi$, where $\sin\theta = 0$. As θ increases beyond 2π, we think of the values of θ as the cumulative arclength swept out by θ as it rotates in a counterclockwise direction. Figure 1.2 also shows that we can admit negative values of θ, corresponding to rotating θ clockwise in Figure 1.1. The units of θ here are called *radians*, which are more commonly used than degrees, and in a sense, are more fundamental than degrees.

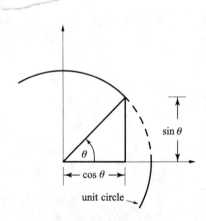

Figure 1.1. A unit circle, illustrating how an angle θ can be expressed in terms of the arclength that it subtends. In this geometry, the heavy vertical line is equal to $\sin\theta$ and the heavy horizontal line is equal to $\cos\theta$. This figure shows why the trigonometric functions are called circular functions.

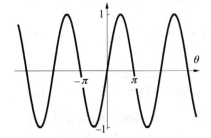

Figure 1.2. A plot of $\sin\theta$ against θ. Note that θ can take on any value. Values of θ greater than 2π mean that θ has gone around the circle in Figure 1.1 more than once in a counterclockwise direction. Negative values of θ mean that θ goes around in a clockwise direction. Note that $\sin\theta$ repeats itself every 2π units. It is periodic with a period 2π, which can be expressed by $\sin(\theta + 2\pi) = \sin\theta$.

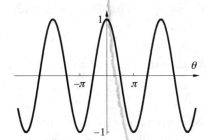

Figure 1.3. A plot of $\cos\theta$ against θ. Note that $\cos\theta$ is periodic with a period 2π, or that $\cos(\theta + 2\pi) = \cos\theta$. Note also that Figures 1.2 and 1.3 are similar, except that one curve is displaced from the other by $\pi/2$ units, or that $\sin(\theta + \pi/2) = \cos\theta$ and $\cos(\theta + \pi/2) = -\sin\theta$.

The heavy horizontal line in Figure 1.1 is $\cos\theta$. As θ sweeps around the unit circle in Figure 1.1 in a counterclockwise direction, the length of the heavy horizontal line is equal to $+1$ at $\theta = 0$, 0 at $\theta = \pi/2$, -1 at $\theta = \pi$, and so on. Figure 1.3 shows $\cos\theta$ plotted against θ. Note that $\cos\theta$ repeats itself as θ varies by $\pm 2\pi$. Functions that repeat, such as $\sin\theta$ and $\cos\theta$, are called *periodic functions*. The *period* of $\sin\theta$ and $\cos\theta$ is 2π because these functions repeat every 2π units. We can express this property as an equation by writing $f(\theta) = f(\theta + 2\pi)$.

A plot of $\tan\theta = \sin\theta/\cos\theta$ is shown in Figure 1.4. Note that $\tan\theta$ diverges (in other words, becomes unbounded) for $\theta = \pm\pi/2$, $\pm 3\pi/2$, ... because $\cos\theta$ is equal to zero at these points. Note that, unlike $\sin\theta$ or $\cos\theta$, $\tan\theta$ has a period of π, so that $\tan\theta = \tan(\theta + \pi)$.

Two important transcendental functions are the exponential function e^x and the logarithmic function $\ln x$. These two functions bear a special relation to each other. If $y = y(x) = e^x$, then $x = x(y) = \ln y$. The two functions, $y(x) = e^x$ and $x(y) = \ln y$, are *inverses* of each other; the function e^x sends x into $y = e^x$ and the function $x = \ln y$ sends y back into x:

$$y = e^x \qquad x = \ln y$$

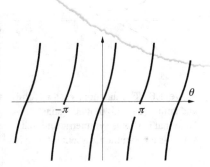

Figure 1.4. A plot of $\tan\theta$ against θ. Note that $\tan\theta$ diverges for $\theta = \pm\pi/2, \pm 3\pi/2, \ldots$ because $\tan\theta = \sin\theta/\cos\theta$ and $\cos\theta = 0$ for $\theta = \pm\pi/2, \pm 3\pi/2, \ldots$. Note that $\tan\theta = \tan(\theta + \pi)$.

Using these two relations, we can see that

$$y = e^{\ln y} \qquad \text{and} \qquad x = \ln e^x \tag{1.1}$$

$$x = \ln y = \ln e^{\ln y} = \ln e^x$$

It's also easy to prove some properties of logarithms using these equations. For example, let $u = e^x$ and $v = e^y$ and write

$$uv = e^x e^y = e^{x+y}$$

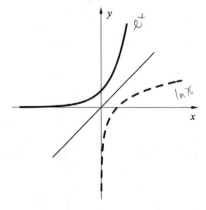

Figure 1.5. The exponential function e^x (solid) and the logarithmic function $\ln x$ (dashed) plotted against x. The two functions are symmetric about the line $y = x$.

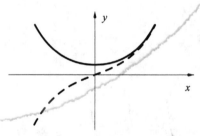

Figure 1.6. The functions $\cosh x$ (solid) and $\sinh x$ (dashed) plotted against x. Note that $\cosh x$ is symmetric and that $\sinh x$ is antisymmetric about the y axis.

$$\tanh(x) = \frac{e^x - e^{-x}}{e^x + e^{-x}}$$ hyperbolic functions

$$\coth(x) = \frac{e^x + e^{-x}}{e^x - e^{-x}}$$

$$\operatorname{csch}(x) = \frac{2}{e^x - e^{-x}}$$

$$\operatorname{sech}(x) = \frac{2}{e^x + e^{-x}}$$

$$\tanh(x) = \frac{\sinh(x)}{\cosh(x)}$$

$$\coth(x) = \frac{\cosh(x)}{\sinh(x)}$$

$$\operatorname{csch}(x) = \frac{1}{\sinh(x)}$$

$$\operatorname{sech}(x) = \frac{1}{\cosh(x)}$$

Taking logarithms of both sides gives

$$\ln uv = x + y = \ln u + \ln v \tag{1.2}$$

Additionally, if $u = e^x$ and $v = e^y$, then

$$\ln \frac{u}{v} = \ln uv^{-1} = x - y = \ln u - \ln v \tag{1.3}$$

We can use this relation to show that $\ln x < 0$ if $x < 1$. To see that this is so, let $x = u/v$ with $u < v$, in which case the right side of Equation 1.3 will be negative (see Figure 1.5). Finally, if we let $u = e^x$, then $u^n = e^{nx} = e^{n \ln u}$ and we see that

$$\ln u^n = n \ln u \tag{1.4}$$

The graphs of $y = e^x$ and $x = \ln y$ are shown in Figure 1.5. Note that one graph can be obtained from the other by simply interchanging the x and y axes, or what amounts to the same thing, by flipping either curve about the line $y = x$.

There are two commonly occurring functions that are defined in terms of e^x, the hyperbolic sine ($\sinh x$) and the hyperbolic cosine ($\cosh x$):

$$\sinh x = \frac{e^x - e^{-x}}{2} \quad \text{and} \quad \cosh x = \frac{e^x + e^{-x}}{2} \tag{1.5}$$

Figure 1.6 shows graphs of $\sinh x$ and $\cosh x$. A cable that is supported at two ends and hanging freely is described by a hyperbolic cosine; the Gateway Arch in St. Louis is an upside-down hyperbolic cosine. Note that $\cosh x$ is symmetric about the y axis and that $\sinh x$ changes sign when it is reflected through the y axis. Analytically, these properties are expressed by $\cosh(-x) = \cosh(x)$ and $\sinh(-x) = -\sinh(x)$. Generally, a function with the property that $f(-x) = f(x)$ is called an *even function* of x and one with the property that $f(-x) = -f(x)$ is called an *odd function* of x.

EXAMPLE 1–1
Show that $\sinh x$ is an odd function of x.

SOLUTION:

$$\sinh(-x) = \frac{e^{-x} - e^x}{2} = -\frac{e^x - e^{-x}}{2} = -\sinh x$$

As a point of interest, the reason that $\sinh x$ and $\cosh x$ are called hyperbolic functions can be seen by comparing the identities

$$\cos^2 u + \sin^2 u = 1 \quad \text{and} \quad \cosh^2 u - \sinh^2 u = 1$$

to the equations of a circle ($x^2 + y^2 = 1$) and a hyperbola ($x^2 - y^2 = 1$).

1.2 Continuity

Most of the functions that we deal with in physical chemistry are continuous. We're going to define just what we mean by continuous in this section, but you probably already have an intuitive idea of continuity, in the sense that the graph of the function can be sketched without lifting your pen from the paper. Although most functions that we deal with are continuous, not all of them are. Figure 1.7 shows the molar entropy of benzene plotted against temperature. The molar entropy is continuous over certain temperature intervals, but there are jumps (discontinuities) at 279 K (the normal melting point of benzene) and at 353 K (the normal boiling point of benzene). From a physical point of view, these points of discontinuity even signify interesting events, phase transitions in this case. Can you think of any other functions in physical chemistry that are discontinuous?

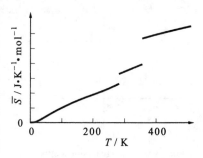

Figure 1.7. The molar entropy of benzene plotted against temperature from 0 K to 500 K.

Before we discuss the idea of continuity, we must introduce the idea of a limit. Consider some function $f(x)$. We say that

$$\lim_{x \to a} f(x) = l \tag{1.6}$$

if the difference between l and the value of $f(x)$ gets arbitrarily small as x gets arbitrarily close to a. Expressed another way, we have

$$|f(x) - l| \longrightarrow 0 \qquad \text{as } x \longrightarrow a \tag{1.7}$$

Equation 1.7 says that we can make $f(x)$ as close as we wish to l as long as we make x sufficiently close to a, *but not equal to a itself.* In fact, $f(x)$ need not even be defined at $x = a$. As an example, let's look at

$$\lim_{x \to 0} \frac{\sqrt{x + 16} - 4}{x}$$

We certainly can't let $x = 0$ because of the x in the denominator, but we can let it be as small as we wish. Figure 1.8 shows $f(x) = (\sqrt{x + 16} - 4)/x$ plotted against x for small values of x, and it seems to go to a finite limit ($1/8$) as $x \to 0$. To see that this is indeed the case, multiply and divide $f(x)$ by $\sqrt{x + 16} + 4$ to get

$$\lim_{x \to 0} \frac{\sqrt{x + 16} - 4}{x} \cdot \frac{\sqrt{x + 16} + 4}{\sqrt{x + 16} + 4} = \lim_{x \to 0} \frac{x}{x(\sqrt{x + 16} + 4)}$$

$$= \lim_{x \to 0} \frac{1}{\sqrt{x + 16} + 4} = \frac{1}{8}$$

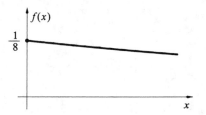

Figure 1.8. The function $f(x) = (\sqrt{x + 16} - 4)/x$ plotted against x for small values of x.

where we use the fact that $x + 16 \to 16$ as $x \to 0$.

In some applications, we are interested in the limit of a function at some point a where x approaches a from one side only. If $f(x)$ has the limit l as x approaches a through positive values of $x - a$ (in other words, from the right), then we express the limit as

$$\lim_{x \to a+} f(x) = l_+ \tag{1.8}$$

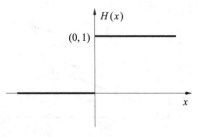

Figure 1.9. The Heaviside step function, $H(x) = 0$ when $x < 0$ and $H(x) = 1$ when $x > 0$.

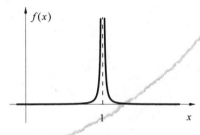

Figure 1.10. The discontinous function $f(x) = 1/(1 - x)^2$ plotted against x near the point $x = 1$.

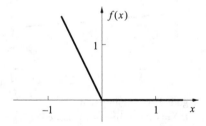

Figure 1.11. The function $f(x) = |x| - x$ plotted against x.

An equivalent way of expressing this limit is

$$\lim_{x \to a+} f(x) = \lim_{\epsilon \to 0} f(a + \epsilon) = l_+ \qquad \text{for } \epsilon > 0 \qquad (1.9)$$

This limit is called the *right-hand limit* of $f(x)$ at $x = a$ and is sometimes denoted by $f(a+)$. Of course, we can have *left-hand limits* also:

$$\lim_{x \to a-} f(x) = \lim_{\epsilon \to 0} f(a - \epsilon) = f(a-) = l_- \qquad \text{for } \epsilon > 0 \qquad (1.10)$$

A good example of a function that has different right-hand and left-hand limits at a point is the Heaviside step function (Figure 1.9), defined by

$$H(x) = \begin{cases} 0 & x < 0 \\ 1 & x > 0 \end{cases} \qquad (1.11)$$

In this case, $H(0+) = 1$ and $H(0-) = 0$. As you might assume, a function that has different right-hand and left-hand limits at a point (0 in the case of the Heaviside step function) must be discontinuous at that point.

In fact, a function is continuous at a point $x = a$ if

$$\lim_{x \to a\pm} f(x) = \lim_{\epsilon \to 0} f(a \pm \epsilon) = f(a) \qquad \text{for } \epsilon > 0 \qquad (1.12)$$

Of course, we are assuming here that $f(a)$ is finite. If Equation 1.12 (along with the requirement that $f(a)$ is finite) is not satisfied, then $f(x)$ is discontinuous at $x = a$. Intuitively, a discontinuity is a jump in the graph of the function. For example, the Heaviside step function (Figure 1.9) has a jump discontinuity of unity at $x = 0$. Another type of discontinuity is displayed by the function $f(x) = 1/(1 - x)^2$ at $x = 1$ (Figure 1.10). In this case, $f(x)$ is not finite at $x = 1$, and so is not continuous there.

Just as we have right-hand and left-hand limits, we have continuity from the right and continuity from the left. For example, we say that $f(x)$ is continuous from the right at $x = a$ if $\lim_{x \to a+} f(x) = f(a)$, or more simply, if $f(a+) = f(a)$. A function for which $f(a-) = f(a+)$ is continuous at $x = a$.

EXAMPLE 1–2
Is $f(x) = |x| - x$ continuous along the x axis?

SOLUTION: $f(x) = -2x$ when $x \le 0$ and $f(x) = 0$ when $x \ge 0$. Thus, $f(x)$ is continuous over the entire x axis (see Figure 1.11).

Although the Heaviside step function is not a continuous function over its entire domain, it is piecewise continuous. A function is said to be *piecewise continuous* in an interval $a \le x \le b$ if the interval can be subdivided into subintervals in each of which the function is continuous and has finite limits at its endpoints. Figure 1.9 shows that the Heaviside step function is piecewise continuous over the subintervals $(-\infty, 0)$ and $(0, \infty)$. Another example of a piecewise continuous function is the molar entropy of benzene, shown in Figure 1.7.

1.3 Differentiation

Recall that the derivative of a function $y(x)$ at a point x is the slope of the straight line tangent to $y(x)$ at x. The derivative of $y(x)$ at a point x is defined by the limiting process

$$y'(x) = \frac{dy}{dx} = \lim_{\Delta x \to 0} \frac{\Delta y}{\Delta x} = \lim_{\Delta x \to 0} \frac{y(x + \Delta x) - y(x)}{\Delta x} \qquad (1.13)$$

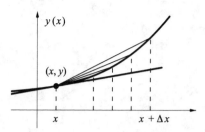

Figure 1.12. An illustration of the limiting process in the definition of the derivative of $y(x)$. As $\Delta x \to 0$, the ratio $[y(x + \Delta x) - y(x)]/\Delta x$ approaches the slope of the tangent line at the point (x, y).

Recall that the limit as $\Delta x \to 0$ means that Δx takes on values as small as you wish them to be, without ever letting Δx be *exactly* zero. Figure 1.12 illustrates this limiting process. In the beginning of your calculus course, you learned how to differentiate a number of functions using the above formulas. For example, if $y(x) = x^2 + 2$, then

$$\begin{aligned} y'(x) = \frac{dy}{dx} &= \lim_{\Delta x \to 0} \frac{(x + \Delta x)^2 + 2 - x^2 - 2}{\Delta x} \\ &= \lim_{\Delta x \to 0} (2x + \Delta x) = 2x \end{aligned}$$

As another example, consider $f(x) = 1/x$. Then

$$\begin{aligned} y'(x) = \frac{dy}{dx} &= \lim_{\Delta x \to 0} \frac{1}{\Delta x} \left(\frac{1}{x + \Delta x} - \frac{1}{x} \right) \\ &= \lim_{\Delta x \to 0} \frac{1}{\Delta x} \left[-\frac{\Delta x}{x(x + \Delta x)} \right] = \lim_{\Delta x \to 0} \left[-\frac{1}{x(x + \Delta x)} \right] = -\frac{1}{x^2} \end{aligned}$$

A little more challenging is $y(x) = \sin x$:

$$\frac{d \sin x}{dx} = \lim_{\Delta x \to 0} \frac{\sin(x + \Delta x) - \sin x}{\Delta x}$$

We now use the trigonometric identity

$$\sin \alpha - \sin \beta = 2 \cos \frac{\alpha + \beta}{2} \sin \frac{\alpha - \beta}{2}$$

to write

$$\begin{aligned} \frac{d \sin x}{dx} &= \lim_{\Delta x \to 0} \frac{2 \cos(x + \Delta x/2) \sin(\Delta x/2)}{\Delta x} \\ &= \lim_{\epsilon \to 0} \frac{\cos(x + \epsilon) \sin \epsilon}{\epsilon} \end{aligned}$$

where $\epsilon = \Delta x/2$. The limit of $\cos(x + \epsilon)$ is just $\cos x$; the limit of $(\sin \epsilon)/\epsilon$ is a so-called indeterminate form $0/0$, and is actually equal to one (Problem 1–30). Therefore, we find that

$$\frac{d \sin x}{dx} = \cos x$$

By the way, a number of websites give tables of trigonometric identities. A particularly good one is *http://www.sosmath.com/trig/Trig5/trig5/trig5.html*.

Typically, in a calculus course, you use Equation 1.13 to differentiate a variety of functions and then use these results and some general rules such as $(uv)' = uv' + u'v$ and $(u/v)' = (vu' - uv')/v^2$ to differentiate just about anything. There are a number of websites that give a table of derivatives; a very good one is the Wikipedia website *http://en.wikipedia.org/wiki/Table_of_derivatives*.

EXAMPLE 1–3
Differentiate $y = f(x) = x^2 e^{-x} \cos x$.

SOLUTION: We'll use a straightforward extension of the derivative of a product, $(uvw)' = u'vw + uv'w + uvw'$. Then

$$y'(x) = 2xe^{-x} \cos x - x^2 e^{-x} \cos x - x^2 e^{-x} \sin x$$
$$= xe^{-x}[2\cos x - x(\cos x + \sin x)]$$

You also learn to differentiate composite functions in a calculus class. A composite function is a function of a function. If $y = f(u)$ and $u = g(x)$, then $y = f(g(x))$ is a composite function of x. Recall that the derivative of y with respect to x is given by the *chain rule*.

$$\frac{dy}{dx} = \frac{dy}{du}\frac{du}{dx} \tag{1.14}$$

For example, if $y(x) = \sin(x^2+2)$, then we have $y = f(u) = \sin u$ and $u = x^2+2$, so

$$\frac{dy}{dx} = (\cos u)2x = 2x\cos(x^2 + 2)$$

EXAMPLE 1–4
Find dy/dx if $y(x) = e^{-(x^2+a^2)^{1/2}}$ where a is a constant.

SOLUTION: We use the chain rule with $y = e^{-u}$ and $u = (x^2 + a^2)^{1/2}$.

$$\frac{dy}{dx} = \frac{dy}{du}\frac{du}{dx} = (-e^{-u})\left[\frac{x}{(x^2 + a^2)^{1/2}}\right]$$
$$= -\frac{xe^{-(x^2+a^2)^{1/2}}}{(x^2 + a^2)^{1/2}}$$

You learn in calculus that if c_1 and c_2 are constants, then

$$\frac{d}{dx}[c_1 f_1(x) + c_2 f_2(x)] = c_1\frac{df_1}{dx} + c_2\frac{df_2}{dx} \tag{1.15}$$

An operation with this property is said to be *linear*. Differentiation is a *linear* operation.

We don't have to stop at first derivatives, do we? We can take sequential derivatives to form second derivatives, third derivatives, and so on. For example, if $y(x) = x^2 e^x$, then

$$y'(x) = \frac{dy}{dx} = (2x + x^2)\,e^x$$

and

$$y''(x) = \frac{d^2 y}{dx^2} = \frac{d}{dx}(2x + x^2)\,e^x = (2 + 4x + x^2)\,e^x$$

and so on.

In all the cases that we have discussed so far, we have had an explicit expression $y = f(x)$. More generally, we might have a function $f(x, y) = c$, where c is a constant, and where we are not able to express y in terms of x in any simple form. For example, consider

$$f(x, y) = x^3 y + xy^3 = 2 \tag{1.16}$$

which is plotted in Figure 1.13. A function that is given by $f(x, y) = c$ is called an *implicit function*. Even though we can't easily solve Equation 1.16 for y in terms of x (or for x in terms of y for that matter), we can still find dy/dx. If we differentiate $f(x, y)$ with respect to x, realizing that $y = y(x)$ (even if we don't have an explicit formula for $y(x)$), then we obtain

$$x^3 \frac{dy}{dx} + 3x^2 y + 3xy^2 \frac{dy}{dx} + y^3 = 0$$

or

$$\frac{dy}{dx} = -\frac{3x^2 y + y^3}{x^3 + 3xy^2} \tag{1.17}$$

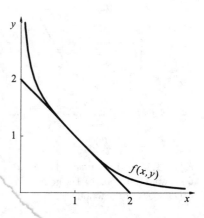

Figure 1.13. The implicit function $f(x, y) = x^3 y + xy^3$ and the tangent line $y = -x + 2$ at the point $(1, 1)$.

This process is called *implicit differentiation*. It may seem strange or unsatisfying that we have dy/dx in terms of both x and y, but it's not a problem. Suppose we want the slope of the curve in Figure 1.13 at the point $(1, 1)$. We simply substitute $x = y = 1$ into Equation 1.17 to obtain $dy/dx = -1$. The equation of the tangent line is $y = -x + b$ and if we let $x = y = 1$, we see that $b = 2$ and that $y = -x + 2$ (see Figure 1.13).

1.4 Extrema

One of the principal applications of derivatives is finding the local maximum and minimum values (extrema) of a function. Points at which $f'(x) = 0$ have a horizontal slope and are called *critical points* of $f(x)$. Although the condition $f'(c) = 0$ defines a critical point, it does not guarantee that $f(x)$ has a local extremum there. The simplest illustration of this is $f(x) = x^3$; in this case $f'(0) = 0$, but $f(x)$ is not an extremum at $x = 0$ (look ahead at Figure 1.15b). The condition $f'(c) = 0$ is a *necessary* condition that $f(c)$ is an extremum, but it is not a sufficient condition.

Let $f(x)$ and $f'(x)$ be continuous for $a \le x \le b$. Realize that if $f'(x) > 0$, then $f(x)$ increases as x increases and that if $f'(x) < 0$, then $f(x)$ decreases as

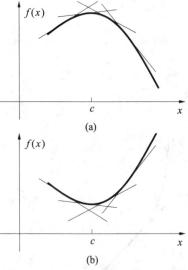

(a)

(b)

Figure 1.14. (a) The graph of a concave downward function. (b) The graph of a concave upward function.

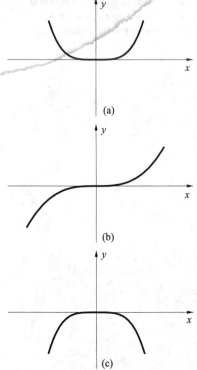

(a)

(b)

(c)

Figure 1.15. The functions (a) $f(x) = x^4$, (b) $g(x) = x^3$, and (c) $h(x) = -x^4$ plotted against x.

x increases. Therefore, if $f'(x) > 0$ for $x < c$ and $f'(x) < 0$ for $x > c$, then $f(x)$ is concave downward at c. A function $f(x)$ is concave downward at $x = c$ if the graph of $f(x)$ in the neighborhood of c lies below the tangent line at $x = c$ (Figure 1.14a). On the other hand, if $f'(x) < 0$ for $x < c$ and $f'(x) > 0$ for $x > c$, then $f(x)$ is concave upward at c. A function $f(x)$ is concave upward at $x = c$ if the graph of $f(x)$ in the neighborhood of c lies above the tangent line at $x = c$ (Figure 1.14b).

The type of concavity is related to the sign of the second derivative. Because $f''(x) = df'(x)/dx$, $f''(x)$ determines how the slope of $f'(x)$ varies with increasing x. If $f''(x) > 0$, then $f'(x)$ increases with x, and so $f(c)$ will be a minimum. If $f''(x) < 0$, then $f'(x)$ decreases with increasing x, and so $f(c)$ will be a maximum. These conditions give us the second derivative test to determine if a critical point is a local extremum or not: If $f'(c) = 0$ and $f''(c)$ exists, then

1. $f(x)$ has a local maximum at $x = c$ if $f''(c) < 0$
2. $f(x)$ has a local minimum at $x = c$ if $f''(c) > 0$
3. no conclusion can be drawn without further analysis if $f''(c) = 0$

A point $x = c$ is called an *inflection point* if the concavity of $f(x)$ changes at c. Consequently, $f''(x) = 0$ at an inflection point. For a point c to be an inflection point, we must have that

1. $f''(c) = 0$
2. $f''(x)$ must change sign at $x = c$

Incidentally, it seems that many students think that the condition $f''(c) = 0$ implies an inflection point at $x = c$, but this is not so. All three functions, $f(x) = x^4$, $g(x) = x^3$, and $h(x) = -x^4$, have first and second derivatives that are equal to zero at $x = 0$, yet $f(x)$ has a minimum, $g(x)$ has an inflection point, and $h(x)$ has a maximum at $x = 0$. (See Figure 1.15.) If $f''(x) = 0$ at a critical point, you must investigate higher-order derivatives to determine its nature.

EXAMPLE 1–5
Find the local extrema and any inflection points of $f(x) = 3x^5 - 20x^3$ over the entire x axis.

SOLUTION: The equation

$$f'(x) = 15x^4 - 60x^2 = 0$$

shows that there are critical points at $x = 0$ and ± 2. The second derivative is

$$f''(x) = 60x^3 - 120x$$

The fact that $f''(2) = 240 > 0$ and $f''(-2) = -240 < 0$ says that the critical point $x = 2$ is a minimum and the critical point $x = -2$ is a maximum. The fact that $f''(0) = 0$ says that $x = 0$ is a possible inflection point. The fact that $f''(x)$ changes sign at $x = 0$ confirms that $x = 0$ is an inflection point. (See Figure 1.16.)

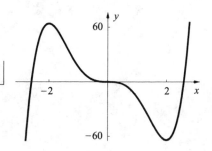

Figure 1.16. The function in Example 1–5, $f(x) = 3x^5 - 20x^3$, plotted against x.

EXAMPLE 1–6
Find the local extrema and any inflection points of $f(x) = x^2(1 - x)^2$ over the entire x axis.

SOLUTION: The equation

$$f'(x) = 2x(1 - x)^2 - 2x^2(1 - x) = 2x(1 - x)(1 - 2x) = 0$$

shows that there are critical points at $x = 0$, $x = 1/2$, and $x = 1$. The second derivative is

$$f''(x) = 12x^2 - 12x + 2$$

The fact that $f''(0) > 0$ and $f''(1) > 0$ and that $f''(1/2) < 0$ tells us that the critical points $x = 0$ and $x = 1$ are local minima and that $x = 1/2$ is a local maximum. The possible inflection points are given by $f''(x) = 0$, or at $x = (3 \pm \sqrt{3})/6$, or at $x = 0.211 \ldots$ and $0.787 \ldots$. The following table shows that both points are inflection points, as can be seen in Figure 1.17.

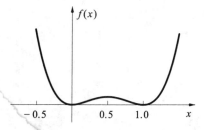

Figure 1.17. The function in Example 1–6, $f(x) = x^2(1 - x)^2$, plotted against x.

   Critical point
$0.211 \ldots$ $f''(0.20) = 0.08 > 0$ $f''(0.22) = -0.059 < 0$
$0.787 \ldots$ $f''(0.78) = -0.059 < 0$ $f''(0.79) = 0.009 > 0$

As a final twist to consider, let's look at $f(x) = x^{2/3}$ for $-1 < x < 1$. In this case, $f'(x) = 2/(3x^{1/3})$, which diverges to ∞ as $x \to 0$ through positive values and to $-\infty$ as $x \to 0$ through negative values. Thus, $f'(x)$ is not continuous at $x = 0$, yet $f(x)$ has a minimum value there (see Figure 1.18). This illustrates the fact that $f'(x)$ must *exist* and be continuous throughout the interval in order to use the above criteria.

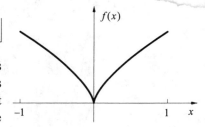

Figure 1.18. The function $f(x) = x^{2/3}$ plotted against x for $-1 < x < 1$.

Problems

1–1.   Plot the functions (a) $y = |x|$, (b) $y = 1/(x - 1)$, and (c) $y = x^{2/3}$ for $-3 \le x \le 3$.

1–2.   Plot the function $x/|x|$ for $-5 < x < 5$.

1–3.   Plot $y(x) = x - |x|$ for $-5 < x < 5$.

1–4.   Classify each of the following functions as even, odd, or neither.

     (a) $\tanh x$ (b) $e^x \sin x$ (c) $\dfrac{e^x}{(e^x + 1)^2}$ (d) $\cos x + \sin x$

1–5.   Plot the function $y(x) = H(x) - H(x - 1)$ for $0 < x < 4$, where $H(x)$ is the Heaviside step function, defined by Equation 1.11.

1–6. Plot the function defined by

$$f(t) = t - 2(t-2)H(t-2) + 2(t-4)H(t-4) - 2(t-6)H(t-6) + \cdots$$

for $t \geq 0$, where $H(x)$ is the Heaviside step function defined by Equation 1.11. Describe the function $f(t)$ in words.

1–7. Which of the following functions are periodic? What are their periods?
(a) $\sin x + 2\cos x$ (b) $\tan 2x$ (c) $|\cos x|$ (d) $\dfrac{\sin x}{x}$

1–8. Find the values of x for which
(a) $x^2 + 2 < 3x$ (b) $-2 < \dfrac{3x+1}{x+1} < 2$ (c) $0 \leq \sin x \leq 1$

1–9. Why do the graphs of $\sinh x$ and $\cosh x$ in Figure 1.6 merge as x gets large?

1–10. Plot $\sin x$ and $\sin^{-1} x$ on the same graph. Are the two curves symmetric about the line $y = x$? Is $\sin^{-1} x$ a single-valued function of x?

1–11. Evaluate (a) $\displaystyle\lim_{x \to 0+} \frac{1}{1 - e^{-1/x}}$ and (b) $\displaystyle\lim_{x \to 0-} \frac{1}{1 - e^{-1/x}}$.

1–12. Find α and β such that $f(x)$ is continuous for $0 < x < 2\pi$:

$$f(x) = \begin{cases} -\sin x & 0 < x < \pi/2 \\ \alpha \sin x + \beta & \pi/2 < x < 3\pi/2 \\ \left(x - \dfrac{3\pi}{2}\right)^2 & 3\pi/2 < x < 2\pi \end{cases}$$

1–13. Differentiate (a) $(2+x)e^{-x^2}$; (b) $\dfrac{\sin x}{x}$; (c) $x^2 \tan 2x$; (d) $e^{-\sin x}$

1–14. Does $f(x) = |x|$ have a derivative at $x = 0$?

1–15. Find $y''(x)$ for
(a) $y(x) = x^2 \cos x$ (b) $y(x) = e^{-x} \sin x$ (c) $y(x) = x^2 \ln x$

1–16. Find $y'(x)$ and $y''(x)$ for
(a) $y(x) = e^{-x^2}$ (b) $y(x) = \sin e^{-x}$ (c) $y(x) = e^{-\tan x}$

1–17. Find dy/dx at the point $(1, 1)$ for $f(x, y) = x^4 y + x^2 y^3 = 2$. What is the equation of the tangent line?

1–18. Find the local extrema and inflection points of $f(x) = x^3 - 3x + 1$ over the entire x axis.

1–19. Find the local extrema and inflection points of $f(x) = x^5 + 2x$ over the entire x axis.

1–20. Find the local extrema and inflection points of $f(x) = 2x^3 - 3x^2 - 12x + 3$ over the entire x axis.

1–21. Find the local extrema and inflection points of $f(x) = 3x^4 - 4x^3 - 24x^2 + 48x - 20$ over the entire x axis.

1–22. Show that $\dfrac{d}{dx} u^v = vu^{v-1}\dfrac{du}{dx} + (\ln u)u^v \dfrac{dv}{dx}$. *Hint*: Let $y = u^v$ and differentiate $\ln y$. Use this result to find $\dfrac{d}{dx} x^x$.

1–23. The height of a body that is shot vertically upward is given as a function of time by $h(t) = 40(128t - 32t^2)$. How high will it go?

1–24. Show that the rectangle of largest possible area for a given perimeter is a square.

1–25. Which points on the curve $xy^2 = 1$ are closest to the origin? *Hint*: Consider the function $D = (x^2 + y^2)^{1/2}$.

1–26. The blackbody radiation law is given by

$$\rho(\lambda, T) = \frac{8\pi hc}{\lambda^5} \frac{1}{e^{hc/\lambda k_B T} - 1}$$

where $\rho(\lambda, t)\, d\lambda$ is the energy between λ and $\lambda + d\lambda$, λ is the wavelength of the radiation, h is the Planck constant, k_B is the Boltzmann constant, c is the speed of light, and T is the kelvin temperature. The Wien displacement law says that $\lambda_{max} T = $ constant where λ_{max} is the value of λ at which $\rho(\lambda, t)$ is a maximum. Derive the Wien displacement law from the blackbody radiation law. Show that "constant" $= hc/4.965 k_B$.

1–27. Show that

$$\frac{d^2 uv}{dx^2} = v\frac{d^2 u}{dx^2} + 2\frac{du}{dx}\frac{dv}{dx} + u\frac{d^2 v}{dx^2}$$

Notice that the numerical coefficients here are 1, 2, and 1, the same as those of $(x + y)^2 = x^2 + 2xy + y^2$. Now do the same for higher derivatives of uv and compare the numerical coefficients to those of the expansion of $(x + y)^n$.

1–28. This problem illustrates why e is such a "natural" number in calculus. Consider the exponential function $y(x) = b^x$, where b is a constant. Show that

$$\frac{dy}{dx} = b^x \lim_{h \to 0} \frac{(b^h - 1)}{h} \tag{1}$$

Whatever this limit is, denote it by c, so that equation 1 reads

$$\frac{dy}{dx} = cb^x = cy(x)$$

If $c = 1$, then $y(x)$ would be a function whose derivative is equal to itself. Show that this will be the case if

$$b = (1 + h)^{1/h} \qquad \text{as } h \longrightarrow 0$$

or, equivalently,

$$b = \left(1 + \frac{1}{n}\right)^n \qquad \text{as } n \longrightarrow \infty$$

Use either a hand calculator or a computer to show that $b \to 2.71828\ldots = e$.

1–29. Here is another example of how the number e arises naturally. Suppose that we deposit money in the bank at an interest r, which is paid annually. Argue that after n years, your money will grow by a factor of $(1 + r)^n$. Now suppose that the interest is paid semiannually. Show that the growth factor in this case is $(1 + r/2)^{2n}$. Continue this argument to show that the growth factor is $(1 + r/365)^{365n}$ if the interest is compounded daily. After 5 years at 6% interest, these factors are equal to 1.3382 (annually), 1.3439 (semiannually), and 1.3498 (daily). Finally, show that the growth factor is $e^{rn} = 1.3499$ if the interest is compounded continuously.

1–30. Use Figure 1.19 to show that $\lim\limits_{x \to 0} \dfrac{\sin x}{x} = 1$. *Hint*: Use the fact that area of $\triangle oab \leq$ area of sector $oad \leq$ area of $\triangle ocd$.

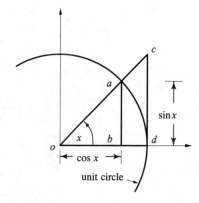

Figure 1.19. Geometry associated with the proof that $\lim\limits_{x \to 0} \dfrac{\sin x}{x} = 1$.

CHAPTER 2

Functions of a Single Variable: Integration

The first two chapters of this book constitute a review of your calculus course. The first chapter reviewed differentiation, and in this chapter we shall review integration. The two operations are related by what is called the fundamental theorem of calculus, which shows that an integral amounts to antidifferentiation, in the sense that the process of integration undoes the process of differentiation. We first present Riemann's definition of an integral as a limit of a sum of rectangles that approximates the area under a curve, and then discuss the fundamental theorem of calculus. We then very briefly discuss some methods that you can use to evaluate integrals, and in the final section we discuss what are called improper integrals, where at least one of the integration limits is infinity. We present a simple test of convergence of integrals of this type.

2.1 Definition of an Integral

The idea of an integral was originally developed to calculate the area bounded by given curves, but nowadays an integral is defined by a limiting process. Consider the situation in Figure 2.1. The interval (a, b) is subdivided into n subintervals, $(a, x_1), (x_1, x_2), \ldots, (x_{n-1}, b)$, and the point, ξ_j, is located anywhere within the jth interval for $j = 1, 2, \ldots, n$. We now form the sum

$$S_n = \sum_{j=1}^{n} f(\xi_j)(x_j - x_{j-1}) = \sum_{j=1}^{n} f(\xi_j)\Delta x_j$$

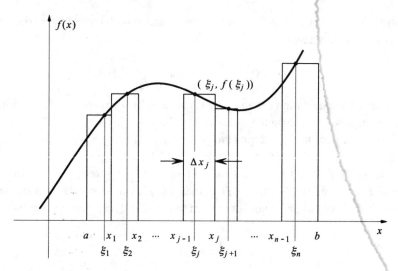

Figure 2.1. The construction associated with a Riemann sum.

where $x_0 = a$, $x_n = b$, and $\Delta x_j = x_j - x_{j-1}$. Geometrically, this sum represents the sum of the areas of the rectangles in Figure 2.1 and is called a *Riemann sum*. If we subdivide (a, b) into more and more subintervals with smaller and smaller widths, we approach a limit, called the *Riemann integral* of $f(x)$, denoted by

$$\int_a^b f(x)dx = \lim_{h \to 0} \sum_{j=1}^n f(\xi_j)\Delta x_j \qquad (2.1)$$

where h is the width of the largest subinterval. If all the intervals are taken to be the same size, then $h = (b-a)/n$, and so $n \to \infty$ as $h \to 0$ in Equation 2.1. This limit exists if $f(x)$ is continuous (or even piecewise continuous) for $a \leq x \leq b$. Geometrically, the integral of $f(x)$ from a to b represents the area between $f(x)$ and the x axis and the vertical lines at $x = a$ and $x = b$ if $f(x)$ is positive everywhere between a and b. Otherwise, it represents the net area, with areas above the x axis treated as positive and areas below the x axis treated as negative.

We certainly don't use Equation 2.1 to evaluate integrals. As you know, and we show below, integration and differentiation are inverse operations, so we often evaluate integrals by working backward from tables of differentiation formulas. Before we do that, however, there are a few properties of integrals that we should point out here that follow from the definition given in Equation 2.1. If c_1 and c_2 are constants, then

$$\int_a^b [c_1 f_1(x) + c_2 f_2(x)]\, dx = c_1 \int_a^b f_1(x)\, dx + c_2 \int_a^b f_2(x)\, dx \qquad (2.2)$$

Integration is a linear operation. Equation 2.2 is similar to Equation 1.15 for differentiation. Another property that follows from Equation 2.1 is

$$\int_a^b f(x)\, dx = \int_a^c f(x)\, dx + \int_c^b f(x)\, dx \qquad (2.3)$$

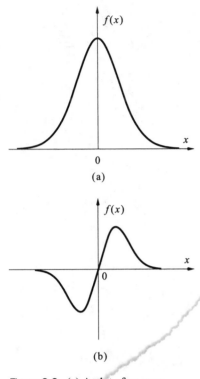

(a)

(b)

Figure 2.2. (a) A plot of an even function of x and (b) a plot of an odd function of x. An even function is symmetric about the y axis and an odd function changes sign.

where $a \leq c \leq b$. Because there is nothing special about labeling the horizontal axis by x, we also have

$$\int_a^b f(x)\,dx = \int_a^b f(u)\,du = \int_a^b f(t)\,dt$$

and so on. The integration variable here is called a *dummy variable*. The resulting integral depends only upon the values of a and b.

We can appeal to the interpretation of an integral being equal to the net area under the curve to illustrate a special property of the integrals of even and odd functions. We said in Chapter 1 that if $f(-x) = f(x)$, then $f(x)$ is called an even function, and if $f(-x) = -f(x)$, then $f(x)$ is called an odd function (see Figure 2.2). Consider an integral of an even function, $f_{even}(x)$, between symmetric limits

$$I = \int_{-a}^a f_{even}(x)\,dx$$

By referring to Figure 2.2a, you can see that the area from $-a$ to 0 is equal to the area from 0 to a, so I is equal to

$$I = \int_{-a}^a f_{even}(x)\,dx = 2\int_0^a f_{even}(x)\,dx \qquad (2.4)$$

Similarly, we have

$$\int_{-a}^a f_{odd}(x)\,dx = 0 \qquad (2.5)$$

2.2 The Fundamental Theorem of Calculus

The integrals that we have been discussing so far have been between fixed limits and are called *definite integrals*. If the upper limit is replaced by x, then the integral defines a function of x and is called an *indefinite integral*. We express this as

$$F(x) = \int_a^x f(u)\,du \qquad (2.6)$$

Notice, incidentally, that we are using u and not x as our variable of integration so that we can distinguish between the variable of integration and the upper limit. The designation of the integration variable is arbitrary since it is a dummy variable. In other words,

$$F(x) = \int_a^x f(u)\,du = \int_a^x f(z)\,dz = \int_a^x f(t)\,dt$$

are completely equivalent. However, writing

$$F(x) = \int_a^x f(x)\,dx$$

is poor practice and should be avoided because the x in the upper limit and the x in the integrand represent different quantities.

The *fundamental theorem of calculus* says that if $f(x)$ is continuous in the interval (a, b) and if

$$F(x) = \int_a^x f(u)\, du + c \tag{2.7}$$

where c is a constant, then $F(x)$ is an *antiderivative* of $f(x)$; in other words, $F'(x) = f(x)$ in the interval (a, b). The fundamental theorem of calculus gives us the inverse relation between differentiation and integration.

To obtain $f(x)$ from Equation 2.7, we differentiate with respect to the upper limit, x. Write

$$F(x+\Delta x) - F(x) = \int_a^{x+\Delta x} f(u)\, du - \int_a^x f(u)\, du = \int_x^{x+\Delta x} f(u)\, du \tag{2.8}$$

If $f(u)$ is continuous in the interval $(x, x + \Delta x)$, then the value of $f(u)$ will be essentially constant and equal to $f(x)$ for very small values of Δx. If we divide Equation 2.8 by Δx and then take the limit $\Delta x \rightarrow 0$, we can take $f(x)$ from under the integral sign and write

$$\frac{dF}{dx} = F'(x) = f(x) \tag{2.9}$$

Equations 2.7 and 2.9 summarize the fundamental theorem of calculus.

We can use Equation 2.7 along with its interpretation as the net area between $f(x)$ and the x axis to display the standard formula for a definite integral. If we first let $x = \beta$ and then let $x = \alpha$ and then form $F(\beta) - F(\alpha)$, we see that

$$F(\beta) - F(\alpha) = \int_\alpha^\beta f(x)\, dx = \left[F(x) \right]_\alpha^\beta \tag{2.10}$$

where the last term here is common notation for $F(\beta) - F(\alpha)$.

Equation 2.10 can be used to verify Equation 2.3 because

$$\int_\alpha^\beta f(x)\, dx = \int_\alpha^c f(x)\, dx + \int_c^\beta f(x)\, dx = F(c) - F(\alpha) + F(\beta) - F(c)$$
$$= F(\beta) - F(\alpha)$$

We can also use Equation 2.10 to derive the important result

$$\int_\alpha^\beta f(x)\, dx = -\int_\beta^\alpha f(x)\, dx = -[F(\alpha) - F(\beta)] = F(\beta) - F(\alpha) \tag{2.11}$$

Equations 2.7 and 2.10 show us that you can evaluate an integral by finding the function whose derivative is $f(x)$. For example, to evaluate $\int_0^{\pi/2} \cos x\, dx$, we can look through a table of derivatives to find $d \sin x / dx = \cos x$, and then write

$$\int_0^{\pi/2} \cos x\, dx = \left[\sin x \right]_0^{\pi/2} = \sin(\pi/2) - \sin 0 = 1$$

EXAMPLE 2–1

Evaluate $\int_0^1 x^3\,dx$.

SOLUTION: We use the fact that $dx^4/dx = 4x^3$ to write

$$\int_0^1 x^3\,dx = \frac{1}{4}\Big[x^4\Big]_0^1 = \frac{1}{4}$$

These two examples admittedly have been pretty easy. You'd be lucky to find exactly the derivative of $F(x)$ in a table of derivatives, and you often have to manipulate the integral into a form that you can use.

2.3 Methods of Integration

You spent weeks in your calculus course learning how to evaluate various integrals. There is a bewildering array of techniques or tricks that can be used to manipulate expressions into forms that can be recognized as antiderivatives. For example, we have integration by parts, trigonometric substitutions, partial fractions, and "miscellaneous" substitutions. With enough practice, most students can become pretty proficient in these techniques, but integration is still somewhat of an art. Of these methods, integration by parts is probably the most useful. Recall that integration by parts uses the formula

$$\int u\,dv = uv - \int v\,du \qquad\qquad (2.12)$$

which is obtained by integrating the formula for the derivative of a product (Problem 2–4).

Let's use integration by parts to evaluate

$$F(x) = \int x \cos x\,dx$$

In this case, we let "u" $= x$ and "dv" $= \cos x\,dx$. Therefore, "du" $= dx$ and "v" $= \sin x$ and Equation 2.12 gives

$$\int x \cos x\,dx = x \sin x - \int \sin x\,dx = x \sin x + \cos x + c$$

where c is an integration constant. You can check this result by differentiating $x \sin x + \cos x + c$ to get $x \cos x$.

EXAMPLE 2–2

Often a substitution can transform an integral into a more transparent form. Evaluate

$$I = \int_0^2 x\,e^{-x^2}\,dx$$

SOLUTION: We know how to evaluate integrals involving $e^{\pm x}$ because $de^{\pm x}/dx = \pm e^{\pm x}$, so let's try the substitution $x^2 = u$. Because $du = 2x\,dx$, we have

$$I = \int_0^2 x\,e^{-x^2}\,dx = \frac{1}{2}\int_0^4 e^{-u}\,du = \frac{1}{2}\Big[-e^{-u}\Big]_0^4 = \frac{1-e^{-4}}{2}$$

Notice that we had to change the limits from 0 to 2 for the x integration to 0 to 4 for the u integration because $u = 0$ when $x = 0$ and $u = 4$ when $x = 2$.

EXAMPLE 2–3
Evaluate the integral

$$I = \int_0^\pi \cos^2\theta \sin\theta\,d\theta$$

which occurs frequently in quantum mechanics.

SOLUTION: Let $x = \cos\theta$, in which case $dx = -\sin\theta\,d\theta$, and so

$$I = \int_0^\pi \cos^2\theta \sin\theta\,d\theta = -\int_1^{-1} x^2\,dx = \int_{-1}^1 x^2\,dx = \frac{1}{3}\Big[x^3\Big]_{-1}^1 = \frac{2}{3}$$

We could go on and on to evaluate many types of integrals, but you've already done this once.

Certainly, having access to a good table of integrals goes a long way to being able to evaluate many integrals. The *CRC Standard Mathematical Tables and Formulae*, which is a standard "hard copy" reference and is also available online in many universities, lists over 50 pages of integrals. The most comprehensive tables (over 1000 pages!) are the *Tables of Integrals, Series, and Products* by Gradshteyn and Ryzhik, which are indispensable for anyone who works in applied mathematics. There are also a number of websites that give tables of integrals; one good one is the Wikipedia website *http://en.wikipedia.org/wiki/table_of_integrals*. In addition to the above references, there are a number of commercial packages, such as *Mathematica*, *MathCad*, and *Maple*, that can be used to evaluate derivatives and indefinite and definite integrals. These programs not only can provide numerical answers but also can perform algebraic manipulations, and so are called computer algebra systems (CAS). For example, the one line in *Mathematica*

Integrate [x^3 Cos [a x], x]

gives the indefinite integral

$$\int x^3 \cos ax\,dx = \frac{3(a^2x^2 - 2)\cos ax}{a^4} + \frac{x(a^2x^2 - 6)\sin ax}{a^3}$$

and

$$\text{Integrate} \, [\, x \, \text{Log} \, [\, a \, x + b \,], \{x, 0, 1\} \,]$$

gives the definite integral

$$\int_0^1 x \ln(ax+b)\,dx = \frac{2ab - a^2 + 2b^2 \ln b + 2(a^2 - b^2)\ln(a+b)}{4a^2}$$

Other CAS, such as *Maple* and *MathCad*, have the same capabilities. Almost every academic science or engineering department or industrial research laboratory owns a license for at least one of these CAS. Learning to use any one of these programs will not only save you hours of algebra, along with its concomitant errors, but will also allow you to concentrate on the central aspects of a problem rather than on algebraic drudgery.

A function does not have to be continuous in order to have a well-defined definite integral; it need only be piecewise continuous. (See the last paragraph of Section 1.2 for the definition of piecewise continuous.) Figure 2.3 shows a simple piecewise continuous function. Suppose that $f(x)$ is piecewise continuous for $a \le x \le b$ with a jump discontinuity at $x = c$. Then

$$I = \int_a^b f(x)\,dx = \lim_{\epsilon \to 0} \int_a^{c-\epsilon} f(x)\,dx + \lim_{\epsilon \to 0} \int_{c+\epsilon}^b f(x)\,dx \qquad (2.13)$$

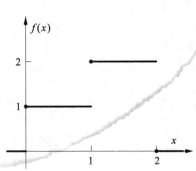

Figure 2.3. The function $f(x) = 0$ for $x < 0$; $f(x) = 1$ for $0 \le x < 1$; $f(x) = 2$ for $1 \le x < 2$; and $f(x) = 0$ for $x > 2$.

EXAMPLE 2–4

Find the area between the x axis and (Figure 2.3)

$$f(x) = \begin{cases} 0 & x < 0 \\ 1 & 0 \le x < 1 \\ 2 & 1 \le x < 2 \\ 0 & x > 2 \end{cases}$$

SOLUTION:

$$I = \lim_{\epsilon \to 0} \int_0^{1-\epsilon} dx + \lim_{\epsilon,\epsilon' \to 0} \int_{1+\epsilon}^{2-\epsilon'} 2\,dx$$

$$= \lim_{\epsilon \to 0}(1 - \epsilon) + 2 \lim_{\epsilon,\epsilon' \to 0}(2 - \epsilon' - 1 - \epsilon) = 3$$

You can also see geometrically that the area under $f(x)$ is equal to 3 because it consists of two rectangles of area 1 and 2.

The next Example shows that you don't always have to go through the limiting processes in Equation 2.13 to evaluate integrals of discontinuous functions.

EXAMPLE 2–5

The square-well potential for the interaction of two spherically symmetric molecules separated by a distance r is given by (see Figure 2.4)

$$u(r) = \begin{cases} \infty & 0 < r < \sigma \\ -\varepsilon & \sigma < r < \lambda\sigma \\ 0 & r > \lambda\sigma \end{cases}$$

where σ, λ, and ε are constants that are characteristic of the molecule. The second virial coefficient of imperfect gas theory is given by

$$B(T) = -2\pi \int_0^\infty [e^{-u(r)/k_BT} - 1] r^2 dr$$

where k_B is the Boltzmann constant and T is the kelvin temperature. Derive an expression for $B(T)$ for the square-well potential.

SOLUTION:

$$B(T) = -2\pi \int_0^\sigma (-1) r^2 \, dr - 2\pi \int_\sigma^{\lambda\sigma} (e^{\epsilon/k_BT} - 1) r^2 \, dr$$

$$= \frac{2\pi\sigma^3}{3} [1 + (\lambda^3 - 1)(e^{\epsilon/k_BT} - 1)]$$

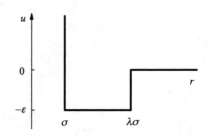

Figure 2.4. The square-well potential for the interaction of two spherically symmetric molecules.

We have three jump discontinuities in Example 2–4 and two in Example 2–5, but the generalization of Equation 2.11 to more than one jump discontinuity is apparent. Also, the values of $f(x)$ at the endpoints of each interval are irrelevant. Whether we write $f(x) = 1$ for $0 \leq x < 1$, or $0 < x < 1$, or $0 \leq x \leq 1$, or $0 < x \leq 1$, for example, in Example 2–4 makes no difference at all.

Although our derivative formulas allow us to evaluate derivatives of any (differentiable) function, there are many integrals that cannot be expressed in terms of known functions. We'll see in Chapter 4 that a number of well-known functions are actually *defined* in terms of integrals. For example, we'll see that the error function, $\text{erf}(x)$, is defined by

$$\text{erf}(x) = \frac{2}{\sqrt{\pi}} \int_0^x e^{-u^2} du$$

To evaluate $\text{erf}(1.5)$, for example, we determine the area between the u axis and the graph of e^{-u^2} from 0 to 1.5 as shown in Figure 2.5. There are many numerical routines to do this. You may remember learning the trapezoidal approximation or Simpson's approximation in your calculus course. Basically, these methods provide numerical approximations for areas (see Chapter 23). The commercially available computer programs, *Mathematica, MathCad,* and *Maple*, can be used to evaluate integrals numerically. The one-line command in *Mathematica*

NIntegrate [Exp [-x ^ 2], {x, 0, 3/2}]

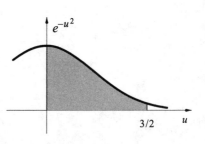

Figure 2.5. The area between the u axis and the graph of e^{-u^2} from 0 to 3/2 is equal to $\int_0^{3/2} e^{-u^2} du$.

gives

$$\int_0^{3/2} e^{-u^2}\, du = 0.856\ 188$$

2.4 Improper Integrals

Integrals that have infinite limits are called *improper integrals* and are defined by

$$\int_a^\infty f(x)\, dx = \lim_{c\to\infty} \int_a^c f(x)\, dx \qquad (2.14)$$

and

$$\int_{-\infty}^b f(x)\, dx = \lim_{c\to-\infty} \int_c^b f(x)\, dx \qquad (2.15)$$

When the limiting process defining an improper integral exists, the integral is said to converge or to be *convergent*. Otherwise, the integral is said to diverge or to be *divergent*.

EXAMPLE 2–6

Examine the convergence of $\int_1^\infty dx/x^p$ as a function of p.

SOLUTION:

$$\int_1^\infty \frac{dx}{x^p} = \lim_{b\to\infty} \int_1^b \frac{dx}{x^p} = \lim_{b\to\infty} \left[\frac{1}{1-p}\frac{1}{x^{p-1}}\right]_1^b$$

If $p > 1$, then we have

$$\lim_{b\to\infty} \frac{1}{1-p}\left(\frac{1}{b^{p-1}} - 1\right) = \frac{1}{1-p}(0-1) = \frac{1}{p-1}$$

If $p < 1$, then

$$\lim_{b\to\infty} \frac{1}{1-p}(b^{1-p} - 1) = \infty$$

If $p = 1$, then

$$\int_1^\infty \frac{dx}{x^p} = \lim_{b\to\infty} \int_1^b \frac{dx}{x} = \lim_{b\to\infty} \ln b = \infty$$

So we see that $\int_1^\infty dx/x^p$ converges if $p > 1$ and diverges if $p \leq 1$. This result is worth remembering.

EXAMPLE 2–7

Examine the convergence of $\int_a^\infty e^{-sx} dx$ as a function of s.

SOLUTION:

$$\int_a^\infty e^{-sx} dx = \lim_{b\to\infty} \int_a^b e^{-sx} dx$$

$$= \lim_{b\to\infty} \frac{e^{-sa} - e^{-sb}}{s} = \frac{e^{-sa}}{s} \quad (s > 0)$$

Thus, the integral converges if $s > 0$ and diverges if $s \leq 0$.

It is useful to be able to determine if an integral converges or not without having to evaluate it. Here is a test, called the *p test*, that is based on the result in Example 2–6:

If $\lim_{x\to\infty} x^p f(x) = K$, then

1. $\int_a^\infty f(x) dx$ converges if $p > 1$ and K is finite, and

2. $\int_a^\infty f(x) dx$ diverges if $p \leq 1$ and $K \neq 0$.

Let's investigate the convergence of

$$I = \int_0^\infty f(x) dx = \int_0^\infty \frac{e^{-x}}{(1+x)^2} dx$$

We see that $\lim_{x\to\infty} x^2 f(x) = K = 0$; thus, $p > 1$ and K is finite, and so the integral converges.

EXAMPLE 2–8

Use the *p* test to investigate the convergence of

$$I = \int_1^\infty f(x) dx = \int_1^\infty \frac{x^2}{(x^6 + 1)^{1/2}} dx$$

SOLUTION: We see that $\lim_{x\to\infty} xf(x) = K = 1$. Thus, $p = 1$ and $K \neq 0$, and so the integral diverges.

Problems

2–1. Find the antiderivatives of the following:

(a) x^n (b) $\dfrac{1}{x}$ (c) e^{-x} (d) $\sin x$ (e) $\dfrac{1}{1+x}$

2–2. Evaluate the following integrals:

(a) $\displaystyle\int_{1}^{2} \frac{dT}{T}$ (b) $\displaystyle\int_{0}^{\pi} \sin u \, du$ (c) $\displaystyle\int_{0}^{1} e^{-z} \, dz$ (d) $\displaystyle\int_{T_1}^{T_2} \frac{dT}{T^2}$

2–3. Use a substitution to evaluate the following integrals:

(a) $\displaystyle\int_{0}^{\pi/4} \tan u \, du$ (b) $\displaystyle\int_{0}^{1} \frac{t \, dt}{1+t^2}$ (c) $\displaystyle\int_{0}^{2} \frac{w \, dw}{1+w^2}$

2–4. Derive the formula for integration by parts.

2–5. Use integration by parts to evaluate

(a) $\displaystyle\int xe^{-x} dx$ (b) $\displaystyle\int x \sin x \, dx$ (c) $\displaystyle\int \ln x \, dx$ (d) $\displaystyle\int x^2 \ln x \, dx$

2–6. Use integration by parts to evaluate

(a) $\displaystyle\int_{0}^{\pi} x^2 \cos x \, dx$ (b) $\displaystyle\int_{1}^{2} x \ln x \, dx$ (c) $\displaystyle\int_{0}^{\infty} x^2 e^{-x} \, dx$

(d) $\displaystyle\int_{0}^{\pi/2} x^2 \sin x \, dx$

2–7. Evaluate

(a) $\displaystyle\int_{-\pi}^{\pi} \sin^3 \theta \, d\theta$ (b) $\displaystyle\int_{-\infty}^{\infty} xe^{-x^2} dx$ (c) $\displaystyle\int_{-2}^{2} \frac{t^3 \, dt}{1+t^2}$

2–8. Evaluate

(a) $\displaystyle\int_{0}^{\pi} \sin 2x \, dx$ (b) $\displaystyle\int_{0}^{\infty} xe^{-x^2} dx$ (c) $\displaystyle\int_{0}^{\pi/2} x \sin x^2 \, dx$

(d) $\displaystyle\int_{0}^{\pi} \cos \theta \sin^2 \theta \, d\theta$

2–9. Find the area bounded by $y = 2x$ and $y = x^2$ from $x = 0$ to 2.

2–10. Find the area between the curve $f(x) = \sqrt{a^2 - x^2}$, $-a \le x \le a$, and the x axis.

2–11. Show explicitly that $\int_{-A}^{A} f(x) \, dx = 2 \int_{0}^{A} f(x) \, dx$ if $f(x) = f(-x)$ and that $\int_{-A}^{A} f(x) \, dx = 0$ if $f(x) = -f(x)$.

2–12. This problem illustrates an important and practical manipulation of integrals. The Stefan–Boltzmann law says that the integral of the blackbody radiation law (Problem 1–26) over all wavelengths has a temperature dependence of T^4. Can you show this without integrating?

2–13. Show that $\int_{0}^{\infty} e^{-\beta x^2} \, dx/(1 + \beta x^2)^2$ is equal to a constant$/\beta^{1/2}$ without evaluating the integral.

2–14. Show that $\int_{0}^{\infty} x^n e^{-\alpha x} \, dx$ is equal to constant$/\alpha^{n+1}$ without evaluating the integral.

2–15. The integral

$$I = \int_{0}^{\pi} (3\cos^2 \theta - 1)^2 \sin \theta \, d\theta$$

occurs in the quantum-mechanical treatment of a rigid rotator or a hydrogen atom. Let $x = \cos \theta$ to show that

$$I = \int_{-1}^{1} (3x^2 - 1)^2 \, dx = \frac{8}{5}$$

2–16.   Show that the indefinite integral of the function described in Example 2–4 is given by

$$F(x) = \begin{cases} 0 & x < 0 \\ x & 0 < x < 1 \\ 1 + 2(x-1) & 1 < x < 2 \\ 3 & x > 2 \end{cases}$$

Show that $F(x)$ is continuous everywhere. Note that integration of the discontinuous function in Example 2–4 is a continuous function. This result illustrates that integration is a smoothing operation.

2–17.   Find the area between the x axis and

$$f(x) = \begin{cases} 0 & x < 0 \\ x & 0 < x < 1 \\ 1 & 1 < x < 2 \\ 3 - x & 2 < x < 3 \\ 0 & x > 3 \end{cases}$$

by integration. Now plot $f(x)$ and determine the area geometrically and compare your results.

2–18.   Evaluate the indefinite integral $F(x)$ of the function given in the previous problem. Plot it and show that it is continuous (see Problem 2–16).

2–19.   Use the p test to show that $\int_1^\infty \dfrac{x^2 + 1}{(x^6 + 1)^{1/2}}\,dx$ diverges.

2–20.   Use the p test to show that $\int_1^\infty \dfrac{x^2\,dx}{x^4 + 1}$ converges.

2–21.   Show that $\int_a^b \dfrac{dx}{(x-a)^p}$ converges if $p < 1$ and diverges if $p \geq 1$. (Take $b > a$.)

2–22.   Show that $\int_0^\infty e^{-x^2 + 6x}\,dx$ converges.

2–23.   Use the definition of a derivative to show that if $G(x) = \displaystyle\int_a^{u(x)} f(z)\,dz$, then

$$\frac{dG}{dx} = f(u(x))\frac{du}{dx}$$

This is called *Leibniz's rule*. What is the result if the lower limit is also a function of x, say, $v(x)$?

CHAPTER 3

Series and Limits

We often need to investigate the behavior of an equation for small values (or perhaps large values) of one of the variables in the equation. For example, we might want to investigate the low-frequency behavior of the Planck blackbody distribution law

$$\rho_v(T)\,dv = \frac{8\pi h}{c^3}\frac{v^3 dv}{e^{\beta h v}-1}$$

where v is the frequency, $\beta = 1/k_B T$ where T is the kelvin temperature and k_B is the Boltzmann constant, h is the Planck constant, and c is the speed of light. To do this, we use the fact that e^x can be written as the infinite series (that is, a series containing an unending number of terms)

$$e^x = 1 + x + \frac{x^2}{2!} + \frac{x^3}{3!} + \cdots$$

If we apply this result to $\rho_v(T)$, then, we have

$$\rho_v(T)\,dv = \frac{8\pi h}{c^3}\frac{v^3 dv}{[1+\beta h v + (\beta h v)^2/2 + \cdots] - 1}$$

$$\approx \frac{8\pi h}{c^3}\frac{v^3 dv}{\beta h v} = \frac{8\pi k_B T}{c^3}v^2 dv$$

for small values of v. Thus, we see that $\rho_v(T)$ goes as v^2 for small values of v. In this chapter, we shall review some useful series and apply them to some physical problems.

3.1 Convergence and Divergence of Infinite Series

An *infinite series* is an expression of the form

$$\sum_{n=1}^{\infty} u_n = u_1 + u_2 + u_3 + \cdots \tag{3.1}$$

We emphasize at the outset that the n in the summation in Equation 3.1 is a dummy variable; it simply runs over the values 1, 2, 3, ... Therefore, we can write

$$\sum_{n=1}^{\infty} u_n = \sum_{j=1}^{\infty} u_j = \sum_{l=1}^{\infty} u_l$$

and so on. We define the Nth partial sum of the series in Equation 3.1 by

$$S_N = \sum_{n=1}^{N} u_n$$

If

$$\lim_{N \to \infty} S_N = S$$

where S is finite, then the series is convergent and S is called the sum of the infinite series; otherwise, the series is divergent.

The standard example of an infinite series is the geometric series, whose Nth partial sum is

$$S_N = \sum_{j=0}^{N} x^j = 1 + x + x^2 + \cdots + x^N \tag{3.2}$$

Note that we start the summation with a $j = 0$ term in this case. It turns out that it is possible (and easy) to obtain a closed-form expression for S_N. Multiply S_N by x and subtract the result from S_N to get

$$S_N - x S_N = 1 - x^{N+1}$$

or

$$S_N = \frac{1 - x^{N+1}}{1 - x} \tag{3.3}$$

Thus, we see that

$$\lim_{N \to \infty} S_N = \begin{cases} \dfrac{1}{1-x} & |x| < 1 \\ \infty & |x| > 1 \end{cases}$$

You can see directly from Equation 3.2 that $S_N = N + 1$ when $x = 1$ and that it oscillates between 1 and 0 if $x = -1$. Thus, the limit of the partial sums diverges for $x = 1$ and has no unique value for $x = -1$, and so we see that the geometric series converges if $|x| < 1$ and diverges if $|x| \geq 1$. Figure 3.1 shows the partial

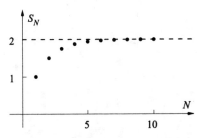

Figure 3.1. A plot of $2[1 - (1/2)^{N+1}]$, the partial sums of the geometric series for $x = 1/2$, against N. The limiting value is shown as a dashed line.

sums plotted against N for $x = 1/2$. The geometric series is often written as

$$\sum_{n=0}^{\infty} x^n = \frac{1}{1-x} = 1 + x + x^2 + x^3 + \cdots \qquad |x| < 1 \qquad (3.4)$$

EXAMPLE 3–1

The partition function of a diatomic molecule modeled as a quantum-mechanical harmonic oscillator is

$$q(T) = \sum_{n=0}^{\infty} e^{-(n+\frac{1}{2})h\nu/k_B T} \qquad (1)$$

where h is the Planck constant, ν is the frequency of the oscillator, k_B is the Boltzmann constant, and T is the kelvin temperature. Express $q(T)$ in closed form.

SOLUTION: Factor $e^{-h\nu/2k_B T}$ from the sum and let $r = e^{-h\nu/k_B T} < 1$. Then, using the geometric series

$$q(T) = e^{-h\nu/2k_B T} \sum_{n=0}^{\infty} r^n$$

$$= \frac{e^{-h\nu/2k_B T}}{1 - e^{-h\nu/k_B T}}$$

We say that $q(T)$ has been evaluated in closed form because its numerical evaluation requires only a finite number of steps, in contrast to equation 1, which would require an infinite number of steps.

It's obvious that it is necessary that $u_n \to 0$ as $n \to \infty$ if a series converges, but it is interesting and important to note that although this requirement is a *necessary* condition for convergence, it is *not* sufficient. The classic example of a series whose nth term goes to zero but for which the series does not converge is the *harmonic series*

$$S = \sum_{n=1}^{\infty} \frac{1}{n}$$

Problem 3–5 leads you through the standard proof that $S_N \to \infty$ as $N \to \infty$.

The convergence or lack of convergence of an infinite series is not affected by the addition or deletion of a finite number of terms. For example, if we were to add 100 terms (which add up to c) to the beginning of a convergent series, then its partial sums would be $S_N + c$ instead of S_N, and the sum, S_+, of the augmented series would be

$$S_+ = \lim_{N \to \infty} (S_N + c) = S + c$$

We can use the geometric series to show that any recurring decimal expression must be a rational number. Consider, for example, the decimal $a = 0.0909090909\cdots$. We can write a in the form of a geometric series:

$$a = \frac{9}{10^2} + \frac{9}{10^4} + \frac{9}{10^6} + \cdots$$

$$= 9\sum_{n=1}^{\infty} \frac{1}{10^{2n}} = 9\sum_{n=1}^{\infty} \frac{1}{(100)^n} = \frac{9}{100}\sum_{n=0}^{\infty} \frac{1}{(100)^n} = \frac{9/100}{1 - 1/100} = \frac{9}{99} = \frac{1}{11}$$

Deriving Equation 3.4 from Equation 3.3 brings us to an important point regarding infinite series: Equation 3.4 converges for $x < 1$ and diverges for $|x| \geq 1$. How can we tell whether a given infinite series converges or diverges? There are a number of so-called convergence tests, but one simple and useful one is the *ratio test*. To apply the ratio test, we form the absolute value of the ratio of the $(n + 1)$th term, u_{n+1}, to the nth term, u_n, and then let n become very large:

$$r = \lim_{n\to\infty} \left| \frac{u_{n+1}}{u_n} \right| \tag{3.5}$$

If $r < 1$, the series converges; if $r > 1$, the series diverges; and if $r = 1$, the test is inconclusive. Let's apply this test to the geometric series (Equation 3.4):

$$r = \lim_{n\to\infty} \left| \frac{x^{n+1}}{x^n} \right| = |x|$$

Thus, we see that the series converges if $|x| < 1$ and diverges if $|x| > 1$. It actually diverges at $x = 1$, but the ratio test does not tell us that. We would have to use some other method to determine the behavior at $x = 1$.

EXAMPLE 3–2

We claimed in the introduction to this chapter and we shall show in the next section that e^x can be written as

$$e^x = \sum_{n=0}^{\infty} \frac{x^n}{n!} = 1 + x + \frac{x^2}{2!} + \frac{x^3}{3!} + \cdots \tag{3.6}$$

Determine the values of x for which this series converges.

SOLUTION: The ratio test gives us

$$r = \lim_{n\to\infty} \left| \frac{x^{n+1}/(n+1)!}{x^n/n!} \right| = \lim_{n\to\infty} \left| \frac{x}{n+1} \right| = 0$$

for any fixed value of x. Thus, we conclude that the exponential series converges for all values of x.

Another test for the convergence of a series that is useful in practice is the *limit comparison test*.

Let $\sum u_n$ and $\sum v_n$ be series consisting of positive terms. If the limit

$$t = \lim_{n \to \infty} \frac{u_n}{v_n}$$

exists and $0 < t < \infty$, then either both series converge or both series diverge.

In order to apply this test, you have to know some series that you can use for comparisons. Two series that are very useful and cover many cases are the geometric series

$$\sum_{n=0}^{\infty} t^n = 1 + t + t^2 + t^3 + \cdots \qquad |t| < 1$$

and the series

$$\sum_{n=1}^{\infty} \frac{1}{n^s} = \frac{1}{1^s} + \frac{1}{2^s} + \frac{1}{3^s} + \cdots \qquad s > 1 \qquad (3.7)$$

(For $s = 1$, this is the harmonic series, which diverges.)

Let's use the limit comparison test to investigate the convergence of $\sum_{n=1}^{\infty} u_n = \sum_{n=1}^{\infty} (n^2 + n)/(n^4 + n + 6)$. Because this test involves the limit where $n \to \infty$, we may let n be very large. In this case, $n^2 + n \to n^2$ and $n^4 + n + 6 \to n^4$, and so $u_n = (n^2 + n)/(n^4 + n + 6) \to 1/n^2$ as $n \to \infty$. Thus, if we choose v_n to be $1/n^2$ in the limit comparison test, we see that

$$t = \lim_{n \to \infty} \frac{u_n}{v_n} = 1$$

Because the series $\sum_{n=1}^{\infty} v_n = \sum_{n=1}^{\infty} 1/n^2$ converges (see Equation 3.7), the series $\sum_{n=1}^{\infty} (n^2 + n)/(n^4 + n + 6)$ converges.

EXAMPLE 3–3

Investigate the convergence of $\sum_{n=1}^{\infty} u_n = \sum_{n=1}^{\infty} (2n^2 - 1)/(n^3 + n + 2)$.

SOLUTION: For large values of n, $2n^2 - 1 \to 2n^2$ and $n^3 + n + 2 \to n^3$, and so $u_n \to 2/n$ as $n \to \infty$. If we choose $v_n = 1/n$, then $t = \lim_{n \to \infty} u_n/v_v = 2$. Because the series $\sum_{n=1}^{\infty} 1/n$ diverges, the series $\sum_{n=1}^{\infty} (2n^2 - 1)/(n^3 + n + 2)$ diverges.

Both of the above examples would have been inconclusive with the ratio test. Problem 3–28 gives another test for convergence.

3.2 Power Series

An infinite series of the form

$$S(x) = \sum_{n=0}^{\infty} a_n x^n = a_0 + a_1 x + a_2 x^2 + \cdots \qquad (3.8)$$

is called a *power series* in x. Let's apply the ratio test to determine the convergence of $S(x)$ in Equation 3.8.

$$\lim_{n\to\infty} \left| \frac{a_{n+1} x^{n+1}}{a_n x^n} \right| = |x| \lim_{n\to\infty} \left| \frac{a_{n+1}}{a_n} \right| = |x| l$$

where $l = \lim_{n\to\infty} |a_{n+1}/a_n|$. Therefore, we see that the series converges for $|x| < 1/l = R$ and diverges for $|x| > 1/l = R$. The range of x for which the series converges, $-R < x < R$, is called the *interval of convergence* of the series.

EXAMPLE 3–4

Find the interval of convergence of

$$S(x) = \sum_{n=1}^{\infty} \frac{x^n}{n \cdot 2^n}$$

SOLUTION:

$$\lim_{n\to\infty} \left| \frac{a_{n+1} x^{n+1}}{a_n x^n} \right| = |x| \lim_{n\to\infty} \left| \frac{n \cdot 2^n}{(n+1) 2^{n+1}} \right| = \frac{|x|}{2}$$

The series converges if $|x| < 2$ and diverges if $|x| > 2$. The interval of convergence is $-2 < x < 2$.

Power series have the following important properties:

1. If $f(x) = \sum_{n=0}^{\infty} a_n x^n$ converges in the interval $-R < x < R$, then $f(x)$ is continuous in that interval.

2. If $f(x) = \sum_{n=0}^{\infty} a_n x^n$ converges in the interval $-R < x < R$, then

$$\int_0^x f(t)\, dt = \sum_{n=0}^{\infty} \frac{a_n x^{n+1}}{n+1}$$

converges in that interval.

3. If $f(x) = \sum_{n=0}^{\infty} a_n x^n$ converges in the interval $-R < x < R$, then

$$f'(x) = \sum_{n=1}^{\infty} n a_n x^{n-1}$$

converges in that interval.

In essence, then, we can differentiate or integrate a power series term by term.

For example, the geometric (power) series converges for $|t| < 1$.

$$\frac{1}{1-t} = \sum_{n=0}^{\infty} t^n = 1 + t + t^2 + \cdots \quad |t| < 1$$

If we integrate both sides of this equation from $t = 0$ to $t = x$ with $|x| < 1$, we obtain

$$\int_0^x \frac{dt}{1-t} = -\ln(1-x) = \sum_{n=1}^{\infty} \frac{x^n}{n} = x + \frac{x^2}{2} + \frac{x^3}{3} + \cdots \quad -1 \leq x < 1 \quad (3.9)$$

We wrote $-1 \leq x < 1$ here because an analysis of the endpoints, $x = \pm 1$, shows that this series converges for $x = -1$ and diverges for $x = 1$. We can also differentiate the geometric series term by term to obtain

$$\frac{1}{(1-x)^2} = \sum_{n=1}^{\infty} nx^{n-1} = 1 + 2x + 3x^2 + \cdots \quad -1 < x < 1 \quad (3.10)$$

EXAMPLE 3–5

Derive a closed expression for

$$f(x) = \sum_{n=0}^{\infty} nx^n$$

SOLUTION: We first note that the interval of convergence of this series is $-1 < x < 1$ since

$$\lim_{n \to \infty} \left| \frac{(n+1)x^{n+1}}{nx^n} \right| = |x| \lim_{n \to \infty} \left| \frac{n+1}{n} \right| = |x|$$

Now notice that $f(x)$ is similar to a geometric series, but with n in front of x^n. You can arrive at this form, however, by differentiating x^n (see Equation 3.10) to get nx^{n-1} and then multiplying by x. So if we start with

$$\frac{1}{1-x} = \sum_{n=0}^{\infty} x^n = 1 + x + x^2 + \cdots \quad |x| < 1$$

and differentiate both sides and then multiply by x, we get

$$\frac{x}{(1-x)^2} = x + 2x^2 + 3x^3 \cdots = \sum_{n=1}^{\infty} nx^n \quad |x| < 1$$

This series is used to calculate the average vibrational energy of a quantum-mechanical harmonic oscillator (Problem 3–17).

In this section we have derived power series for $\ln(1-x)$ and $(1-x)^{-2}$, each of which were derived from the geometric series. In the next section, we shall learn about a general method to derive power series for any (suitably well-behaved) function.

3.3 Maclaurin Series

A practical question that arises is how we find the infinite series that corresponds to a given function. For example, how do we derive Equation 3.6? First, assume that the function $f(x)$ can be expressed as a power series (i.e., a series in powers of x):

$$f(x) = c_0 + c_1 x + c_2 x^2 + c_3 x^3 + \cdots$$

where the c_j are to be determined. Then let $x = 0$ and find that $c_0 = f(0)$. Now differentiate once with respect to x,

$$\frac{df}{dx} = c_1 + 2c_2 x + 3c_3 x^2 + \cdots$$

and let $x = 0$ to find that $c_1 = (df/dx)_{x=0}$. Differentiate again,

$$\frac{d^2 f}{dx^2} = 2c_2 + 3 \cdot 2c_3 x + \cdots$$

and let $x = 0$ to get $c_2 = (d^2 f/dx^2)_{x=0}/2$. Differentiate once more,

$$\frac{d^3 f}{dx^3} = 3 \cdot 2c_3 + 4 \cdot 3 \cdot 2x + \cdots$$

and let $x = 0$ to get $c_3 = (d^3 f/dx^3)_{x=0}/3!$. The general result is

$$c_n = \frac{1}{n!} \left(\frac{d^n f}{dx^n} \right)_{x=0} \tag{3.11}$$

so we can write

$$f(x) = f(0) + \left(\frac{df}{dx} \right)_{x=0} x + \frac{1}{2!} \left(\frac{d^2 f}{dx^2} \right)_{x=0} x^2 + \frac{1}{3!} \left(\frac{d^3 f}{dx^3} \right)_{x=0} x^3 + \cdots \tag{3.12}$$

Equation 3.12 is called the Maclaurin series of $f(x)$.

Before we go on to apply Equation 3.12 to some simple functions, we should consider its validity. You might ask just under what conditions does the infinite series on the right of Equation 3.12 equal the function $f(x)$. Surely there must be some conditions on $f(x)$ for the expansion to be valid. Our derivation of Equation 3.12 certainly requires that all the derivatives of $f(x)$ be finite at $x = 0$. (Unfortunately, however, this is a *necessary* condition, but not a sufficient condition. Problem 3–33 shows a function whose derivatives are all finite at $x = 0$, yet for which Equation 3.12 is not valid.) The requirements that $f(x)$ must satisfy are a little too involved to go into here, but they are discussed in calculus books.

Although almost any function we deal with in physical chemistry will satisfy these requirements, you should at least be aware that Equation 3.12 does impose certain mathematical restrictions on $f(x)$.

If we apply Equation 3.12 to $f(x) = e^x$, we find that

$$\left(\frac{d^n e^x}{dx^n}\right)_{x=0} = 1$$

so

$$e^x = 1 + x + \frac{x^2}{2!} + \frac{x^3}{3!} + \cdots$$

in agreement with Equation 3.6. Some other important Maclaurin series, which can be obtained from a straightforward application of Equation 3.12 (Problem 3–18) are

$$\sin x = x - \frac{x^3}{3!} + \frac{x^5}{5!} - \frac{x^7}{7!} + \cdots \tag{3.13}$$

$$\cos x = 1 - \frac{x^2}{2!} + \frac{x^4}{4!} - \frac{x^6}{6!} + \cdots \tag{3.14}$$

$$\tan x = x + \frac{x^3}{3} + \frac{2x^5}{15} + \cdots \tag{3.15}$$

$$\ln(1+x) = x - \frac{x^2}{2} + \frac{x^3}{3} - \frac{x^4}{4} + \cdots \quad -1 < x \leq 1 \tag{3.16}$$

and

$$(1+x)^n = 1 + nx + \frac{n(n-1)}{2!}x^2 + \frac{n(n-1)(n-2)}{3!}x^3 + \cdots \quad |x| < 1 \tag{3.17}$$

Series 3.13 through 3.15 converge for all values of x, but as indicated, Series 3.16 converges only for $-1 < x \leq 1$ and Series 3.17 converges only for $|x| < 1$. Note that if n is a positive integer in Series 3.17, the series truncates. For example, if $n = 2$ or 3, we have

$$(1+x)^2 = 1 + 2x + x^2$$

and

$$(1+x)^3 = 1 + 3x + 3x^2 + x^3$$

Equation 3.17 for a positive integer is called the binomial expansion. If n is not a positive integer, the series continues indefinitely, and Equation 3.17 is called the binomial series. For example,

$$(1+x)^{1/2} = 1 + \frac{x}{2} - \frac{1}{8}x^2 + \cdots \tag{3.18}$$

and

$$(1+x)^{-1/2} = 1 - \frac{x}{2} + \frac{3}{8}x^2 + \cdots \tag{3.19}$$

The Wikipedia website *http://en.wikipedia.org/wiki/Calculus.Taylor_Series* lists power series for a number of functions. Problem 3–26 discusses a Taylor series, which is an expansion about a point $x = x_0$ rather than $x = 0$.

3.4 Applications of Power Series

Power series have numerous applications in practice. For example, suppose we want to know if the integral

$$I = \int_0^1 \frac{\sin x}{x^2} dx$$

is finite. The region around $x = 0$ is problematic because the denominator approaches zero. Let's use Equation 3.13 for $\sin x$ and write I as

$$I = \int_0^1 \frac{1}{x^2} \left(x - \frac{x^3}{3!} + \frac{x^5}{5!} + \cdots \right) dx$$

$$= \int_0^1 \frac{dx}{x} - \int_0^1 \frac{x}{3!} dx + \int_0^1 \frac{x^3}{5!} dx + \cdots$$

All the integrals beyond the first are finite, but the first integral diverges because

$$\int_0^1 \frac{dx}{x} = \left[\ln x \right]_0^1 = \infty$$

and so I itself diverges.

We can also use power series to evaluate integrals. The integral

$$I = \int_0^\infty \frac{x^3}{e^x - 1} dx$$

occurs in the theory of blackbody radiation. We can evaluate this integral by multiplying the numerator and denominator of the integrand by e^{-x} and expanding the denominator in powers of e^{-x} using the geometric series.

$$I = \int_0^\infty \frac{x^3 e^{-x} dx}{1 - e^{-x}} = \int_0^\infty x^3 e^{-x} \left[\sum_{n=0}^\infty (e^{-x})^n \right] dx$$

The series here is a convergent power series (in e^{-x}) for $x > 0$ and so we can integrate term by term to write

$$I = \sum_{n=0}^\infty \int_0^\infty x^3 e^{-x(n+1)} dx = \sum_{n=0}^\infty \frac{1}{(n+1)^4} \int_0^\infty u^3 e^{-u} du$$

$$= 6 \sum_{n=0}^\infty \frac{1}{(n+1)^4} = 6 \sum_{n=1}^\infty \frac{1}{n^4}$$

The summation here is in most handbooks and is equal to $\pi^4/90$. Thus, we see that $I = \pi^4/15$.

This is probably a good place to introduce a frequently used notation that is very useful when working with power series. We write $O(x^2)$ to indicate terms in x^2 and higher. For example, we write $\cos x = 1 - x^2/2 + O(x^4)$ because $\cos x = 1 - x^2/2! + x^4/4! + \cdots$. This notation, in a bookkeeping sense, is used to keep track of powers of x that we neglect when truncating series expansions, as we shall see in the next Example.

EXAMPLE 3–6

One statistical-mechanical theory of solutions of strong electrolytes (such as an aqueous solution of sodium chloride) gives the thermodynamic energy of the solution as

$$U(\kappa, a) = -\frac{x^2 + x - x(1 + 2x)^{1/2}}{4\pi\beta a^3}$$

where $\beta = 1/k_B T$, a is the average radius of the positive and negative ions, and $x = \kappa a$, where κ is a known parameter that is a measure of the concentration of the solution. Show that U goes as κ^3 as $\kappa \to 0$, in other words, for small concentration.

SOLUTION: We need to write U as a power series in x. Using the binomial series

$$-(4\pi\beta a^3)U = x^2 + x - x(1 + 2x)^{1/2}$$

$$= x^2 + x - x\left[1 + \frac{2x}{2} - \frac{(2x)^2}{8} + O(x^3)\right]$$

$$= x^2 + x - x - x^2 + \frac{x^3}{2} + O(x^4)$$

$$= \frac{x^3}{2} + O(x^4)$$

or, using the fact that $x = \kappa a$,

$$U = -\frac{\kappa^3}{8\pi\beta} + O(\kappa^4)$$

Another example from the theory of electrolyte solutions is the following: One theory expresses the osmotic pressure of a solution of a strong electrolyte in terms of a function σ given by

$$\sigma(x) = \frac{3}{x^3}\left[1 + x - \frac{1}{1 + x} - 2\ln(1 + x)\right] \tag{3.20}$$

where once again $x = \kappa a$. The small-kappa (dilute solution) behavior of $\sigma(x)$ is given by

$$\sigma(x) = \frac{3}{x^3} \left\{ 1 + x - [1 - x + x^2 - x^3 + x^4 + O(x^5)] \right.$$

$$\left. - 2 \left[x - \frac{x^2}{2} + \frac{x^3}{3} - \frac{x^4}{4} + O(x^5) \right] \right\}$$

$$= \frac{3}{x^3} \left[\frac{x^3}{3} - \frac{x^4}{2} + O(x^5) \right] = 1 - \frac{3}{2}x + O(x^2)$$

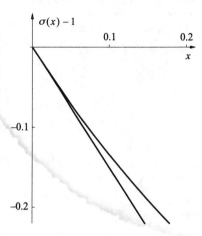

Figure 3.2 shows that $\sigma(x) - 1 = -3x/2 + O(x^2)$ for small values of x. Note that we keep track of the powers of x in the expansions of $1/(1 + x)$ and $\ln(1 + x)$ by writing $O(x^5)$ to indicate the first power of x neglected in each expansion. Without this notation to remind us, it would be easy to make the mistake of neglecting some power of x in one term but not the other. This is particularly important when a number of lower powers cancel, as they do (the x^0, x^1, and x^2 powers cancel) in the above two cases.

We can use the series presented in this chapter to derive a number of results used in physical chemistry. For example, consider the limit

$$l = \lim_{x \to 0} \frac{\sin x}{x} \qquad (3.21)$$

Because both $\sin x$ and x approach 0 as $x \to 0$, the ratio is called an *indeterminate form*. In cases like this, we can use *l'Hôpital's rule*, which says that

$$\lim \frac{f(x)}{g(x)} = \lim \frac{f'(x)}{g'(x)} \qquad (3.22)$$

Figure 3.2. A comparison of the plots of $\sigma(x) - 1$ given by Equation 3.20 and $\sigma(x) - 1 = -3x/2$ against x for small values of x, showing that $\sigma(x) - 1 = -3x/2 + O(x^2)$ as $x \to 0$.

If $\lim f'(x)/g'(x)$ turns out to be indeterminate, you use second derivatives instead of first derivatives, and so on, in Equation 3.22. Applying l'Hôpital's rule to Equation 3.21 gives

$$\lim_{x \to 0} \frac{\sin x}{x} = \lim_{x \to 0} \frac{\dfrac{d \sin x}{dx}}{\dfrac{dx}{dx}} = \lim_{x \to 0} \cos x = 1$$

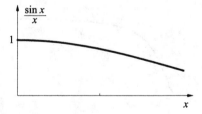

Figure 3.3 shows $\sin x/x$ plotted against x, showing that $\sin x/x \to 1$ as $x \to 0$. (Problem 1–30 offers a geometric proof of this result.) We could derive the same result by dividing Equation 3.13 by x and then letting $x \to 0$. (These two methods are really equivalent. See Problem 3–23.)

Figure 3.3. A plot of $\sin x/x$ against x, showing that $\sin x/x \to 1$ as $x \to 0$.

EXAMPLE 3–7
Evaluate

$$\lim_{x \to 0} \frac{1 - \cos x}{x^2}$$

using l'Hôpital's rule and the Maclaurin expansion of $\cos x$.

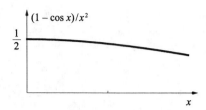

Figure 3.4. A plot of $(1 - \cos x)/x^2$ against x, showing that $(1 - \cos x)/x^2 \to 1/2$ as $x \to 0$.

SOLUTION: L'Hôpital's rule gives

$$\lim_{x \to 0} \frac{1 - \cos x}{x^2} = \lim_{x \to 0} \frac{\sin x}{2x} = \lim_{x \to 0} \frac{\cos x}{2} = \frac{1}{2}$$

The Maclaurin expansion of $\cos x$ gives

$$\lim_{x \to 0} \frac{1 - \cos x}{x^2} = \lim_{x \to 0} \frac{1 - \left(1 - \dfrac{x^2}{2} + O(x^4)\right)}{x^2} = \frac{1}{2}$$

(See Figure 3.4.)

We shall do one final example involving series and limits. Einstein's theory of the temperature dependence of the molar heat capacity of a crystal is given by

$$C_V = 3R \left(\frac{\Theta_E}{T}\right)^2 \frac{e^{-\Theta_E/T}}{(1 - e^{-\Theta_E/T})^2} \tag{3.23}$$

where R is the molar gas constant and Θ_E is a constant, called the Einstein temperature, that is characteristic of the solid. We'll now show that this equation gives the Dulong and Petit limit ($C_V \to 3R$) at high temperatures. First, let $x = \Theta_E/T$ in Equation 3.23 to obtain

$$C_V = 3Rx^2 \frac{e^{-x}}{(1 - e^{-x})^2} \tag{3.24}$$

When T is large, x is small, and so we shall use

$$e^{-x} = 1 - x + O(x^2)$$

Equation 3.24 becomes

$$C_V = 3Rx^2 \frac{1 - x + O(x^2)}{[x + O(x^2)]^2} \longrightarrow 3R$$

as $x \to 0$ ($T \to \infty$). This result is called the law of Dulong and Petit; the molar heat capacity of a crystal becomes $3R = 24.9$ J·K^{-1}·mol^{-1} for a monatomic crystal at high temperatures. By "high temperatures" we actually mean that $T \gg \Theta_E$, which for many substances is around 500 K. Figure 3.5 shows $C_V/3R$ plotted against T/Θ_E.

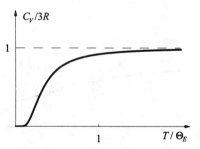

Figure 3.5. A plot of $C_V/3R$ in Equation 3.23 plotted against T/Θ_E, showing that $C_V \to 3R$ as $T \to \infty$.

Problems

3–1.   Evaluate the series $S = \sum_{n=0}^{\infty} 1/3^n$.

3–2.   Evaluate the series $S = \sum_{n=1}^{\infty} (-1)^{n+1}/2^n$.

3–3.   Express the recurring decimal $0.27272727\cdots$ as a fraction.

3–4.   Show that $0.142\,142\cdots$ is a rational number.

3–5. Here is a standard proof that the harmonic series diverges. Show that

$$S_2 = 1 + \frac{1}{2} = \frac{3}{2} \qquad S_4 > 1 + \frac{2}{2} \qquad S_8 > 1 + \frac{3}{2} \qquad S_{16} > 1 + \frac{4}{2}$$

and so on, and then argue that S_N is unbounded as $N \to \infty$.

3–6. Evaluate the series $S = \dfrac{1}{2^5} + \dfrac{1}{2^7} + \dfrac{1}{2^9} + \dfrac{1}{2^{11}} + \cdots$.

3–7. Test the following series for convergence:

(a) $\displaystyle\sum_{n=1}^{\infty} \frac{1}{n!}$ (b) $\displaystyle\sum_{n=1}^{\infty} \frac{n}{n^3 + 3}$ (c) $\displaystyle\sum_{n=1}^{\infty} \frac{\ln n}{n}$ (d) $\displaystyle\sum_{n=1}^{\infty} \frac{2^n}{3^n + n}$

3–8. Test the following series for convergence:

(a) $\displaystyle\sum_{n=0}^{\infty} \frac{3^n}{n!}$ (b) $\displaystyle\sum_{n=1}^{\infty} \frac{1}{n^2 + \ln n}$ (c) $\displaystyle\sum_{n=1}^{\infty} \frac{1}{4n + 1}$ (d) $\displaystyle\sum_{n=1}^{\infty} \frac{n}{3^n}$

3–9. Find the values of x for which each of the following series converge:

(a) $\displaystyle\sum_{n=0}^{\infty} (2x)^n$ (b) $\displaystyle\sum_{n=0}^{\infty} (x - 1)^n$ (c) $\displaystyle\sum_{n=0}^{\infty} \left(\frac{2x - 1}{3} \right)^n$ (d) $\displaystyle\sum_{n=0}^{\infty} e^{nx}$

3–10. Does $\sum_{n=1}^{\infty} \sin(n\pi/2)$ converge?

3–11. Calculate the percentage difference between e^x and $1 + x$ for $x = 0.0050$, $0.0100,\ 0.0150, \ldots,\ 0.1000$. Comment on the magnitude of the difference as x gets smaller.

3–12. Calculate the percentage difference between $\ln(1 + x)$ and x for $x = 0.0050$, $0.0100,\ 0.0150, \ldots,\ 0.1000$. Comment on the magnitude of the difference as x gets smaller.

3–13. Determine the values of x for which the following series converge:

(a) $\displaystyle\sum_{n=1}^{\infty} \frac{x^n}{n(n + 1)}$ (b) $\displaystyle\sum_{n=1}^{\infty} \frac{x^n}{n^n}$ (c) $\displaystyle\sum_{n=1}^{\infty} \frac{x^{2n+1}}{(2n + 1)!}$ (d) $\displaystyle\sum_{n=1}^{\infty} \frac{x^n}{2n - 1}$

3–14. Show that $\dfrac{1}{(1 - x)^3} = \dfrac{1}{2} \displaystyle\sum_{n=1}^{\infty} n(n + 1)x^{n-1} \qquad |x| < 1$.

3–15. Show that $\dfrac{1}{2} \ln \dfrac{1 + x}{1 - x} = \displaystyle\sum_{n=0}^{\infty} \dfrac{x^{2n+1}}{2n + 1} \qquad |x| < 1$.

3–16. Evaluate $\sum_{n=1}^{\infty} n^2 x^n$ in closed form.

3–17. The energy of a quantum-mechanical harmonic oscillator is given by

$\varepsilon_n = \left(n + \frac{1}{2} \right) h\nu, n = 0,\ 1,\ 2, \ldots$, where h is the Planck constant and ν is the fundamental frequency of the oscillator. The average vibrational energy of a harmonic oscillator is given by

$$\varepsilon_{\text{vib}} = (1 - e^{-h\nu/k_B T}) \sum_{n=0}^{\infty} \varepsilon_n e^{-nh\nu/k_B T}$$

where k_B is the Boltzmann constant and T is the kelvin temperature. Show that

$$\varepsilon_{\text{vib}} = \frac{h\nu}{2} + \frac{h\nu e^{-h\nu/k_B T}}{1 - e^{-h\nu/k_B T}}$$

3–18. Use Equation 3.12 to derive Equations 3.13 and 3.14.

3–19. Show that Equations 3.6, 3.13, and 3.14 are consistent with the relation
$e^{ix} = \cos x + i \sin x$.

3–20. Use Equation 3.6 and the definitions $\sinh x = (e^x - e^{-x})/2$ and
$\cosh x = (e^x + e^{-x})/2$ to show that

$$\sinh x = x + \frac{x^3}{3!} + \frac{x^5}{5!} + \cdots \quad \text{and} \quad \cosh x = 1 + \frac{x^2}{2!} + \frac{x^4}{4!} + \cdots$$

3–21. Show that Equations 3.13 and 3.14 and the results of the previous problem are consistent with the relations

$$\sin ix = i \sinh x \qquad \cos ix = \cosh x$$

$$\sinh ix = i \sin x \qquad \cosh ix = \cos x$$

3–22. Evaluate the limits of the following functions as $x \to 0$:

(a) $f(x) = \dfrac{e^{-x} \sin^2 x}{x^2}$ \qquad (b) $f(x) = x \ln x$ \qquad (c) $f(x) = \dfrac{1 - \cosh x}{x^2}$

3–23. Show that l'Hôpital's rule amounts to forming a Maclaurin expansion of both the numerator and the denominator. Evaluate the limit

$$\lim_{x \to 0} \frac{\ln(1+x) - x}{x^2}$$

both ways.

3–24. There are two limits that don't necessarily arise in the first course in physical chemistry, but do arise in more advanced courses. These two limits are
$\lim\limits_{x \to \infty} x^n e^{-x} = 0$ for any value of n, no matter how large n is, and $\lim\limits_{x \to 0} x^\alpha \ln x = 0$
for $\alpha > 0$, no matter how small α is. Use l'Hôpital's rule to verify both of these limits. These two limits are worth remembering.

3–25. Evaluate the integral

$$I = \int_0^a x^2 e^{-x} \cos^2 x \, dx$$

through terms of order a^5 by expanding I in powers of a.

3–26. A Maclaurin series is an expansion about the point $x = 0$. A series of the form

$$f(x) = c_0 + c_1(x - x_0) + c_2(x - x_0)^2 + \cdots$$

is an expansion about the point x_0 and is called a *Taylor series*. First show that $c_0 = f(x_0)$. Now differentiate both sides of the above expansion with respect to x and then let $x = x_0$ to show that $c_1 = (df/dx)_{x=x_0}$. Now show that

$$c_n = \frac{1}{n!} \left(\frac{d^n f}{dx^n} \right)_{x=x_0}$$

and so

$$f(x) = f(x_0) + \left(\frac{df}{dx} \right)_{x=x_0} (x - x_0) + \frac{1}{2} \left(\frac{d^2 f}{dx^2} \right)_{x=x_0} (x - x_0)^2 + \cdots$$

3–27. Start with

$$\frac{1}{1-x} = 1 + x + x^2 + \cdots$$

Now let $x = 1/x$ to write

$$\frac{1}{1 - \frac{1}{x}} = \frac{x}{x-1} = 1 + \frac{1}{x} + \frac{1}{x^2} + \cdots$$

Now add these two expressions to get

$$1 = \cdots + \frac{1}{x^2} + \frac{1}{x} + 2 + x + x^2 + \cdots$$

Does this make sense? What went wrong?

3–28. We have a p test for the convergence of integrals in Section 2.4. In this problem we give a p test for the convergence of an infinite series. If $\lim_{n \to \infty} n^p u_n = l$, then $\sum u_n$ converges if $p > 1$ and l is finite (even zero) and $\sum u_n$ diverges if $p < 1$ and $l \neq 0$ (but may be infinity). Test each of the following series for convergence:

(a) $\displaystyle\sum_{n=0}^{\infty} n e^{-n^2}$ (b) $\displaystyle\sum_{n=3}^{\infty} \frac{1}{n^2 + 3}$ (c) $\displaystyle\sum_{n=1}^{\infty} \frac{1}{n^{3/2}}$ (d) $\displaystyle\sum_{n=1}^{\infty} \frac{1}{\sqrt{2n + 1}}$

3–29. The Debye theory of the molar heat capacity of a crystal says that

$$C_V = 9R \left(\frac{T}{\Theta_D} \right)^3 \int_0^{\Theta_D/T} \frac{x^4 e^x}{(e^x - 1)^2} \, dx$$

where R is the molar gas constant, T is the kelvin temperature, and Θ_D, called the Debye temperature, is a parameter whose value depends on the crystal. First show that $C_V \to 3R$ (the law of Dulong and Petit) at high temperatures. Now show that $C_V \to \text{constant} \times T^3$ for low temperatures. (You don't need to evaluate any integral to show this.) This last result is the famous T^3 law of Debye.

3–30. Show that $\int_0^\infty e^{-ax} \sinh bx \, dx = b/(a^2 - b^2)$, $|b| < a$, by expanding $\sinh bx$ in a power series and integrating term by term. Can you evaluate this integral another way?

3–31. Does $\displaystyle\int_0^1 \frac{e^{-x} - 1 + x}{x^2} \, dx$ converge?

3–32. Does $\displaystyle\int_0^1 \frac{\cos x}{x} \, dx$ converge?

3–33. Expand e^{-1/x^2} in a Maclaurin series. This is an example of a function where all its derivatives exist at $x = 0$, but for which Equation 3.12 is not valid.

CHAPTER 4

Functions Defined by Integrals

As we mentioned in Chapter 2, there are many functions that do not possess an antiderivative. In other words, many indefinite integrals cannot be expressed in terms of elementary functions. By elementary functions we mean algebraic functions and the trigonometric functions, the exponential function, and the logarithm function that we discussed in Chapter 1. There are many other functions that are used extensively in physical problems that are not considered elementary, although we say at the outset here that they are not necessarily difficult in any sense. A great number of these functions are defined in terms of integrals. For example, we shall see in this chapter that the function of x defined by

$$f(x) = \int_0^x e^{-u^2} \, du$$

arises frequently in physical problems. In fact, it arises so frequently that we give it a name and embrace it as a standard mathematical function, giving it a place alongside our so-called elementary functions. In this chapter, we shall discuss four such functions: the gamma function, the beta function, the error function, and the Dirac delta function.

The standard reference for the various functions that are used in applied mathematics is the 1045-page *Handbook of Mathematical Functions with Formulas, Graphs, and Mathematical Tables* edited by Milton Abramowitz and Irene Stegun. This book, which was first published in 1964 by what was then called the National Bureau of Standards (now called the National Institute of Science and Technology) and referred to as *Applied Mathematical Series 55* (AMS55), is available online in its entirety at *http://www.convertit.com/Go/Convertit/Reference/AMS55.ASP*. For anyone who applies mathematics to physical problems, this book is a treasure.

4.1 The Gamma Function

The expression $n!$, equal to $1 \cdot 2 \cdot 3 \cdots n$, occurs when we enumerate permutations and combinations of things, such as the number of ways that N molecules can be distributed over n molecular quantum states (see Chapter 21). In the 1700s, Euler introduced a function that yields $n!$ when n is a positive integer, but is also well defined when n is not a positive integer. The function is called the *gamma function* and is defined by the integral expression

$$\Gamma(x) = \int_0^\infty z^{x-1} e^{-z} dz \qquad x > 0 \tag{4.1}$$

Notice that the integrand is a function of x and z and that the resulting integral is a function of x upon integrating over z. If x is a positive integer greater than or equal to 2, we can integrate $\Gamma(x)$ by parts. Letting $e^{-z} dz$ be "dv" and z^{x-1} be "u," we obtain

$$\Gamma(x) = \left[-z^{x-1} e^{-z} \right]_0^\infty + (x-1) \int_0^\infty z^{x-2} e^{-z} dz = (x-1) \int_0^\infty z^{x-2} e^{-z} dz$$

If we compare the last integral here to $\Gamma(x)$ in Equation 4.1, we see that it is equal to $\Gamma(x-1)$, so we have

$$\Gamma(x) = (x-1)\Gamma(x-1) \tag{4.2}$$

We can now substitute $\Gamma(x-1) = (x-2)\Gamma(x-2)$ and so on into Equation 4.2 to get

$$\Gamma(x) = (x-1)(x-2) \cdots \Gamma(1) \tag{4.3}$$

where

$$\Gamma(1) = \int_0^\infty e^{-z} dz = 1$$

Therefore, Equation 4.3 reads

$$\Gamma(x) = (x-1)(x-2) \cdots (1) = (x-1)! \qquad x = 2, \ 3, \ \cdots \tag{4.4}$$

Up to this point, Equation 4.4 is restricted to integer values of $x \geq 2$, but we can use it to *define* factorials for other values of x. If we let $x = 1$ in Equation 4.4, we have $\Gamma(1) = 0!$. But Equation 4.1 is perfectly well defined for $x = 1$ and yields $\Gamma(1) = 1$. Therefore, we can say that $0! = \Gamma(1) = 1$, a relation that you may have come across before. The fact that $0! = 1$ shouldn't bother you; the familiar relation $n! = 1 \cdot 2 \cdot 3 \cdots n$ is true only for *positive* integers.

We can use Equation 4.4 to define a factorial for nonintegers. This extension, or generalization, of a factorial turns out to be very convenient in practice. We can let $x = 1/2$ in Equation 4.4 to write $\Gamma(1/2) = (-1/2)!$. Now we can use Equation 4.1 to evaluate $\Gamma(1/2)$. Letting $z = u^2$, Equation 4.1 becomes

$$\Gamma(1/2) = \int_0^\infty z^{-1/2} e^{-z} dz = 2 \int_0^\infty e^{-u^2} du = \sqrt{\pi}$$

so that $\Gamma(1/2) = (-1/2)! = \sqrt{\pi}$ (Problem 4–5). What about $\Gamma(3/2) = (1/2)!$? Simply use Equation 4.2 to write

$$\Gamma(3/2) = (1/2)\Gamma(1/2) = \frac{\sqrt{\pi}}{2}$$

Similarly,

$$\Gamma\left(\frac{5}{2}\right) = \frac{3}{2}\Gamma\left(\frac{3}{2}\right) = \frac{3}{2} \cdot \frac{1}{2} \cdot \Gamma\left(\frac{1}{2}\right) = \frac{3\sqrt{\pi}}{4}$$

You might be wondering why we would want to consider factorials of numbers other than positive integers. The definitions of many functions that occur naturally in physical problems involve quantities such as $\Gamma(1/2)$ and $\Gamma(1/3)$. The following Example and Examples 4–3 and 4–4 that follow give another reason.

EXAMPLE 4–1

Evaluate $\int_0^\infty xe^{-ax^4}dx$ in terms of a gamma function ($a > 0$).

SOLUTION: Let $u = ax^4$; then $x = (u/a)^{1/4}$, $dx = u^{-3/4}du/4a^{1/4}$, and

$$\int_0^\infty xe^{-ax^4}dx = \frac{1}{4a^{1/2}} \int_0^\infty \frac{e^{-u}}{u^{1/2}}\,du = \frac{\Gamma(1/2)}{4a^{1/2}} = \frac{1}{4}\left(\frac{\pi}{a}\right)^{1/2}$$

Chapter 6 of Abramowitz and Stegun gives many other properties of the gamma function.

4.2 The Beta Function

In this section we shall discuss another useful function introduced by Euler. It is the *beta function*, defined by

$$B(x, y) = \int_0^1 z^{x-1}(1 - z)^{y-1}dz \qquad 0 < x,\ 0 < y \tag{4.5}$$

By letting $1 - z = u$, we find that $B(x, y) = B(y, x)$ (Problem 4–11); in other words, the beta function is a symmetric function in x and y. It turns out that $B(x, y)$ is related to the gamma function by the relation

$$B(x, y) = \frac{\Gamma(x)\Gamma(y)}{\Gamma(x + y)} \tag{4.6}$$

which also shows that $B(x, y) = B(y, x)$. The proof of Equation 4.6 is outlined in Problem 4–12.

The integral in Equation 4.5 occurs in a variational treatment of the quantum-mechanical problem of a particle in a box. In that case, x and y take on the integral values 2, 3, ...

EXAMPLE 4–2
Evaluate the integral $\int_0^1 z^4(1-z)^4 dz$.

SOLUTION: We simply use Equations 4.5 and 4.6 with $x = y = 5$ to get

$$\int_0^1 z^4(1-z)^4 dz = \frac{\Gamma(5)\Gamma(5)}{\Gamma(10)} = \frac{(4!)^2}{9!} = \frac{4\cdot3\cdot2}{9\cdot8\cdot7\cdot6\cdot5} = \frac{1}{630}$$

We could have evaluated the integral in Example 4–2 by expanding $(1-z)^4$ using the binomial formula and then integrating term by term, but that approach gets tedious pretty quickly.

The beta function can be transformed into another useful form by a simple change of variable. If we let $z = \sin^2\theta$ in Equation 4.5, then (Problem 4–13)

$$B(x, y) = 2\int_0^{\pi/2} (\sin\theta)^{2x-1}(\cos\theta)^{2y-1} d\theta \qquad 0 < x,\ 0 < y \qquad (4.7)$$

Equations 4.5, 4.6, and 4.7 can be used to evaluate a host of integrals. These are best illustrated by examples.

EXAMPLE 4–3
Evaluate $\int_0^2 u^3(4-u^2)^{3/2} du$.

SOLUTION: First let $u^2 = 4z$, so that $(4-u^2)^{3/2} = 8(1-z)^{3/2}, u^3 = 8z^{3/2}$, and $du = dz/z^{1/2}$. Then

$$\int_0^2 u^3(4-u^2)^{3/2} du = 64\int_0^1 z(1-z)^{3/2} dz$$

$$= 64\frac{\Gamma(2)\Gamma(5/2)}{\Gamma(9/2)} = 64\frac{\frac{3}{2}\cdot\frac{1}{2}\cdot\Gamma\left(\frac{1}{2}\right)}{\frac{7}{2}\cdot\frac{5}{2}\cdot\frac{3}{2}\cdot\frac{1}{2}\cdot\Gamma\left(\frac{1}{2}\right)} = \frac{256}{35}$$

where we used Equation 4.2 repeatedly to evaluate $\Gamma(5/2)$ and $\Gamma(9/2)$.

EXAMPLE 4–4
Evaluate $\int_0^\pi \cos^6\theta\, d\theta$.

SOLUTION: We could use Equation 4.7 with $x = 1/2$ and $y = 7/2$ if the limits were 0 to $\pi/2$. Note, however, that $\cos^6\theta$ is symmetric about

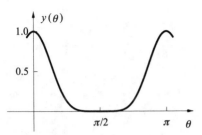

Figure 4.1. The function $y(\theta) = \cos^6 \theta$ is symmetric about $\theta = \pi/2$.

$\theta = \pi/2$ (Figure 4.1), so that $\int_0^\pi = 2 \int_0^{\pi/2}$. Now we use Equation 4.7 to write

$$
\int_0^\pi \cos^6 \theta \, d\theta = 2 \int_0^{\pi/2} \cos^6 \theta \, d\theta = \frac{\Gamma(1/2)\Gamma(7/2)}{\Gamma(4)}
$$

$$
= \frac{\sqrt{\pi} \cdot \dfrac{5}{2} \cdot \dfrac{3}{2} \cdot \dfrac{1}{2} \cdot \sqrt{\pi}}{3 \cdot 2} = \frac{5\pi}{16}
$$

With a little practice and experience, you can cast many integrals in terms of a beta function. Chapter 6 of Abramowitz and Stegun gives many other properties of the beta function.

4.3 The Error Function

One of the most commonly occurring and important integrals that cannot be expressed in terms of elementary functions is of the form $\int_0^x e^{-u^2} du$. This type of integral occurs so frequently that the function of x that it defines is a standard function of applied mathematics. We *define* the *error function* by the integral

$$
\mathrm{erf}\,(x) = \frac{2}{\sqrt{\pi}} \int_0^x e^{-u^2} du \qquad -\infty < x < \infty \tag{4.8}
$$

Even though we cannot express $\mathrm{erf}\,(x)$ in terms of simpler functions, it is a perfectly well-defined function of x and can be evaluated by numerical integration (see Chapter 23).

The $2/\sqrt{\pi}$ factor in Equation 4.8 is chosen so that $\mathrm{erf}\,(\infty) = 1$. Furthermore, it's easy to show that (Problem 4–16)

$$
\mathrm{erf}\,(-x) = -\mathrm{erf}\,(x) \tag{4.9}
$$

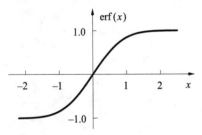

Figure 4.2. The error function $\mathrm{erf}\,(x)$ plotted against x.

so that $\mathrm{erf}\,(-\infty) = -1$. Tables of the error function are available at many websites. (Either go to page 310 of the Abramowitz–Stegun website cited in the introduction to this chapter or just put "error function tables" into Google.) Figure 4.2 shows $\mathrm{erf}\,(x)$ plotted against x.

EXAMPLE 4–5
The error function occurs frequently in the kinetic theory of gases. The fraction of molecules that have a component (x, y, or z) of velocity between v and $v + dv$ is given by

$$
f(v)\,dv = \left(\frac{m}{2\pi k_{\mathrm{B}} T} \right)^{1/2} e^{-mv^2/2k_{\mathrm{B}}T}\,dv
$$

where m is the mass of a molecule, T is the kelvin temperature, and k_B is the Boltzmann constant. Calculate the fraction of molecules with $-(2k_BT/m)^{1/2} \leq v \leq (2k_BT/m)^{1/2}$.

SOLUTION: Let $(2k_BT/m)^{1/2} = v_0$. The fraction is given by

$$F = \int_{-v_0}^{v_0} f(v)\, dv = 2\int_0^{v_0} f(v)\, dv = \frac{2}{v_0\sqrt{\pi}} \int_0^{v_0} e^{-mv^2/2k_BT}\, dv$$

because the integrand is an even function of v. Now let $mv^2/2k_BT = v^2/v_0^2 = u^2$ to write

$$F = \frac{2}{\sqrt{\pi}} \int_0^1 e^{-u^2}\, du = \mathrm{erf}\,(1) = 0.84270$$

Thus, about 84% of the molecules have a component of velocity whose magnitude is less than $(2k_BT/m)^{1/2}$ or, equivalently, a one-dimensional kinetic energy, $mv_0^2/2$, less than k_BT.

Suppose we want to calculate the fraction of molecules in a gas whose x component of velocity is greater in magnitude than v_0 (perhaps to determine the fraction of particularly energetic molecules). We would calculate the following (see Figure 4.3):

$$F = \int_{-\infty}^{-v_0} f(v)\, dv + \int_{v_0}^{\infty} f(v)\, dv = 2\left(\frac{m}{2\pi k_BT}\right)^{1/2} \int_{|v_0|}^{\infty} e^{-mv^2/2k_BT}\, dv$$

We can evaluate this integral just like we did in Example 4–5 to obtain

$$F = \frac{2}{\sqrt{\pi}} \int_1^{\infty} e^{-u^2}\, du$$

which is a special case of

$$F(x) = \frac{2}{\sqrt{\pi}} \int_x^{\infty} e^{-u^2}\, du$$

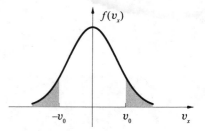

Figure 4.3. The shaded area represents the fraction of molecules in a gas with an x component of the velocity that exceeds some value v_0 in magnitude.

Equation 4.8 says that $F(1) = 1 - \mathrm{erf}(1)$, or that about 16% of the molecules have a component of velocity whose magnitude exceeds $(2k_BT/m)^{1/2}$. Of course, we should expect this result and the result of Example 4–5 to add to one.

Although $F(x) = 1 - \mathrm{erf}\,(x)$, this integral occurs frequently enough that it is used to define the *complementary error function*, erfc (x),

$$\mathrm{erfc}\,(x) = 1 - \mathrm{erf}(x) = \frac{2}{\sqrt{\pi}} \int_x^{\infty} e^{-u^2}\, du \qquad (4.10)$$

Note that erfc $(-\infty) = 2$, erfc $(0) = 1$, and erfc $(\infty) = 0$. (See Figure 4.4.)

A number of definite integrals can be expressed in terms of the error function or its related functions. For example, a useful general integral is (Problem 4–20)

$$\int_0^{\infty} e^{-(au^2+2bu+c)}\, du = \left(\frac{\pi}{4a}\right)^{1/2} e^{(b^2-ac)/a} \,\mathrm{erfc}\,\left(\frac{b}{\sqrt{a}}\right)$$

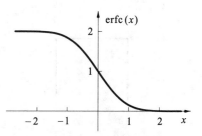

Figure 4.4. The complementary error function erfc (x) plotted against x.

Chapter 7 of Abramowitz and Stegun lists many integrals like this one, along with many other properties of the error function.

4.4 The Dirac Delta Function

In this section, we shall introduce a famous function that is not a function. Consider the function defined by (Figure 4.5)

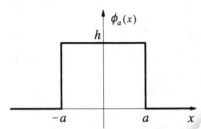

$$\phi_a(x) = \begin{cases} 0 & x < -a \\ h & -a < x < a \\ 0 & x > a \end{cases} \tag{4.11}$$

where h and a are related such that

$$\int_{-\infty}^{\infty} \phi_a(x)\, dx = 2ah = 1 \tag{4.12}$$

Figure 4.5. The function $\phi_a(x)$ defined by Equation 4.11 plotted against x.

In other words, the area under the curve is unity. Now let's form the integral

$$I = \int_{-\infty}^{\infty} \phi_a(x) f(x)\, dx$$

where $f(x)$ is a continuous function. Because of the definition of $\phi_a(x)$,

$$I = \int_{-a}^{a} \phi_a(x) f(x)\, dx$$

We are interested in the limit of this integral as $a \to 0$. In this case, the function $f(x)$ does not differ much from $f(0)$ as x varies from $-a$ to $+a$ if $f(x)$ is continuous, and so we can remove $f(0)$ from under the integral sign. Therefore, as $a \to 0$, I becomes

$$\lim_{a \to 0} I = f(0) \lim_{a \to 0} \int_{-a}^{a} \phi_a(x)\, dx = f(0) \lim_{a \to 0} \int_{-a}^{a} h\, dx = f(0) \tag{4.13}$$

By multiplying $f(x)$ by $\phi_a(x)$ and then integrating, letting $a \to 0$ and $h \to \infty$, such that $2ah = 1$, we have sifted out the value of $f(x)$ at $x = 0$.

We can use a construction like $\phi_a(x)$ to sift out $f(x)$ at any value of x. To isolate $f(x_0)$, let

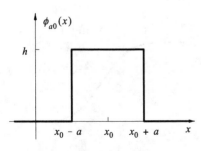

$$\phi_{a0}(x) = \begin{cases} 0 & x < x_0 - a \\ h & x_0 - a < x < x_0 + a \\ 0 & x_0 + a < x \end{cases} \tag{4.14}$$

with $2ah = 1$ (Figure 4.6). Equation 4.14 simply defines $\phi_a(x)$ to be centered at x_0 rather than at $x = 0$. Clearly,

$$\lim_{a \to 0} \int_{-\infty}^{\infty} \phi_{a0}(x) f(x)\, dx = f(x_0) \tag{4.15}$$

Figure 4.6. The function $\phi_{a0}(x)$ defined by Equation 4.14 plotted against x.

In the limiting process in Equation 4.15, $\phi_{a0}(x)$ is getting increasingly narrow and increasingly tall with the area of the rectangle $2ah = 1$. We denote this limiting "function" by $\delta(x - x_0)$ and write

$$\delta(x - x_0) = \begin{cases} 0 & x \neq x_0 \\ \infty & x = x_0 \end{cases} \qquad (4.16)$$

always keeping in mind that Equation 4.16 is a shorthand notation for $\phi_{a0}(x)$ as $a \to 0$ and $h \to \infty$ such that $2ah = 1$. Physically, Equation 4.16 represents a voltage spike at $x = x_0$ or an impulsive force if x represents time. In terms of $\delta(x - x_0)$, we write

$$\int_{-\infty}^{\infty} \delta(x - x_0)\, dx = 1 \qquad (4.17)$$

and

$$\int_{-\infty}^{\infty} \delta(x - x_0) f(x)\, dx = f(x_0) \qquad (4.18)$$

where $f(x)$ is a continuous function. Equations 4.17 and 4.18 serve to define the *Dirac delta function*, which was introduced by the British theoretical physicist Paul Dirac in 1927. Equation 4.18 illustrates the *sifting property* of the delta function.

EXAMPLE 4–6
Evaluate $I = \int_{-\infty}^{\infty} \delta(x - x_0)\cos x\, dx$.

SOLUTION: The function $\cos x$ is continuous, so we simply replace x by x_0 in $\cos x$ to get

$$I = \cos x_0$$

Another function that can act as a delta function in a limiting process is the *Gaussian distribution function:*

$$p_\sigma(x) = \frac{1}{(2\pi\sigma^2)^{1/2}} e^{-(x-x_0)^2/2\sigma^2} \qquad (4.19)$$

The Gaussian distribution is the famous bell-shaped curve. Figure 4.7 shows $p_\sigma(x)$ plotted against x for several values of σ. The function is centered at $x = x_0$, and the width about x_0 is governed by the value of σ; the smaller σ is, the narrower and more peaked is $p_\sigma(x)$. The factor of $(2\pi\sigma^2)^{-1/2}$ in front assures that

$$\int_{-\infty}^{\infty} p_\sigma(x)\, dx = 1$$

The curves in Figure 4.7 get both narrower and taller with decreasing values of σ because the areas under the curves are all the same (equal to 1).

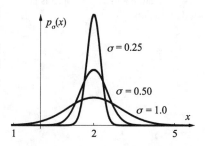

Figure 4.7. The Gaussian distribution in Equation 4.19 plotted against x for $\sigma = 1.0$, 0.50, and 0.25 and $x_0 = 2$.

If we multiply a continuous function $f(x)$ by $p_\sigma(x)$ in Equation 4.19 and let $\sigma \to 0$, then $p_\sigma(x)$ approaches $\delta(x - x_0)$ and

$$\lim_{\sigma \to 0} (2\pi \sigma^2)^{-1/2} \int_{-\infty}^{\infty} f(x) e^{-(x-x_0)^2/2\sigma^2} dx = f(x_0)$$

EXAMPLE 4–7
Evaluate

$$I(\sigma) = (2\pi\sigma^2)^{-1/2} \int_{-\infty}^{\infty} e^{-(x-x_0)^2/2\sigma^2} \cos x \, dx$$

explicitly and then let $\sigma \to 0$ to show that $\lim_{\sigma \to 0} I(\sigma) = \cos x_0$.

SOLUTION:

$$I(\sigma) = (2\pi\sigma^2)^{-1/2} \int_{-\infty}^{\infty} e^{-z^2/2\sigma^2} \cos(z + x_0) \, dz$$

Using the relation $\cos(z + x_0) = \cos x_0 \cos z - \sin x_0 \sin z$, we have

$$I(\sigma) = (2\pi\sigma^2)^{-1/2} \left[\cos x_0 \int_{-\infty}^{\infty} e^{-z^2/2\sigma^2} \cos z \, dz \right.$$
$$\left. - \sin x_0 \int_{-\infty}^{\infty} e^{-z^2/2\sigma^2} \sin z \, dz \right]$$

The first integral is equal to $2(\pi/2)^{1/2} \sigma e^{-\sigma^2/2}$, and the second integral is equal to zero because the integrand is an odd function of z (Figure 4.8). Thus,

$$I(\sigma) = e^{-\sigma^2/2} \cos x_0$$

and so

$$\lim_{\sigma \to 0} I(\sigma) = \cos x_0$$

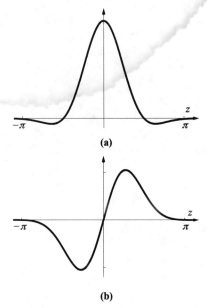

(a)

(b)

Figure 4.8. (a) A plot of $e^{-z^2/2\sigma^2} \cos z$ and (b) $e^{-z^2/2\sigma^2} \sin z$ against z for $\sigma = 1$, showing that $e^{-z^2/2\sigma^2} \cos z$ is an even function and $e^{-z^2/2\sigma^2} \sin z$ is an odd function.

The delta function is not a function in the strict sense, but has meaning only if it occurs multiplying a continuous function under an integral sign. For example, we can assign a meaning to $x\delta(x)$ by multiplying by a continuous function $f(x)$ and integrating to get

$$\int_{-\infty}^{\infty} f(x) x \, \delta(x) \, dx = f(0) \cdot 0 \cdot 1 = 0 \qquad (4.20)$$

and so one often sees the expression

$$x\delta(x) = 0 \qquad (4.21)$$

Keep in mind, however, that Equation 4.21 is a shorthand notation for Equation 4.20. Similarly, we write (Problem 4–24)

$$x\delta'(x) = -\delta(x) \tag{4.22}$$

and

$$\delta(ax) = \frac{1}{|a|}\delta(x) \tag{4.23}$$

where, once again, we must be aware of the meaning of these relations.

Problems

4–1. Evaluate $\int_0^\infty e^{-au}u^{3/2}du$, where $a > 0$.

4–2. Evaluate (a) $3\,\Gamma(5/2)/2\,\Gamma(1/2)$ and (b) $3\Gamma(5/4)/2\Gamma(1/4)$.

4–3. Evaluate $\int_0^\infty xe^{-ax^2}\,dx$, where $a > 0$, in terms of a gamma function.

4–4. Evaluate $\int_0^1 (\ln x)^n dx$, where $n > -1$, in terms of a gamma function. *Hint*: Let $\ln x = -z$.

4–5. This problem involves plane polar coordinates, which we discuss in Chapter 14. In this problem, we shall use a trick to prove that $I = \int_0^\infty e^{-x^2}dx = \sqrt{\pi}/2$. First square I and write it as

$$I^2 = \int_0^\infty e^{-x^2}dx \int_0^\infty e^{-y^2}dy = \int_0^\infty \int_0^\infty e^{-(x^2+y^2)}dxdy$$

Now convert to plane polar coordinates ($x = r\cos\theta$, $y = r\sin\theta$), and use the fact that the differential area in cartesian coordinates is $dxdy$ while that in plane polar coordinates is $rdrd\theta$ to show that $I^2 = \pi/4$, or that $I = \sqrt{\pi}/2$.

4–6. A factorial notation that is sometimes used is $n!! = n(n-2)(n-4)\cdots$, where the product terminates at $n = 1$ if n is odd and at $n = 2$ if n is even. Evaluate (a) $10!!$ and (b) $7!!$.

4–7. Evaluate $\int_0^\infty x^m e^{-x^n}\,dx$, where m and n are positive integers, in terms of a gamma function.

4–8. Show that $\int_0^\infty x^{2n}e^{-ax^2}dx = \Gamma\left(n + \tfrac{1}{2}\right)/2a^{n+1/2}$, where $a > 0$.

4–9. Many calculations of average values for the hydrogen atom involve integrals of the form

$$I_n = \int_0^\infty r^n e^{-\beta r}\,dr$$

where $\beta > 0$. Show that

$$I_n = \frac{n!}{\beta^{n+1}}$$

4–10. The logarithm function $\ln x$ can be *defined* by the integral $\ln x = \int_1^x du/u$. Using only this definition, show that $\ln ab = \ln a + \ln b$ and that $\ln a^b = b\ln a$. *Hint*: For the first part, write $\int_1^{ab} = \int_1^a + \int_a^{ab}$ and then make a substitution in the second integral.

4–11.   Show that $B(x, y) = B(y, x)$.

4–12.   This problem involves plane polar coordinates, which we discuss in
        Chapter 14. You can derive Equation 4.6 in the following way. Start with
        $\Gamma(m) = \int_0^\infty z^{m-1} e^{-z} dz$ and let $z = x^2$. Do the same for $\Gamma(n)$. Now form
        $\Gamma(m)\Gamma(n)$ as a double integral, transform to plane polar coordinates ($x = r \cos\theta$,
        $y = r \sin\theta$), use the fact that the differential area in cartesian coordinates is $dxdy$
        while that in plane polar coordinates is $r\, dr d\theta$, and finally use Equation 4.7.

4–13.   Derive Equation 4.7 by letting $z = \sin^2\theta$ in Equation 4.5.

4–14.   Evaluate $\int_0^{2\pi} \cos^6\theta d\theta$. *Hint*: Plot $\cos^6\theta$ first.

4–15.   Evaluate $\int_0^2 u^2 \sqrt[3]{8 - u^3}\, du$.

4–16.   Prove that $\mathrm{erf}\,(-x) = -\mathrm{erf}\,(x)$.

4–17.   Referring to Example 4–5, show that $\mathrm{Prob}\,\{-v_{x0} \le v_x \le v_{x0}\} =$
        $\mathrm{erf}\,[(m/2k_BT)^{1/2}v_{x0}]$.

4–18.   The probability that the x component of the velocity of a gas molecule exceeds a
        value of $v_{x0} > 0$ is given by

$$\mathrm{Prob}\,\{v_x > v_{x0}\} = \left(\frac{m}{2\pi k_B T}\right)^{1/2} 2 \int_{v_{x0}}^\infty e^{-mv_x^2/2k_B T} dv_x$$

        Show that this probability is given by $\mathrm{erfc}\,(u_0)$, where $u_0 = +(m/2k_BT)^{1/2}v_{x0}$.

4–19.   The kinetic theory of gases provides the following formula for the probability that
        the *speed* of a gas molecule exceeds a value c_0:

$$\mathrm{Prob}\,\{c \ge c_0\} = 4\pi \left(\frac{m}{2\pi k_B T}\right)^{3/2} \int_{c_0}^\infty c^2 e^{-mc^2/2k_B T} dc$$

        where $c = (v_x^2 + v_y^2 + v_z^2)^{1/2} \ge 0$ is the speed of the molecule. Show that

$$\mathrm{Prob}\,\{c \ge c_0\} = \frac{2}{\sqrt{\pi}} \left\{ \left(\frac{m}{2k_B T}\right)^{1/2} c_0 e^{-mc_0^2/2k_B T} + \frac{\sqrt{\pi}}{2} \mathrm{erfc}\left[\left(\frac{m}{2k_B T}\right)^{1/2} c_0\right] \right\}$$

4–20.   Show that $\int_0^\infty e^{-(ax^2 + 2bx + c)} dx = (\pi/4a)^{1/2} e^{(b^2 - ac)/a} \mathrm{erfc}\left(b/\sqrt{a}\right)$, where $a > 0$.

4–21.   Show that $\displaystyle\int_0^\infty \frac{e^{-at} dt}{\sqrt{t + x^2}} = \left(\frac{\pi}{a}\right)^{1/2} e^{ax^2} \mathrm{erfc}\,(\sqrt{a}x)$, where $a > 0$.

4–22.   The complementary error function also occurs in the quantum-mechanical
        discussion of a harmonic oscillator. A harmonic oscillator can be used as a model
        for the oscillatory behavior of two masses connected by a spring (a model of a
        diatomic molecule). If x is the displacement of the spring from its equilibrium
        position, then the two masses oscillate sinusoidally about $x = 0$ and the potential
        energy of the system is given by $V(x) = kx^2/2$, where k is the force constant of
        the spring. Show that the maximum displacement of the two masses is given by
        $x_{\max} = (2E/k)^{1/2}$, where E is the total energy. The quantity $x_{\max}$ is called the
        classically allowed amplitude. One of the many strange results of quantum
        mechanics is that there is a nonzero probability that the displacement of the
        oscillator will exceed its classically allowed amplitude, even though it doesn't

have sufficient energy. For a quantum-mechanical harmonic oscillator in its lowest energy state, this probability is given by

$$\text{Prob} = 2 \left(\frac{\alpha}{\pi} \right)^{1/2} \int_{\alpha^{-1/2}}^{\infty} e^{-\alpha x^2} dx$$

where α is a positive constant that is characteristic of the oscillator. Show that this probability is given by $\text{Prob} = \text{erfc}\,(1) = 1 - \text{erf}\,(1) = 0.15730$. Thus, we see that there is almost a 16% chance that the displacement will exceed its classically allowed amplitude.

4–23. Write out the actual meanings of Equations 4.21 and 4.22.

4–24. Verify Equation 4.23.

4–25. Evaluate $I(\sigma) = (2\pi\sigma^2)^{-1/2} \int_{-\infty}^{\infty} e^{-(x-x_0)^2/2\sigma^2} \sin x\, dx$ explicitly and then let $\sigma \to 0$ to show that $\lim_{\sigma \to 0} I(\sigma) = \sin x_0$.

4–26. Verify Equation 4.13 explicitly by evaluating the integral
$I = h \int_{-a}^{a} \cos(x + b)\, dx$ and then letting $a \to 0$ such that $2ah = 1$.

4–27. Argue that

$$\int_{-\infty}^{\infty} \delta(x - a)\, \delta(x - b)\, dx = \delta(a - b)$$

for $a \neq b$.

4–28. Show that

$$\delta(f(x)) = \frac{1}{|f'(a)|} \delta(x - a)$$

where $f'(a)$ is the derivative of $f(x)$ and $f(a) = 0$. *Hint*: Use the fact that $f(x) \approx f(a)(x - a)$ for $x \approx a$ and then use Equation 4.23.

4–29. Another representation of a delta function is

$$\delta(x) = \frac{1}{\pi} \lim_{\epsilon \to 0} \frac{\epsilon}{x^2 + \epsilon^2}$$

First show that $\int_{-\infty}^{\infty} \delta(x)\, dx = 1$. Now verify Equation 4.18 by showing that

$$\frac{1}{\pi} \lim_{\epsilon \to 0} \int_{-\infty}^{\infty} \frac{\epsilon \cos(x + b)}{x^2 + \epsilon^2}\, dx = \cos b$$

4–30. Derive the Maclaurin expansion of $\text{erf}\,(x)$. *Hint*: Do not use Equation 3.12; expand e^{-u^2} in Equation 4.8 and integrate term by term.

4–31. Show that $(2n)!! = 2^n n!$ and that $(2n + 1)!! = (2n + 1)!/2^n n!$.

CHAPTER 5

Complex Numbers

Complex numbers are usually introduced by considering a quadratic equation of the type $x^2 - x + 1 = 0$, where the quadratic formula gives

$$x = \frac{1}{2} \pm \frac{\sqrt{-3}}{2} = \frac{1}{2} \pm i\frac{\sqrt{3}}{2}$$

where $i = \sqrt{-1}$ is the imaginary unit. A number of the form $a + ib$, where a and b are real numbers, is called a *complex number*. If $a = 0$, then $x = ib$ is called an *imaginary number*. The message here is that we must introduce imaginary numbers in order to be able to solve quadratic equations in general. It shouldn't be surprising that initially there was a great resistance to the introduction of complex numbers and that it took many years for them to be accepted as legitimate members of our number system. The very name "imaginary number" seems to convey a certain degree of mysticism to these numbers.

If complex numbers had arisen only with quadratic equations, then it might have been easy to reject them by asserting that the equation $x^2 - x + 1 = 0$ has no solutions. After all, we're probably comfortable saying that $\sin x = 2$ has no solution for real values of x. Historically, imaginary numbers were most puzzling in the study of the solutions to cubic equations. Consider the cubic equation $x^3 + 2x^2 - x - 2 = 0$. You can verify by inspection that this equation has three real roots, ± 1 and -2. Yet when you solve this equation using the standard (fairly messy) formula for calculating the three roots, square roots of negative numbers occur at several intermediate steps. The final results are the three real roots, so it is apparent that the occurrence of imaginary numbers doesn't invalidate any of the formulas. Eventually, mathematicians came not only to tolerate imaginary numbers but to embrace them fully.

You might wonder if more complicated polynomial equations (such as 17th-degree equations) require the introduction of types of numbers "beyond" complex numbers. It turns out that they do not. There is a remarkable theorem of algebra called nothing less than the *fundamental theorem of algebra*, which says that every Nth-degree polynomial equation, $a_N x^N + a_{N-1} x^{N-1} + \cdots + a_1 x + a_0 = 0$, even with complex numbers as coefficients, has exactly N roots over the complex numbers. In other words, complex numbers are sufficient to solve *all* polynomial equations.

5.1 Complex Numbers and the Complex Plane

As we said in the introduction, a number of the form $a + ib$, where a and b are real numbers and $i^2 = -1$, is called a complex number. Usually, we write a complex number as

$$z = x + iy \tag{5.1}$$

where x is the real part of z and y is the imaginary part, which we express as

$$x = \text{Re}(z) \qquad y = \text{Im}(z) \tag{5.2}$$

We add or subtract complex numbers by adding or subtracting their real and imaginary parts separately. For example, if $z_1 = 2 + 3i$ and $z_2 = 1 - 4i$, then

$$z_1 - z_2 = (2 - 1) + [3 - (-4)]i = 1 + 7i$$

Furthermore, we can write

$$2z_1 + 3z_2 = 2(2 + 3i) + 3(1 - 4i) = 4 + 6i + 3 - 12i = 7 - 6i$$

To multiply complex numbers together, we simply multiply the two quantities as binomials and use the fact that $i^2 = -1$. For example,

$$(2 - i)(-3 + 2i) = -6 + 3i + 4i - 2i^2$$
$$= -4 + 7i$$

To divide complex numbers, it is convenient to introduce the complex conjugate of z, which we denote by z^* and form by replacing i by $-i$. For example, if $z = x + iy$, then $z^* = x - iy$. Note that a complex number multiplied by its complex conjugate is a real quantity:

$$zz^* = (x + iy)(x - iy) = x^2 - i^2 y^2 = x^2 + y^2 \tag{5.3}$$

The square root of zz^* is called the magnitude or the absolute value of z, and is denoted by $|z|$. Consider now the quotient of two complex numbers:

$$z = \frac{2 + i}{1 + 2i}$$

This ratio can be written in the form $x + iy$ if we multiply both the numerator and the denominator by $1 - 2i$, the complex conjugate of the denominator:

$$z = \frac{2+i}{1+2i}\left(\frac{1-2i}{1-2i}\right) = \frac{4-3i}{5} = \frac{4}{5} - \frac{3}{5}i$$

EXAMPLE 5–1
Show that

$$z^{-1} = \frac{x}{x^2+y^2} - \frac{iy}{x^2+y^2}$$

SOLUTION:

$$z^{-1} = \frac{1}{z} = \frac{1}{x+iy} = \frac{1}{x+iy}\left(\frac{x-iy}{x-iy}\right) = \frac{x-iy}{x^2+y^2}$$

$$= \frac{x}{x^2+y^2} - \frac{iy}{x^2+y^2}$$

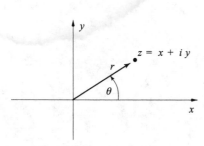

Figure 5.1. Representation of a complex number $z = x + iy$ as a point in a two-dimensional coordinate system. The plane of this figure is called the complex plane.

Because complex numbers consist of two parts, a real part and an imaginary part, we can represent a complex number by a point in a two-dimensional coordinate system where the real part is plotted along the horizontal (x) axis and the imaginary part is plotted along the vertical (y) axis, as shown in Figure 5.1. The plane of such a figure is called the *complex plane*. If we draw a vector $\mathbf{r}$ from the origin of this figure to the point $z = (x, y)$, then the length of the vector, $r = (x^2 + y^2)^{1/2}$, is the *magnitude* or the *absolute value* of z. The angle θ that the vector $\mathbf{r}$ makes with the x axis is the *argument* of z.

EXAMPLE 5–2
Determine the curve in the complex plane that is described by $|z - 1| = 2$.

SOLUTION:

$$|z - 1| = |(x - 1) + iy| = [(x - 1)^2 + y^2]^{1/2}$$

and so $|z - 1| = 2$ corresponds to

$$(x - 1)^2 + y^2 = 4$$

which is a circle of radius 2 centered at $x = 1$, $y = 0$ (Figure 5.2).

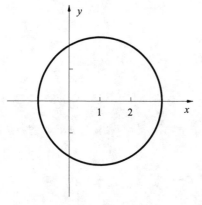

Figure 5.2. The graph of $|z - 1| = 2$, or $(x - 1)^2 + y^2 = 4$, in the complex plane.

The addition of two numbers in the complex plane has a nice geometrical interpretation. Figure 5.3 illustrates the addition of $z_1 = x_1 + iy_1$ and $z_2 = x_2 + iy_2$. The point $z_1 + z_2$ completes the parallelogram whose legs are z_1 and z_2. We say that the addition of z_1 and z_2 satisfies the *parallelogram law*.

If we refer to Figure 5.1, we see that we can represent a complex number z in *polar form*:

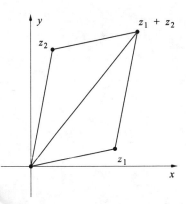

$$x = r\cos\theta \qquad y = r\sin\theta \tag{5.4}$$

and

$$z = r\cos\theta + ir\sin\theta \tag{5.5}$$

where

$$r = (x^2 + y^2)^{1/2} \tag{5.6}$$

is the distance from the origin to the point (x, y) and

$$\tan\theta = \frac{y}{x} \tag{5.7}$$

Figure 5.3. A geometrical interpretation of the addition of two complex numbers, z_1 and z_2.

As we said above, the angle θ is called the argument of z and r is its magnitude. We often denote these two quantities by $\theta = \arg z$ and $r = |z|$. Equation 5.5 is called the *polar form* of z.

EXAMPLE 5–3
Express $z = -1 + i$ in polar form.

SOLUTION: The magnitude of z is $\sqrt{2}$ and $\tan\theta = \frac{1}{-1} = -1$. As Figure 5.4 shows, the angle θ lies in the second quadrant, so $\theta = 3\pi/4$. Thus, the polar form of z is

$$z = \sqrt{2}\left(\cos\frac{3\pi}{4} + i\sin\frac{3\pi}{4}\right)$$

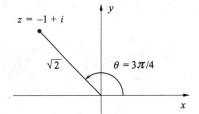

Figure 5.4. The complex number $z = -1 + i$ in polar form.

The Example above illustrates the fact that you must be aware in which quadrant the angle θ lies. If you use a hand calculator to evaluate $\tan^{-1}(-1)$, you'll get $-\pi/4$. The reason for this is that $\tan^{-1}\theta$ is a multivalued function (Problem 5–3).

5.2 Euler's Formula and the Polar Form of Complex Numbers

Another way to express $z = x + iy$ in terms of r and θ is by using *Euler's formula*,

$$e^{i\theta} = \cos\theta + i\sin\theta \tag{5.8}$$

which is derived in Problem 5–14. Using Equation 5.8, we can express z as

$$z = r(\cos\theta + i\sin\theta) = re^{i\theta} \tag{5.9}$$

Equation 5.9 shows that the polar form of z and Euler's formula are equivalent. Note that $|z| = (zz^*)^{1/2} = (re^{i\theta}re^{-i\theta})^{1/2} = r$.

EXAMPLE 5–4

Show that $e^{-i\theta} = \cos\theta - i\sin\theta$ and use this result and the polar representation of z to show that $|e^{i\theta}| = 1$.

SOLUTION: To prove that $e^{-i\theta} = \cos\theta - i\sin\theta$, we use Equation 5.8 and the fact that $\cos\theta$ is an even function of θ [$\cos(-\theta) = \cos\theta$] and that $\sin\theta$ is an odd function of θ [$\sin(-\theta) = -\sin\theta$]. Therefore,

$$e^{-i\theta} = \cos(-\theta) + i\sin(-\theta) = \cos\theta - i\sin\theta$$

Furthermore,

$$|e^{i\theta}| = [(\cos\theta + i\sin\theta)(\cos\theta - i\sin\theta)]^{1/2}$$
$$= (\cos^2\theta + \sin^2\theta)^{1/2} = 1$$

EXAMPLE 5–5

Express $z_1 = 1 + i$ and $z_2 = -1 - i$ in terms of Euler's formula.

SOLUTION: In both cases, $r = \sqrt{2}$. The point $z = 1 + i$ lies in the first quadrant, so $\tan^{-1}(1) = \pi/4$, and

$$z_1 = \sqrt{2}\,e^{i\pi/4}$$

The point $z = -1 - i$ lies in the third quadrant, so $\tan^{-1}(1) = 5\pi/4$, and

$$z_2 = \sqrt{2}\,e^{5\pi i/4}$$

Note that both z_1 and z_2 have the same value of y/x. Once again, we see that you must be aware in which quadrant the angle θ lies.

Multiplying and dividing complex numbers is easy in polar form:

$$z_1 z_2 = (r_1 e^{i\theta_1})(r_2 e^{i\theta_2}) = r_1 r_2 e^{i(\theta_1 + \theta_2)}$$

$$\frac{z_1}{z_2} = \frac{r_1}{r_2} e^{i(\theta_1 - \theta_2)}$$

For example, the product of z_1 and z_2 in Example 5–5 is $z_1 z_2 = 2e^{6\pi i/4} = 2e^{3\pi i/2} = -2i$ and their ratio is $z_1/z_2 = e^{-i\pi} = -1$.

We can use the polar form of complex numbers to derive many trigonometric identities. For example, start with

$$e^{i\alpha}e^{i\beta} = e^{i(\alpha+\beta)}$$

and write

$$(\cos\alpha + i\sin\alpha)(\cos\beta + i\sin\beta) = \cos(\alpha+\beta) + i\sin(\alpha+\beta)$$

Expand the left side and equate the real and imaginary parts of both sides to get

$$\cos\alpha\cos\beta - \sin\alpha\sin\beta = \cos(\alpha+\beta)$$

$$\sin\alpha\cos\beta + \cos\alpha\sin\beta = \sin(\alpha+\beta)$$

The expression $e^{i\omega t}$ occurs frequently in physical applications. Because $e^{i\omega t} = \cos\omega t + i\sin\omega t$, this expression has the physical interpretation of a unit vector (a vector whose length is one) that rotates about the origin in the complex plane in a counterclockwise direction with a frequency of ω radians per second (Figure 5.5). Given this interpretation, $e^{-i\omega t}$ can be viewed as a unit vector rotating in a clockwise direction.

We can also use Euler's formula to derive the formulas (Problem 5–10)

$$\sin\theta = \frac{e^{i\theta} - e^{-i\theta}}{2i} \quad \text{and} \quad \cos\theta = \frac{e^{i\theta} + e^{-i\theta}}{2} \quad (5.10)$$

These two formulas have a nice geometrical interpretation. Consider $\cos\omega t = (e^{i\omega t} + e^{-i\omega t})/2$. As we said above, $e^{i\omega t}/2$ may be viewed as a vector rotating in the complex plane in a counterclockwise direction and $e^{-i\omega t}/2$ may be viewed as one rotating in a clockwise direction, as shown in Figure 5.6. At $t = 0$, the sum of the two vectors points a unit length along the positive real axis. As t evolves, the two vectors rotate in opposite directions in such a way that their vertical components cancel and the sum of their horizontal components oscillates back and forth between $+1$ and -1 with a frequency of ω radians per second, according to $\cos\omega t$.

We can use Equations 5.10 to derive trigonometric identities. For example,

$$\begin{aligned}
\sin\alpha\cos\beta &= \frac{(e^{i\alpha} - e^{-i\alpha})}{2i}\frac{(e^{i\beta} + e^{-i\beta})}{2} \\
&= \frac{e^{i(\alpha+\beta)} - e^{-i(\alpha+\beta)}}{4i} + \frac{e^{i(\alpha-\beta)} - e^{-i(\alpha-\beta)}}{4i} \\
&= \frac{1}{2}\sin(\alpha+\beta) + \frac{1}{2}\sin(\alpha-\beta)
\end{aligned}$$

We can also use Equations 5.10 to evaluate integrals involving $\sin x$ or $\cos x$.

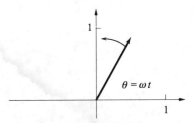

Figure 5.5. The expression $e^{i\omega t}$ has the physical interpretation of a unit vector that is rotating about the origin in a counterclockwise direction at a frequency of ω radians per second.

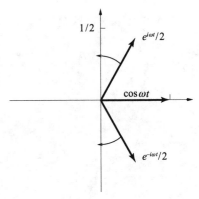

Figure 5.6. Geometrical interpretation of the expression $\cos\omega t = (e^{i\omega t} + e^{-i\omega t})/2$.

EXAMPLE 5–6
Evaluate

$$I = \int_0^\infty e^{-\alpha t}\sin t\, dt \quad (\alpha > 0)$$

by using Equation 5.10.

SOLUTION:

$$I = \frac{1}{2i} \int_0^\infty e^{-(\alpha - i)t} dt - \frac{1}{2i} \int_0^\infty e^{-(\alpha + i)t} dt$$

$$= \frac{1}{2i} \left(\frac{1}{\alpha - i} - \frac{1}{\alpha + i} \right) = \frac{2i}{2i(\alpha^2 + 1)} = \frac{1}{\alpha^2 + 1}$$

We can evaluate I in Example 5–6 another way. Because $e^{it} = \cos t + i \sin t$, we can write I as

$$I = \int_0^\infty e^{-\alpha t} \sin t \, dt = \text{Im} \int_0^\infty e^{-(\alpha - i)t} dt$$

$$= \text{Im} \left(\frac{1}{\alpha - i} \right) = \frac{1}{\alpha^2 + 1}$$

This procedure gives us

$$\int_0^\infty e^{-\alpha t} \cos t \, dt = \text{Re} \int_0^\infty e^{-(\alpha - i)t} dt = \text{Re} \left(\frac{1}{\alpha - i} \right) = \frac{\alpha}{\alpha^2 + 1}$$

as a by-product.

EXAMPLE 5–7
Summations such as

$$S(\theta) = \sum_{n=0}^{N} \cos n\theta$$

occur in group theory, crystallography, and optics. Derive a closed expression for $S(\theta)$.

SOLUTION: We express $\cos n\theta$ as $(e^{in\theta} + e^{-in\theta})/2$ and use Equation 3.2 with $x = e^{\pm i\theta}$. Therefore, we have

$$S(\theta) = \frac{1}{2} \sum_{n=0}^{N} e^{in\theta} + \frac{1}{2} \sum_{n=0}^{N} e^{-in\theta}$$

$$= \frac{1}{2} \left[\frac{1 - e^{i(N+1)\theta}}{1 - e^{i\theta}} \right] + \frac{1}{2} \left[\frac{1 - e^{-i(N+1)\theta}}{1 - e^{-i\theta}} \right]$$

$$= \frac{1 - \cos\theta + \cos N\theta - \cos(N + 1)\theta}{2(1 - \cos\theta)}$$

where in getting to the last line we combined the two terms and used Equation 5.10 several times (see Problem 5–24).

Problems

5–1. Find the real and imaginary parts of the following quantities:
(a) $(2-i)^3$ (b) $e^{\pi i/2}$ (c) $e^{-2+i\pi/2}$ (d) $(\sqrt{2}+2i)e^{-i\pi/2}$

5–2. If $z = x + 2iy$, then find
(a) $\mathrm{Re}\,(z^*)$ (b) $\mathrm{Re}\,(z^2)$ (c) $\mathrm{Im}\,(z^2)$ (d) $\mathrm{Re}\,(zz^*)$ (e) $\mathrm{Im}\,(zz^*)$

5–3. Determine the value of $\tan^{-1}\theta$ for the following complex numbers:
(a) $-1-i$ (b) $-1+i$ (c) $1-i$ (d) $-i$

5–4. Express the following numbers in the form $re^{i\theta}$:
(a) $6i$ (b) $4-\sqrt{2}\,i$ (c) $-1-2i$ (d) $1+i$

5–5. Express the following complex numbers in the form $x+iy$:
(a) $e^{-i\pi/4}$ (b) $6e^{2\pi i/3}$ (c) $e^{-(\pi/4)i+\ln 2}$ (d) $e^{-2\pi i}+e^{4\pi i}$

5–6. Discuss the statement that multiplying a complex number by i has a geometric interpretation of rotating the number by $90°$ counterclockwise in the complex plane.

5–7. Prove that $e^{i\pi} = -1$. Comment on the nature of the numbers in this relation.

5–8. Show that $\mathrm{Re}\,(z) = (z+z^*)/2$ and that $\mathrm{Im}\,(z) = (z-z^*)/2i$.

5–9. Determine the region in the complex plane described by $1 \le |z+i| \le 3$.

5–10. Show that $\cos\theta = (e^{i\theta}+e^{-i\theta})/2$ and that $\sin\theta = (e^{i\theta}-e^{-i\theta})/2i$.

5–11. Use Equation 5.8 to derive the formula of de Moivre,

$$\cos n\theta + i\sin n\theta = (\cos\theta + i\sin\theta)^n$$

Use the formula of de Moivre to derive the trigonometric identities

$$\cos 2\theta = \cos^2\theta - \sin^2\theta$$
$$\sin 2\theta = 2\sin\theta\cos\theta$$
$$\cos 3\theta = \cos^3\theta - 3\cos\theta\sin^2\theta = 4\cos^3\theta - 3\cos\theta$$
$$\sin 3\theta = 3\cos^2\theta\sin\theta - \sin^3\theta = 3\sin\theta - 4\sin^3\theta$$

5–12. Evaluate (a) $(1+i)^{10}$ and (b) $(1-i)^{12}$.

5–13. Consider the set of functions

$$\Phi_m(\phi) = \frac{1}{\sqrt{2\pi}}e^{im\phi} \qquad \begin{matrix} m = 0, \pm1, \pm2, \ldots \\ 0 \le \phi \le 2\pi \end{matrix}$$

First show that

$$\int_0^{2\pi} \Phi_m(\phi)\,d\phi = \begin{cases} 0 & \text{for all values of } m \neq 0 \\ \sqrt{2\pi} & m = 0 \end{cases}$$

Now show that

$$\int_0^{2\pi} \Phi_m^*(\phi)\Phi_n(\phi)\,d\phi = \begin{cases} 0 & m \neq n \\ 1 & m = n \end{cases}$$

5–14. This problem offers a derivation of Euler's formula. Start with

$$f(\theta) = \ln\left(\cos\theta + i\sin\theta\right) \tag{1}$$

Show that

$$\frac{df}{d\theta} = i \tag{2}$$

Now integrate both sides of equation 2 to obtain

$$f(\theta) = \ln\left(\cos\theta + i\sin\theta\right) = i\theta + c \tag{3}$$

where c is a constant of integration. Show that $c = 0$ and then exponentiate equation 3 to obtain Euler's formula.

5–15. We can use Euler's formula and the formulas of de Moivre (Problem 5–11) to evaluate a host of integrals. First show that

$$\int_0^\pi e^{i2n\theta}\,d\theta = 0 \qquad n = \pm 1,\ \pm 2,\ \ldots$$

Use this result to show that

$$\int_0^\pi \sin^2\theta\,d\theta = \frac{\pi}{2} \qquad \text{and} \qquad \int_0^\pi \cos^2\theta\,d\theta = \frac{\pi}{2}$$

Now use the same method to show that

$$\int_0^\pi \cos^4\theta\,d\theta = \int_0^\pi \sin^4\theta\,d\theta = \frac{3\pi}{8}$$

(See the following problem.)

5–16. Here is another way to evaluate $\int_0^\pi \cos^{2n}\theta\,d\theta$ and $\int_0^\pi \sin^{2n}\theta\,d\theta$ (see the previous problem). First write $\cos\theta$ as

$$\cos\theta = \frac{e^{i\theta} + e^{-i\theta}}{2}$$

Now use the binomial theorem in the form

$$(x + y)^{2n} = \sum_{m=0}^{2n} \frac{(2n)!}{m!(2n-m)!} x^m y^{2n-m}$$

with $x = e^{i\theta}$ and $y = e^{-i\theta}$ to show that

$$\int_0^\pi \cos^{2n}\theta\,d\theta = \pi\frac{(2n)!}{2^{2n}(n!)^2} \qquad n = 0,\ 1,\ 2,\ \ldots$$

Use a similar approach to show that

$$\int_0^\pi \sin^{2n}\theta\,d\theta = \pi\frac{(2n)!}{2^{2n}(n!)^2} \qquad n = 0,\ 1,\ 2,\ \ldots$$

5–17. Use Euler's formula to show that

$$\cos ix = \cosh x \qquad \text{and} \qquad \sin ix = i\sinh x$$

Now show that

$$\sinh ix = i\sin x \qquad \text{and} \qquad \cosh ix = \cos x$$

5–18. Use Euler's formula to show that

$$\cos \alpha \cos \beta = \frac{1}{2} \cos(\alpha + \beta) + \frac{1}{2} \cos(\alpha - \beta)$$

$$\sin \alpha \sin \beta = \frac{1}{2} \cos(\alpha - \beta) - \frac{1}{2} \cos(\alpha + \beta)$$

5–19. Evaluate i^i.

5–20. The equation $x^2 = 1$ has two distinct roots, $x = \pm 1$. The equation $x^N = 1$ has N distinct roots, called the N roots of unity. This problem shows how to find the N roots of unity. We shall see that some of the roots turn out to be complex, so let's write the equation as $z^N = 1$. Now let $z = e^{i\theta}$ and obtain $e^{iN\theta} = 1$, or

$$\cos N\theta + i \sin N\theta = 1$$

Now argue that $N\theta = 2\pi n$, where n has the N distinct values 0, 1, 2, ..., $N - 1$ or that the N roots of unity are given by

$$z = e^{2\pi i n/N} \qquad n = 0,\ 1,\ 2,\ \ldots,\ N - 1$$

Show that we obtain $z = 1$ and $z = \pm 1$ for $N = 1$ and $N = 2$, respectively. Now show that

$$z = 1,\ -\frac{1}{2} + i\frac{\sqrt{3}}{2},\ \text{and}\ -\frac{1}{2} - i\frac{\sqrt{3}}{2}$$

for $N = 3$. Show that each of these roots is of unit magnitude. Plot these three roots in the complex plane. Now show that $z = 1,\ i,\ -1,\ \text{and}\ -i$ for $N = 4$ and that

$$z = 1,\ -1,\ \frac{1}{2} \pm i\frac{\sqrt{3}}{2},\ \text{and}\ -\frac{1}{2} \pm i\frac{\sqrt{3}}{2}$$

for $N = 6$. Plot the four roots for $N = 4$ and the six roots for $N = 6$ in the complex plane. Compare the plots for $N = 3$, $N = 4$, and $N = 6$. Do you see a pattern?

5–21. Using the results of the previous problem, find the three distinct roots of $z^3 = 8$.

5–22. The *Schwarz inequality* says that if $z_1 = x_1 + iy_1$ and $z_2 = x_2 + iy_2$, then $x_1 x_2 + y_1 y_2 \leq |z_1| \cdot |z_2|$. To prove this inequality, start with its square

$$(x_1 x_2 + y_1 y_2)^2 \leq |z_1|^2 |z_2|^2 = (x_1^2 + y_1^2)(x_2^2 + y_2^2)$$

Now use the fact that $(x_1 y_2 - x_2 y_1)^2 \geq 0$ to prove the inequality.

5–23. Starting with $\int_0^\infty e^{-a^2 x^2}\, dx = \sqrt{\pi}/2a$, let $a = (1 - i)/\sqrt{2}$ and separate the result into real and imaginary parts to show that

$$\int_0^\infty \cos x^2\, dx = \int_0^\infty \sin x^2\, dx = \frac{1}{2}\left(\frac{\pi}{2}\right)^{1/2}$$

5–24. Show that

$$\sum_{n=0}^{N} \sin n\theta = \frac{\sin \theta - \sin(N + 1)\theta + \sin N\theta}{2(1 - \cos \theta)}$$

CHAPTER 6

Ordinary Differential Equations

Many scientific laws can be expressed in terms of differential equations. In fact, differential equations are the most common and most useful means of formulating these laws. A differential equation describes a function of one or more variables in terms of the derivatives of the function. If the unknown function depends upon only one independent variable, then the equation is called an *ordinary differential equation*. Ordinary differential equations necessarily involve ordinary derivatives. Examples of ordinary differential equations are

$$\text{(a) } \frac{dy}{dx} = 2y^2 \qquad \text{(b) } x^2\frac{d^2y}{dx^2} + x\frac{dy}{dx} + y = e^x \qquad \text{(c) } (x^2 + y^2)\frac{dy}{dx} = xy$$

In each case, there is only one independent variable, x, and one dependent variable, y. If the unknown function depends upon more than one independent variable, then its equation is called a *partial differential equation*. Partial differential equations necessarily involve partial derivatives. Examples are

$$\frac{\partial^2 u}{\partial x^2} + \frac{\partial^2 u}{\partial y^2} = 0 \qquad \text{and} \qquad \frac{\partial u}{\partial t} = \frac{\partial^2 u}{\partial x^2}$$

We shall not discuss partial differential equations until Chapters 15 and 16.

There are several terms that we often use when discussing differential equations. We say that $y(x)$ is a *solution* of an ordinary differential equation when the two sides of the equation are equal when $y(x)$ is substituted into it. The *order* of a differential equation is the order of the highest derivative that occurs in the equation. Equations (a) and (c) above are first-order equations and (b) is a second-order equation.

We shall discuss only linear differential equations in this chapter. A differential equation is said to be linear if the dependent variable (the function to be determined) and all its derivatives occur only to the first power and there are no cross terms involving $y(x)$ and its derivatives. Equation (b) is the only linear differential equation of the three. Equation (a) is nonlinear because of the y^2 term, and Equation (c) is nonlinear because of the $y^2 \, dy/dx$ term.

It turns out that linear differential equations are much easier to solve than nonlinear differential equations. Fortunately, many natural laws can be expressed by linear differential equations to a high degree of accuracy. Consequently, linear differential equations occur frequently in applied problems.

6.1 Linear First-Order Differential Equations

The general form of a first-order linear differential equation is

$$\frac{dy}{dx} + p(x)y = q(x) \tag{6.1}$$

where $p(x)$ and $q(x)$ are known functions. For example, the differential equation

$$\frac{dy}{dx} + \frac{1}{x}y = x^2$$

is of this form. If it were not for the $q(x)$ term on the right side of Equation 6.1, then it would be easy to solve because we could write it as

$$\frac{dy}{y} = -p(x)\,dx \tag{6.2}$$

and then integrate both sides to obtain

$$y(x) = e^{-\int p(x)\,dx}$$

where we are ignoring the constant of integration for the moment. To solve Equation 6.1, we simply *try* a solution of the form

$$y(x) = u(x)e^{-\int p(x)\,dx} \tag{6.3}$$

where $u(x)$ is to be determined. If we substitute Equation 6.3 into Equation 6.1, we obtain

$$\frac{du}{dx}e^{-\int p(x)\,dx} - p(x)u(x)e^{-\int p(x)\,dx} + p(x)u(x)e^{-\int p(x)\,dx} = q(x)$$

or simply

$$\frac{du}{dx}e^{-\int p(x)\,dx} = q(x)$$

This equation is easy to solve for $u(x)$ by separating variables to obtain

$$du = q(x)e^{\int p(x)\,dx}\,dx$$

and integrating both sides:

$$u(x) = \int q(x)e^{\int p(x)\,dx}\,dx + \beta \tag{6.4}$$

where β is a constant of integration. Substituting Equation 6.4 into Equation 6.3 gives

$$y(x) = e^{-\int p(x)\,dx}\left[\int q(x)e^{\int p(x)\,dx}\,dx + \beta\right] \tag{6.5}$$

as the solution to Equation 6.1. Notice that Equation 6.5 contains one constant of integration. Solving a first-order differential equation is equivalent to performing a single integration, producing one integration constant.

Before we go on, we should comment on how we obtained the solution to Equation 6.1. We let the term not involving $y(x)$ or its derivative equal zero, readily found a solution to the simplified equation, and then assumed that the solution to the original equation was the product of an unknown function and the solution to the simplified equation. We then were able to determine the unknown function by substituting the product back into the original equation. This may appear to be trickery, and it is. Much of the study of differential equations may seem to the beginner to be various tricks or divine substitutions to find solutions. As in many things, however, these methods become much less inspirational with experience.

Let's use Equation 6.5 to solve

$$x\frac{dy}{dx} + 2y = x^3$$

First divide by x to put this equation into the form of Equation 6.1:

$$\frac{dy}{dx} + \frac{2}{x}y = x^2$$

Then $\int p(x)\,dx = \int 2\,dx/x = 2\ln x = \ln x^2$ and so $e^{\int p\,dx} = e^{\ln x^2} = x^2$. Therefore, Equation 6.5 gives

$$y(x) = \frac{1}{x^2}\left[\int x^4\,dx + \beta\right] = \frac{1}{x^2}\left[\frac{x^5}{5} + \beta\right] = \frac{x^3}{5} + \frac{\beta}{x^2}$$

as you can verify by direct substitution.

EXAMPLE 6–1
Solve

$$x\frac{dy}{dx} - y = x$$

with the condition that $y(1) = 3$.

SOLUTION: First divide by x to put the equation into the form of Equation 6.1.

$$\frac{dy}{dx} - \frac{1}{x}y = 1$$

Then $\int p \, dx = -\ln x$ and so $e^{\int p \, dx} = e^{-\ln x} = 1/x$. Therefore, Equation 6.5 gives

$$y(x) = x \left[\int \frac{dx}{x} + \beta \right] = x[\ln x + \beta] = x \ln x + \beta x$$

The condition $y(1) = 3$ gives $3 = \beta$, and so

$$y(x) = x \ln x + 3x$$

As another example, consider the two-step kinetic process

$$A \xrightarrow{k_1} B \xrightarrow{k_2} C$$

This kinetic process might represent a radioactive decay sequence or a chemical reaction. The differential equations describing this scheme are

$$\frac{dA}{dt} = -k_1 A$$

$$\frac{dB}{dt} = k_1 A - k_2 B \qquad (6.6)$$

$$\frac{dC}{dt} = k_2 B$$

Let's solve the equations for the initial conditions $A(0) = A_0$ and $B(0) = C(0) = 0$. We can solve the first equation for $A(t)$ by writing it as

$$\frac{dA}{A} = -k_1 \, dt$$

and integrating both sides to obtain

$$\ln A(t) = -k_1 t + c$$

where c is the integration constant. Using the fact that $A(0) = A_0$, we have $\ln A(t) = -k_1 t + \ln A_0$, or

$$A(t) = A_0 e^{-k_1 t}$$

where A_0 is the initial value of A. We now substitute this result into the equation for $B(t)$ to obtain

$$\frac{dB}{dt} + k_2 B = k_1 A_0 e^{-k_1 t}$$

This equation is now of the form of Equation 6.1. In this case, "$e^{\int p(x) \, dx}$" $= e^{k_2 t}$, and so

$$B(t) = e^{-k_2 t} \left[k_1 A_0 \int e^{(k_2 - k_1) t} \, dt + \beta \right] = \frac{k_1 A_0 e^{-k_1 t}}{k_2 - k_1} + \beta e^{-k_2 t}$$

We can determine β by setting $B(0) = 0$, which gives $\beta = -k_1 A_0/(k_2 - k_1)$ and

$$B(t) = \frac{k_1 A_0}{k_2 - k_1} (e^{-k_1 t} - e^{-k_2 t}) \qquad (6.7)$$

We can determine $C(t)$ from the mass balance condition $A(t) + B(t) + C(t) = A(0) + B(0) + C(0) = A_0$. The solutions to Equations 6.6 are plotted for $A_0 = 1$,

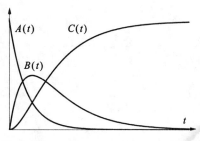

Figure 6.1. A plot of the solutions to Equations 6.6 for $A_0 = 1$, $k_1 = 2$, and $k_2 = 1$.

$k_1 = 2$, and $k_2 = 1$ in Figure 6.1. Note that $A(t)$ decays monotonically, that $B(t)$ rises initially and then decays, and that $C(t)$ increases monotonically. A problem like this one, where time is the independent variable and we are given values of the quantities at $t = 0$, is called an *initial value problem*.

Linear first-order differential equations are used to solve a class of problems called dilution problems. Here is an example. A 100-liter container initially contains 50 liters of a 3.00 molar solution of NaCl. A 1.00 molar solution of NaCl flows into the container at a rate of 8 liters per minute, and the resulting solution (assumed to be homogeneous) flows out of the container at a rate of 6 liters per minute. Let's determine the concentration of the solution when the container becomes full. First we note that the volume of the solution in the container at any time t before it is full is given by

$$V(t) = 50\,\text{L} + (8\,\text{L}\cdot\text{min}^{-1} - 6\,\text{L}\cdot\text{min}^{-1})(t\,\text{min})$$
$$= 50\,\text{L} + (2\,\text{L}\cdot\text{min}^{-1})(t\,\text{min})$$

Note that the container will be full (100 L) in 25 min. If we let $n(t)$ be the number of moles of salt at time t, then the concentration will be given by $n(t)/V(t) = n(t)/[\,50\,\text{L} + (2\,\text{L}\cdot\text{min}^{-1})(t\,\text{min})\,]$. The change in the number of moles of NaCl in the time interval t to $t + \Delta t$ is given by the difference between the inflow and the outflow:

$$\Delta n(t) = (8\,\text{L}\cdot\text{min}^{-1})(1.00\,\text{mol}\cdot\text{L}^{-1})\Delta t - (6\,\text{L}\cdot\text{min}^{-1})\frac{n(t)}{50\,\text{L} + (2\,\text{L}\cdot\text{min}^{-1})(t\,\text{min})}\Delta t$$

The corresponding differential equation (without all the units displayed) is obtained by dividing Δt and taking the limit $\Delta t \to 0$:

$$\frac{dn}{dt} + \frac{6n}{50 + 2t} = 8$$

In this case, "$\int p(x)\,dx$" $= 6\int dt/(50 + 2t) = 3\ln(50 + 2t)$ and so "$e^{\int p\,dx}$" $= (50 + 2t)^3$. Equation 6.5 says that

$$n(t) = \frac{1}{(50 + 2t)^3}\left[8\int (50 + 2t)^3\,dt + \beta\right]$$
$$= \frac{1}{(50 + 2t)^3}\left[64\int (25 + t)^3\,dt + \beta\right]$$

To evaluate the integral here, let $u = 25 + t$ to get

$$n(t) = \frac{1}{(50 + 2t)^3}\left[64\int u^3\,du + \beta\right]$$
$$= \frac{1}{(50 + 2t)^3}[16\,u^4 + \beta]$$
$$= \frac{1}{(50 + 2t)^3}[16\,(25 + t)^4 + \beta]$$
$$= \frac{(50 + 2t)^4 + \beta}{(50 + 2t)^3} \tag{6.8}$$

Using the fact that $n(0) = (50\text{ L})(3.00\text{ mol·L}^{-1}) = 150$ mol, we find that $\beta + (50)^4 = 150(50)^3$, or $\beta = 2(50)^4$. The concentration as a function of time is $c(t) = n(t)/V(t)$, or

$$c(t) = \frac{(50 + 2t)^4 + 2(50)^4}{(50 + 2t)^4}$$

The container is full after 25 minutes, and so the concentration at that time is $c(25) = 9/8$. Figure 6.2 shows $c(t)$ plotted against t.

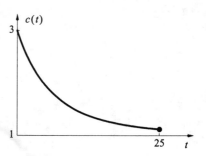

Figure 6.2. A plot of $c(t)$ against time.

6.2 Homogeneous Linear Differential Equations with Constant Coefficients

In the previous section, we found the general solution of a first-order linear differential equation. This is not possible for general higher-order linear differential equations. We'll see in this section, however, that we can solve higher-order linear differential equations if the coefficients are constants. Fortunately, a great many of the differential equations that occur in physical applications are linear and have constant coefficients. We'll discuss some properties of general higher-order linear differential equations first and then spend the rest of this section discussing those with constant coefficients.

A general nth-order linear differential equation can be written in the form

$$a_n(x)\frac{d^n y}{dx^n} + a_{n-1}(x)\frac{d^{n-1}y}{dx^{n-1}} + \cdots + a_1(x)\frac{dy}{dx} + a_0(x)y = f(x) \qquad (6.9)$$

Note that all the terms involving y or its derivatives occur only to the first power and there are no cross terms. If $f(x) = 0$, Equation 6.9 is said to be *homogeneous*; otherwise it is *nonhomogeneous*. It is sometimes convenient to write Equation 6.9 in the abbreviated form,

$$\mathcal{L}y(x) = f(x) \qquad (6.10)$$

where $\mathcal{L}$ is the linear differential operator

$$\mathcal{L} = a_n(x)\frac{d^n}{dx^n} + a_{n-1}(x)\frac{d^{n-1}}{dx^{n-1}} + \cdots + a_1(x)\frac{d}{dx} + a_0(x) \qquad (6.11)$$

In other words, $\mathcal{L}$ acts upon $y(x)$ to give the left side of Equation 6.9. An important property of a homogeneous linear differential equation, $\mathcal{L}y(x) = 0$, is that if $y(x)$ is a solution, then so is $cy(x)$ where c is a constant. Furthermore, if $y_1(x)$ and $y_2(x)$ are solutions to $\mathcal{L}y(x) = 0$, then so is $c_1 y_1(x) + c_2 y_2(x)$, or

$$\mathcal{L}[c_1 y_1(x) + c_2 y_2(x)] = c_1 \mathcal{L}y_1(x) + c_2 \mathcal{L}y_2(x) = 0 \qquad (6.12)$$

because $\mathcal{L}y_1(x) = 0$ and $\mathcal{L}y_2(x) = 0$. We can continue this process and say that if $y_1(x), y_2(x), \ldots, y_n(x)$ are solutions to the homogeneous equation, then so is the linear combination $c_1 y_1(x) + c_2 y_2(x) + \cdots + c_n y_n(x)$.

We'll now spend the rest of the section discussing homogeneous linear differential equations with constant coefficients. These are best discussed by means of examples. Let's start with

$$y''(x) + y'(x) - 6y(x) = 0 \qquad (6.13)$$

This equation can be satisfied by a function whose derviatives are multiples of itself. The function $e^{\alpha x}$ (where α is a constant) is such a function. If we substitute $y = e^{\alpha x}$ into Equation 6.13, we get

$$(\alpha^2 + \alpha - 6)e^x = 0$$

The factor $e^x \neq 0$, and so

$$\alpha^2 + \alpha - 6 = 0 \qquad (6.14)$$

or $\alpha = 2$ and -3. Equation 6.14 is called the *auxiliary equation* of Equation 6.13. Two solutions to Equation 6.13 are e^{2x} and e^{-3x} and according to Equation 6.12

$$y(x) = c_1 e^{2x} + c_2 e^{-3x} \qquad (6.15)$$

Note that this solution to the second-order differential equation, Equation 6.13, contains two arbitrary constants. Because Equation 6.13 is second order, we are essentially integrating twice to arrive at a solution, and so have two integration constants.

Because any multiple of $y(x)$ is a solution if $y(x)$ is a solution, we must distinguish between two solutions that are just multiples of each other and two solutions that are really different. We say that two functions, $y_1(x)$ and $y_2(x)$, are *linearly independent* if the only way that the equation

$$c_1 y_1(x) + c_2 y_2(x) = 0 \qquad (6.16)$$

can be satisfied is by setting $c_1 = c_2 = 0$. Otherwise, $y_1(x)$ and $y_2(x)$ are *linearly dependent*. For example, suppose that $y_1(x) = e^x$ and $y_2(x) = 2e^x$. Clearly, these are linearly dependent since one is a multiple of the other. In Equation 6.15, $c_1 = 1$ and $c_2 = -1/2$ satisfy this equation. The two functions $y_1(x) = e^{2x}$ and $y_2(x) = e^{-3x}$ in Equation 6.15 are linearly independent because there are no nonzero values of c_1 and c_2 that will satisfy Equation 6.15. For just two functions, linear independence means that they are not multiples of each other. For a second-order linear differential equation, there can be only two linearly independent solutions. A solution to a second-order linear differential equation of the form $y(x) = c_1 y_1(x) + c_2 y_2(x)$, where $y_1(x)$ and $y_2(x)$ are linearly independent, is called a *general solution*. Equation 6.15 is the general solution of Equation 6.13. For an nth-order linear differential equation, there are n linearly independent solutions. We shall discuss the linear independence of functions more thoroughly in Chapter 20.

EXAMPLE 6–2

Determine the solution of

$$y'' + y' - 2y = 0$$

subject to the conditions $y(0) = 0$ and $y'(0) = 6$.

SOLUTION: The auxiliary equation is $\alpha^2 + \alpha - 2 = 0$, which gives $\alpha = 1$ and -2. The general solution is

$$y(x) = c_1 e^x + c_2 e^{-2x}$$

Applying the conditions $y(0) = 0$ and $y'(0) = 6$ gives $c_1 + c_2 = 0$ and $c_1 - 2c_2 = 6$, or $c_1 = 2$ and $c_2 = -2$. Thus, the *particular solution* is

$$y(x) = 2e^x - 2e^{-2x}$$

If we attempt to solve

$$y''(x) - 2y'(x) + y(x) = 0 \qquad (6.17)$$

the auxiliary equation, $\alpha^2 - 2\alpha + 1 = 0$, gives us only one distinct root, $\alpha = 1$. Thus, $y(x) = e^x$ is one solution, but we need to find another solution in order to have a general solution. It is not uncommon to have one solution in hand and to search for another. This can be done very nicely by a method called *reduction of order*. The solution that we have in hand is $y(x) = ce^x$. To find a second solution, we *assume* that

$$y(x) = u(x)e^x \qquad (6.18)$$

where $u(x)$ is to be determined. Substitute Equation 6.18 into Equation 6.17 to obtain

$$u''(x)e^x + 2u'(x)e^x + u(x)e^x - 2[u'(x)e^x + u(x)e^x] + u(x)e^x = 0$$

or

$$u''(x)e^x = 0$$

The factor $e^x \neq 0$, so $u''(x) = 0$. This gives us $u(x) = c_1 x + c_2$, which, substituted into Equation 6.18, gives

$$y(x) = (c_1 x + c_2)e^x = c_1 x e^x + c_2 e^x \qquad (6.19)$$

This is the general solution to Equation 6.17 because e^x and xe^x are linearly independent. Although we have introduced the method of reduction of order by a specific example, the method is general. (See Problems 6–7b and 6–8b.)

6.3 Oscillatory Solutions

So far, the roots of the auxiliary equation have been real. Let's consider the equation

$$x''(t) + \omega^2 x(t) = 0 \tag{6.20}$$

where ω is a constant. The auxiliary equation is $\alpha^2 + \omega^2 = 0$, so $\alpha = \pm i\omega$. The general solution in this case is

$$x(t) = c_1 e^{i\omega t} + c_2 e^{-i\omega t} \tag{6.21}$$

We can use Euler's formula, $e^{i\theta} = \cos\theta + i\sin\theta$, to write Equation 6.21 as

$$x(t) = c_3 \cos\omega t + c_4 \sin\omega t \tag{6.22}$$

where $c_3 = c_1 + c_2$ and $c_4 = i(c_1 - c_2)$. Equations 6.21 and 6.22 are equivalent and are both general solutions to Equation 6.20. Problem 6–18 shows that Equation 6.22 can be written in the more lucid form

$$x(t) = A\cos(\omega t + \phi) \tag{6.23}$$

where $A = (c_3^2 + c_4^2)^{1/2}$ and $\phi = \tan^{-1}(-c_4/c_3)$.

EXAMPLE 6–3
Show that

$$x(t) = A\cos(\omega t + \phi)$$

is a general solution to Equation 6.20.

SOLUTION: The second derivative of $x(t)$ is

$$x''(t) = -\omega^2 A\cos(\omega t + \phi)$$

and so $x''(t) + \omega^2 x(t) = 0$. We say that it is a general solution because A and ϕ are constants. Their values can be determined by the initial values of $x(t)$ and $x'(t)$, for example.

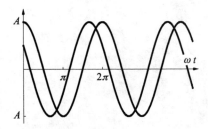

Figure 6.3. The periodic function $x(t) = A\cos(\omega t + \phi)$ plotted against ωt for two values of ϕ. Note that the motion is harmonic with a frequency $\nu = \omega/2\pi$ and an amplitude A.

Figure 6.3 shows $x(t)$ given by Equation 6.23 plotted against ωt; note that $x(t)$ oscillates with a frequency $\nu = \omega/2\pi$ cycles per second and that A, called the amplitude, is the maximum displacement of $x(t)$. The angle ϕ, which dictates the initial value of $x(t)$, is called the phase angle.

Equation 6.20 is the equation of motion of a classical harmonic oscillator. This is an important enough result that we should derive it. Consider a mass m attached to a spring, as shown in Figure 6.4. Suppose there is no gravitational force acting on m so that the only force is from the spring. Let the relaxed or undistorted length of the spring be x_0. Hooke's law says that the force acting on the mass m is $f = -k(x - x_0)$, where k is a constant characteristic of the spring and is called the *force constant* of the spring. The minus sign indicates the direction of the force: to the left in Figure 6.4 if $x > x_0$ (extended) and to the right if $x < x_0$ (compressed). The momentum of the mass is $p = mdx/dt$. Newton's second law says that the rate of change of momentum is equal to the force; the force in this case is $-k(x - x_0)$ and the rate of change of momentum is $dp/dt = md^2x/dt^2$. Consequently, the equation of motion is given by

$$m\frac{d^2x}{dt^2} = -k(x - x_0)$$

Letting $\xi = x - x_0$ be the displacement of the spring from its undistorted length, we have

$$m\frac{d^2\xi}{dt^2} + k\xi = 0$$

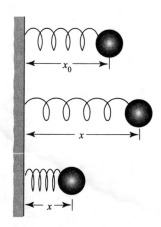

Figure 6.4. A body of mass m connected to a wall by a spring.

If we divide by m and let $\omega^2 = k/m$, we get Equation 6.20. Thus, we see that the angular frequency of the harmonic oscillator is $\omega = (k/m)^{1/2}$. Equation 6.23 represents the oscillatory motion.

Equation 6.20 represents not only the motion of a harmonic oscillator, but also a pendulum swinging through small angles, the electric current in a circuit containing an inductance and capacitance, a quantum-mechanical particle in a box, and numerous others. Let's consider the case of a pendulum swinging in a fixed plane for concreteness. We'll express the equation of motion in terms of the arclength $s(t) = l\theta(t)$, where θ is the angle that the pendulum makes with respect to the vertical (Figure 6.5).

The momentum of the supported mass is given by mds/dt, and so Newton's equation says that

$$m\frac{d^2s}{dt^2} = f \tag{6.24}$$

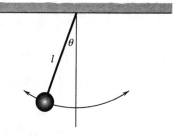

Figure 6.5. A pendulum oscillating in a single plane. The pendulum support is rigid and massless, and supports a mass m.

where f is the force acting on the mass. Using the fact that $s(t) = l\theta(t)$, we can write Equation 6.24 as

$$ml\frac{d^2\theta}{dt^2} = f(\theta) \tag{6.25}$$

We can determine the force from the potential energy, which is given by mgh, where h is the height of the mass above its minimum position. Referring to Figure 6.5, we have

$$V(\theta) = mg(l - l\cos\theta) \tag{6.26}$$

and the force is given by

$$f(\theta) = -\frac{\partial V}{\partial s} = -\frac{1}{l}\frac{\partial V}{\partial \theta} = -mg\sin\theta \tag{6.27}$$

Using the fact that $\sin\theta \approx \theta$ for small values of θ (see Equation 3.13), the force becomes $f = -mg\theta$. Therefore, Equation 6.25 becomes

$$ml\frac{d^2\theta}{dt^2} = -mg\theta$$

or

$$\frac{d^2\theta}{dt^2} + \omega_0^2\theta = 0 \tag{6.28}$$

where $\omega_0 = (g/l)^{1/2}$ is the natural angular frequency of the pendulum. The solution to Equation 6.28 with the initial condition $\theta(0) = \theta_0$ and $d\theta/dt = 0$ at $t = 0$ is $\theta(t) = \theta_0\cos\omega_0 t$. Physically, this solution depicts the back-and-forth motion of the pendulum with an angular frequency $\omega_0 = (g/l)^{1/2}$.

EXAMPLE 6–4

An equation similar to Equation 6.20 occurs in the quantum-mechanical problem of a particle in a one-dimensional box, which is treated in every physical chemistry course. Physically, this system represents a potential-free particle constrained to lie along a line from 0 to a. The Schrödinger equation for this system is

$$\frac{d^2\psi}{dx^2} + \frac{2mE}{\hbar^2}\psi(x) = 0 \qquad 0 \le x \le a \tag{1}$$

where $\psi(x)$ is the wave function of the particle, m is the mass of the particle, $\hbar$ is the Planck constant divided by 2π, and E is the energy of the particle. The wave function must satisfy the conditions $\psi(0) = \psi(a) = 0$, which are called *boundary conditions*. Solving a differential equation subject to boundary conditions is called a *boundary value problem*. Solve the above boundary value problem.

SOLUTION: As usual, we assume that $\psi(x)$ is of the form $e^{\alpha x}$, which leads to the auxiliary equation

$$\alpha^2 + \frac{2mE}{\hbar^2} = 0$$

or to $\alpha = \pm i(2mE/\hbar^2)^{1/2}$. The general solution is given by

$$\psi(x) = A\cos kx + B\sin kx$$

where $k = (2mE/\hbar^2)^{1/2}$. The boundary condition at $x = 0$ requires that $A = 0$; the boundary condition at $x = a$ requires that $B\sin ka = 0$. We could satisfy this condition by letting $B = 0$, but this gives $\psi(x) = 0$ for all x in the interval $(0, a)$, which is called a *trivial solution*. We can also satisfy the boundary condition by requiring that $\sin ka = 0$, which occurs whenever $ka = n\pi$, where $n = 1, 2, 3, \ldots$. (We exclude $n = 0$ because this gives a trivial solution.) Thus, we find that the energy is given by

$$E = \frac{n^2\pi^2\hbar^2}{2ma^2} = \frac{n^2h^2}{8ma^2} \qquad n = 1,\ 2,\ 3,\ \ldots$$

These values are the allowed energies of a particle in a one-dimensional box.

The wave function is given by

$$\psi_n(x) = B \sin \frac{n\pi x}{a} \qquad 0 \le x \le a$$

We can determine the value of B by requiring that $\psi_n(x)$ be normalized, in other words, by requiring that

$$\int_0^a \psi_n^2(x)\,dx = 1 = B^2 \int_0^a \sin^2 \frac{n\pi x}{a}\,dx = \frac{a}{2}B^2$$

Thus, we see that the (normalized) wave functions are

$$\psi_n(x) = \left(\frac{2}{a}\right)^{1/2} \sin \frac{n\pi x}{a} \qquad 0 \le x \le a$$

Let's consider the following equation, where the roots of the auxiliary equation are a complex conjugate pair:

$$x''(t) + 2x'(t) + 10x(t) = 0 \tag{6.29}$$

The auxiliary equation is $\alpha^2 + 2\alpha + 10 = 0$, so $\alpha = -1 \pm 3i$. The general solution is

$$x(t) = c_1 e^{-t} e^{3it} + c_2 e^{-t} e^{-3it}$$
$$= e^{-t}(c_3 \cos 3t + c_4 \sin 3t)$$

where $c_3 = c_1 + c_2$ and $c_4 = i(c_1 - c_2)$. This equation can be written in the equivalent form (Problem 6–18)

$$x(t) = Ae^{-t} \cos(3t + \phi) \tag{6.30}$$

where $A = (c_3^2 + c_4^2)^{1/2}$ and $\phi = \tan^{-1}(-c_4/c_3)$. Thus, in this case, the solution displays damped harmonic behavior, where the displacement decreases with time until the motion stops (Figure 6.6).

As an example, let's consider the pendulum in Figure 6.5 moving through a viscous medium, and thus experiencing a resistance to its motion. We can introduce this resistance in a fairly simple way by saying that the viscous force is proportional to but opposite the motion, $ds/dt = l\,d\theta/dt$. Equation 6.28 then becomes

$$\frac{d^2\theta}{dt^2} + \gamma \frac{d\theta}{dt} + \omega_0^2 \theta = 0 \tag{6.31}$$

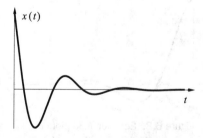

Figure 6.6. The function $x(t)$ given by Equation 6.30 plotted against t for $\phi = 1/2$, showing damped oscillatory behavior.

where γ is a frictional coefficient. Let's solve Equation 6.31 under the initial conditions $\theta(0) = \theta_0$ and $d\theta/dt = 0$. The auxiliary equation is $\alpha^2 + \gamma\alpha + \omega_0^2 = 0$, and yields

$$\alpha = -\frac{\gamma}{2} \pm \frac{1}{2}(\gamma^2 - 4\omega_0^2)^{1/2} \qquad (6.32)$$

You can see that the motion of the pendulum depends upon the relative values of γ^2 and $4\omega_0^2$.

If $\gamma^2 > 4\omega_0^2$, then the α's given in Equation 6.32 will be real and there will be no oscillatory behavior. The pendulum will simply fall to its vertical position and stop, describing motion in a very viscous medium. To see this behavior, let $\gamma = 3$ and $\omega_0^2 = 5/4$, just for concreteness, in which case $\alpha = -\frac{3}{2} \pm 1 = -5/2$ and $-1/2$. For the initial conditions, $\theta(0) = \theta_0$ and $\theta'(0) = 0$, it turns out that (Problem 6–15)

$$\theta(t) = \frac{\theta_0}{4}(5e^{-t/2} - e^{-5t/2}) \qquad (6.33)$$

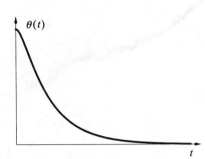

Figure 6.7. Equation 6.33 plotted against t. The resistance to motion is so great in this case that the pendulum simply falls to its vertical position without oscillating.

Equation 6.33 is plotted in Figure 6.7, where you can see that the pendulum simply falls to its vertical position without oscillating.

If $4\omega_0^2 > \gamma^2$, then α is a complex conjugate pair and the solution to Equation 6.31 is

$$\theta(t) = e^{-\gamma t/2}(c_1 \cos \omega t + c_2 \sin \omega t)$$

where $\omega = (4\omega_0^2 - \gamma^2)^{1/2}/2$. For concreteness only, take $\gamma = 1$ and $\omega_0^2 = 5/4$, in which case $\omega = 1$. Applying the initial conditions, $\theta(0) = \theta_0$ and $d\theta/dt = 0$, gives (Problem 6–16)

$$\theta(t) = \theta_0 e^{-t/2}\left(\cos t + \frac{1}{2}\sin t\right) \qquad (6.34)$$

Using the results of Problem 6–18, Equation 6.34 can be written as (Problem 6–19)

$$\theta(t) = \frac{\sqrt{5}\theta_0}{2}e^{-t/2}\cos(t - 0.4636) \qquad (6.35)$$

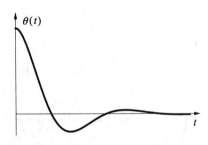

Figure 6.8. Equation 6.35 plotted against t. In this case, the pendulum undergoes a few oscillations before coming to rest at its vertical position.

Equation 6.35 is plotted against t in Figure 6.8. Note that the pendulum undergoes oscillations of decreasing displacement and then stops.

6.4 Two Invaluable Resources for Solutions to Differential Equations

The book *Ordinary Differential Equations and Their Solutions* by George M. Murphy (see the References at the end of the book) is to differential equations what a table of integrals is to integrals. The first half of the book reviews the methods that can be used to solve a great variety of differential equations, and the second half of the book lists over 200 pages of differential equations along with their solutions. The great feature of this list is that it is organized in such a way that any of the many differential equations included can be readily found. The tables include homogeneous and nonhomogeneous equations. Unfortunately, this 1960 book is out of print; nevertheless, it is such a great resource that it is worth checking to see if it is available in your library.

The second invaluable resource is a computer algebra system (CAS) such as *Mathematica*, *Maple*, or *MathCad*. These programs can solve differential equations analytically (often referred to as symbolically). For example, the one-line command

DSolve [y''[x] + 3 y'[x] + 2 y[x] == 12 x Exp[2 x], y[x], x]

in *Mathematica* gives the solution

$$y(x) = e^{2x} \left(-\frac{7}{12} + x \right) + c_1 e^{-2x} + c_2 e^{-x} \qquad (6.36)$$

to the nonhomogeneous differential equation

$$\frac{dy^2}{dx^2} + 3\frac{dy}{dx} + 2y = 12xe^{2x} \qquad (6.37)$$

These programs can also yield analytic solutions to systems of equations. For example, the *Mathematica* command

DSolve [{x'[t] == 4 x[t] - y[t] + Exp [-t], y'[t] == 5 x[t]
- 2 y[t] + 2 Exp[-t], x[0] == 0, y[0] == 0}, {x[t], y[t]}, t]

gives the solution

$$x(t) = -\frac{3}{16}e^{-t} + \frac{3}{16}e^{3t} + \frac{1}{4}te^{-t}$$

$$\qquad (6.38)$$

$$y(t) = -\frac{3}{16}e^{-t} + \frac{3}{16}e^{3t} + \frac{5}{4}te^{-t}$$

to the two simultaneous differential equations

$$\frac{dx}{dt} = 4x - y + e^{-t}$$

$$\qquad (6.39)$$

$$\frac{dy}{dt} = 5x - 2y + 2e^{-t}$$

with $x(0) = 0$ and $y(0) = 0$.

The ability of these CAS to solve differential equations symbolically is somewhat limited, but you should be aware that these systems can also solve differential equations numerically. We're not going to discuss numerical methods for solving differential equations in this book, but they are discussed in most texts on differential equations. (See the reference to Edwards and Penney in the References at the end of the book.) For example, *Mathematica* is unable to provide an analytic solution for the nonlinear equation

$$2y''(x) + y'(x) + 8y^3 = 0 \qquad (6.40)$$

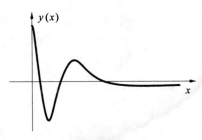

Figure 6.9. A plot of the numerical solution to Equation 6.40 with the initial conditions $y(0) = 1$ and $y'(0) = 0$.

but the command

NDSolve [{ 2 y"[x] + y'[x] + 8 y[x]^3 == 0, y[0] == 1,

y'[0] == 0}, y[x], {x, 0, 20 }]

provides a numerical solution with the initial conditions $y(0) = 1$, $y'(0) = 0$, over the interval $0 \leq x \leq 20$, which is plotted in Figure 6.9.

Neither of these resources is a substitute for an understanding of how solutions are obtained, but they are great supplements.

Problems

6–1. Find the general solutions of

(a) $\dfrac{dy}{dx} = x^2 - 3x^2 y$

(b) $\dfrac{dy}{dx} + \dfrac{2}{x} y = x^2 + 2$

(c) $t\dfrac{ds}{dt} = (3t + 1) s + t^3 e^{3t}$

(d) $(x + y^2)\dfrac{dy}{dx} = 1$

6–2. Find the solution of $x\dfrac{dy}{dx} + y = 2x$, where $y = 2$ when $x = 2$.

6–3. Solve the differential equation

$$\frac{dm}{dt} + \frac{4m}{20 + t} = 2$$

with the initial condition $m(0) = 20$.

6–4. In chemical kinetics and other types of rate processes, you frequently encounter the scheme $A \underset{k_2}{\overset{k_1}{\rightleftharpoons}} B$ representing the interconversion of two species, A and B. The rate equation for this interconversion can be written as $dA/dt = -k_1 A + k_2 B$, where k_1 and k_2 are called *rate constants*. By conservation of mass, $A(t) + B(t) = A_0 + B_0$, where $A_0 = A(0)$ and $B_0 = B(0)$. Solve the above equation for $A(t)$ and $B(t)$. Show that $B_{eq}/A_{eq} = k_1/k_2 = K_{eq}$.

6–5. A large container contains 100 liters of a salt solution whose concentration is 20 grams per liter. A salt solution of concentration 2 grams per liter is added to the container at a rate of 10 liters per minute, and the efflux from the container is 5 liters per minute. Calculate the minimum amount of salt in the container and when it will occur. Assume that the solution is stirred vigorously so that it is maintained at a uniform concentration.

6–6. A large container contains 100 L of a 2.00 molar solution. Pure water is pumped into the container at a rate of $2.00 \text{ L} \cdot \text{s}^{-1}$, and the resulting solution (assumed to be homogeneous) is pumped out at a rate of $1.00 \text{ L} \cdot \text{s}^{-1}$. How long will it take before the solution in the container is less than 0.10 molar?

6–7. Find the general solutions of

(a) $y''(x) - y'(x) - 2y(x) = 0$

(b) $y''(x) - 6y'(x) + 9y(x) = 0$

(c) $y''(x) + 4y'(x) + y(x) = 0$

6–8.   Find the solutions of
(a) $y''(x) - 4y(x) = 0$
(b) $y''(x) + 2y'(x) + 4y(x) = 0$
(c) $y''(x) + 9y(x) = 0$

where $y(0) = 1$ and $y'(0) = 1$ in each case.

6–9.   Find the solutions of
(a) $y''(x) + 6y'(x) = 0$
(b) $y''(x) - 4y'(x) + 3y(x) = 0$
(c) $y''(x) + 4y(x) = 0$

where $y(0) = 0$ and $y'(0) = 1$ in each case.

6–10. Solve
(a) $y''(x) - 4y(x) = 0$ given that $y(0) = 2$ and $y'(0) = 4$
(b) $y''(x) - 5y'(x) + 6y(x) = 0$ given that $y(0) = -1$ and $y'(0) = 0$
(c) $y'(x) - 2y(x) = 0$ given that $y(0) = 2$

6–11. Given that $y = x^2$ satisfies $x^2 y''(x) + x y'(x) - 4y(x) = 0$, use reduction of order to find a second solution.

6–12. Given that $y = x$ satisfies $x^2 y''(x) - x y'(x) + y(x) = 0$, use reduction of order to find a second solution.

6–13. Prove that $x(t) = \cos \omega t$ oscillates with a frequency $\nu = \omega/2\pi$. Prove that $x(t) = A \cos \omega t + B \sin \omega t$ oscillates with the same frequency, $\omega/2\pi$.

6–14. Solve the following initial value problems:

(a) $\dfrac{d^2 x}{dt^2} + \omega^2 x(t) = 0 \quad x(0) = 0; \ x'(0) = v_0$

(b) $\dfrac{d^2 x}{dt^2} + \omega^2 x(t) = 0 \quad x(0) = x_0; \ x'(0) = v_0$

Prove in both cases that $x(t)$ oscillates with frequency $\omega/2\pi$.

6–15. Verify Equation 6.33.

6–16. Verify Equation 6.34.

6–17. Show that $A \cos t + B \sin t$ can be written as $C \sin (t + \phi)$, where $C = (A^2 + B^2)^{1/2}$ and $\phi = \tan^{-1}(A/B)$. *Hint*: Work backward from the trigonometric identity $\sin (\alpha + \beta) = \sin \alpha \cos \beta + \cos \alpha \sin \beta$.

6–18. Show that $A \cos t + B \sin t$ can be written as $C \cos (t + \psi)$, where $C = (A^2 + B^2)^{1/2}$ and $\psi = \tan^{-1}(-B/A)$. *Hint*: Work backward from the trigonometric identity $\cos (\alpha + \beta) = \cos \alpha \cos \beta - \sin \alpha \sin \beta$.

6–19. Use the result of Problem 6–18 to derive Equation 6.35 from Equation 6.34.

6–20. Consider a body falling from a height h, and suppose that it encounters a resistance proportional to its velocity. Show that Newton's equation for this system is (see Figure 6.10)

$$m \frac{d^2 x}{dt^2} = -\gamma \frac{dx}{dt} + mg \qquad \text{or} \qquad \frac{dv}{dt} + \frac{\gamma}{m} v = g$$

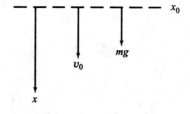

Figure 6.10. The geometry to be used in Problem 6–20.

where γ is a frictional coefficient. Solve this equation for $v(t)$, given that its velocity at $t = 0$ is zero. Show that v attains a limiting velocity given by $v_{\lim} = mg/\gamma$. This limiting velocity is called the *terminal velocity*.

6–21. Repeat the previous problem for a resistance that is proportional to the square of the velocity. Show that

$$v(t) = \left(\frac{mg}{\gamma}\right)^{1/2} \tanh \left(\frac{g\gamma}{m}\right)^{1/2} t$$

and that v attains a limiting velocity of $(mg/\gamma)^{1/2}$.

6–22. There is another way to solve the harmonic oscillator equation, using complex numbers. Show that this second-order equation can be written as two first-order equations:

$$\frac{dp}{dt} = -kx \qquad \text{and} \qquad \frac{dx}{dt} = \frac{p}{m}$$

Now let $\eta = k^{1/2}x + ip/m^{1/2}$, and show that these two equations are equivalent to

$$\frac{d\eta}{dt} = -i\omega\eta$$

where $\omega^2 = k/m$. Solve this equation to get

$$\eta = (k^{1/2}x_0 + im^{1/2}v_0)e^{-i\omega t}$$

or, upon equating real and imaginary parts,

$$x(t) = x_0 \cos \omega t + \frac{v_0}{\omega} \sin \omega t$$

$$p(t) = p_0 \cos \omega t - m\omega x_0 \sin \omega t$$

6–23. Verify that Equation 6.36 is a solution to Equation 6.37.

6–24. Verify that Equations 6.38 are solutions to Equations 6.39.

CHAPTER 7

Power Series Solutions of Differential Equations

In the previous chapter, we learned how to solve linear differential equations with constant coefficients. Generally, however, there is no method that allows us to find an analytic solution, at least in a finite number of steps, to a differential equation of the form

$$a_2(x)\frac{d^2 y}{dx^2} + a_1(x)\frac{dy}{dx} + a_3(x)y(x) = 0$$

whose coefficients are not constants, even if the a_j's are well-behaved functions of x. There *is* a method, however, that does allow us to solve the above equation in terms of a power series in x. In its simplest application, we assume that $y(x)$ is of the form

$$y(x) = \sum_{n=0}^{\infty} a_n x^n$$

then substitute it into the differential equation, and determine all the coefficients, $\{a_n\}$, in the power series. We can then use and manipulate the power series solution much as we would any simple analytic solution.

You may think that resorting to series solutions is not a particularly convenient approach, but it turns out that many of the most important differential equations of the physical sciences can be solved only in terms of infinite series. These equations are often second-order equations with nonconstant coefficients and are named after the mathematicians who introduced them and applied them to significant problems. Thus, we have Legendre's equation with its Legendre polynomials,

Bessel's equation with Bessel functions as its solutions, and a host of others. Even though these "name" functions are formally defined only through power series, it is possible to deduce many of their properties and the relations between them. In time, you can become as comfortable with these functions as you are with the trigonometric functions. In fact, if we were to *define* $\sin x$ and $\cos x$ formally as the odd and even power series solutions to the equation

$$y''(x) + y(x) = 0$$

we would still have our myriad of trigonometric identities.

7.1 The Power Series Method

We shall illustrate the method of solving differential equations by the power series method with a fairly simple example. Consider the equation

$$y''(x) + y(x) = 0 \tag{7.1}$$

We solved this equation several times in the previous chapter to obtain $y(x) = a\cos x + b\sin x$, but let's assume here that we are unable to solve it using the methods of the previous chapter. Now let's assume that $y(x)$ is a power series in x:

$$y(x) = \sum_{n=0}^{\infty} a_n x^n \tag{7.2}$$

Our task then is to determine the a_n in Equation 7.2 such that Equation 7.2 is a solution of Equation 7.1. Substitute Equation 7.2 into Equation 7.1 to obtain

$$\sum_{n=0}^{\infty} n(n-1) a_n x^{n-2} + \sum_{n=0}^{\infty} a_n x^n = 0 \tag{7.3}$$

Notice that the first summation here vanishes if $n = 0$ or $n = 1$, so Equation 7.3 can be written as

$$\sum_{n=2}^{\infty} n(n-1) a_n x^{n-2} + \sum_{n=0}^{\infty} a_n x^n = 0 \tag{7.4}$$

The first summation in Equation 7.4 written out explicitly is

$$2a_2 + 3 \cdot 2a_3 x + 4 \cdot 3a_4 x^2 + \cdots$$

Using this series as a guide, we can change the lower limit in the first summation in Equation 7.4 such that it starts at 0 by changing the summation index from n to $m = n - 2$. In this case, Equation 7.4 becomes

$$\sum_{m=0}^{\infty} (m+2)(m+1) a_{m+2} x^m + \sum_{n=0}^{\infty} a_n x^n = 0$$

Because m and n are just dummy summation indices, this result is equivalent to

$$\sum_{n=0}^{\infty} (n+2)(n+1) a_{n+2} x^n + \sum_{n=0}^{\infty} a_n x^n = 0$$

or

$$\sum_{n=0}^{\infty} [(n+2)(n+1)a_{n+2} + a_n]x^n = 0 \qquad (7.5)$$

(Changing summation indices takes a little practice, and Problem 7–1 gives you some.)

We now use the fact that if a power series vanishes over some interval, then each of the coefficients of the series must equal zero. Therefore, Equation 7.5 gives us

$$a_{n+2} = -\frac{a_n}{(n+2)(n+1)} \qquad n = 0, 1, 2, \ldots \qquad (7.6)$$

Equation 7.6 is a *recursion formula* for a_{n+2} in terms of a_n. As you let $n = 0, 1, 2, \ldots$ in Equation 7.6, you get the even-subscripted a's in terms of a_0 and the odd-subscripted a's in terms of a_1. Thus, we get two separate sets of coefficients. Starting with $n = 0$, we have for even values of n

$$a_2 = -\frac{a_0}{2\cdot 1} \qquad a_4 = -\frac{a_2}{4\cdot 3} = \frac{a_0}{4\cdot 3\cdot 2\cdot 1} = \frac{a_0}{4!} \qquad a_6 = -\frac{a_4}{6\cdot 5} = -\frac{a_0}{6!}$$

The general result is

$$a_{2n} = \frac{(-1)^n}{(2n)!}a_0 \qquad n = 0, 1, 2, \ldots \qquad (7.7)$$

Notice that we have determined only the even-subscripted coefficients (all in terms of a_0). To determine the odd-subscripted coefficients, we start with $n = 1$ and we use odd values of n in Equation 7.6:

$$a_3 = -\frac{a_1}{3\cdot 2} \qquad a_5 = -\frac{a_3}{5\cdot 4} = \frac{a_1}{5\cdot 4\cdot 3\cdot 2} = \frac{a_1}{5!} \qquad a_7 = -\frac{a_5}{7\cdot 6} = -\frac{a_1}{7!}$$

or

$$a_{2n+1} = \frac{(-1)^n}{(2n+1)!}a_1 \qquad n = 0, 1, 2, \ldots \qquad (7.8)$$

If we substitute Equations 7.7 and 7.8 into Equation 7.2, we obtain

$$y(x) = a_0 \sum_{n=0}^{\infty} \frac{(-1)^n}{(2n)!}x^{2n} + a_1 \sum_{n=0}^{\infty} \frac{(-1)^n}{(2n+1)!}x^{2n+1} \qquad (7.9)$$

Notice that the solution has two arbitrary constants as you would expect for the general solution for Equation 7.1. In fact, the two power series in Equation 7.9 are the Maclaurin series of $\cos x$ and $\sin x$ (Equations 3.13 and 3.14), respectively, so Equation 7.9 is

$$y(x) = a_0 \cos x + a_1 \sin x \qquad (7.10)$$

which we could have obtained easily since Equation 7.1 has constant coefficients. Nevertheless, we used the power series method to illustrate the procedure. Usually, you will not be able to identify the resulting power series solution with any known function, but will have to deal with the power series as such.

Suppose now that we did not know that the two solutions in Equation 7.9 were the familiar sine and cosine functions from trigonometry. We might define two functions

$$s(x) = \sum_{n=0}^{\infty} \frac{(-1)^n x^{2n+1}}{(2n+1)!} \quad \text{and} \quad c(x) = \sum_{n=0}^{\infty} \frac{(-1)^n x^{2n}}{(2n)!}$$

The first task might be to evaluate these series numerically as a function of x and plot them. (This is certainly a lot easier nowadays than it was when most of the special functions were first studied.) We then might notice that $s'(x) = c(x)$ and that $c'(x) = -s(x)$. With a little perseverance, you might notice that

$$c(x) \pm i\, s(x) = e^{\pm ix} = \sum_{n=0}^{\infty} \frac{(\pm ix)^n}{n!}$$

and so on. If $s(x)$ and $c(x)$ occurred in a number of different problems, then they would eventually be given names and become part of the mathematical literature.

EXAMPLE 7–1

Solve the equation

$$y''(x) + 3xy'(x) + 3y(x) = 0$$

by the power series method.

SOLUTION: Substitute Equation 7.2 into the above equation to obtain

$$\sum_{n=0}^{\infty} n(n-1)a_n x^{n-2} + 3x \sum_{n=0}^{\infty} n a_n x^{n-1} + 3 \sum_{n=0}^{\infty} a_n x^n = 0$$

Rewrite the first summation as $\sum_{n=0}^{\infty}(n+2)(n+1)a_{n+2}x^n$ and collect terms to get

$$\sum_{n=0}^{\infty} \{(n+2)(n+1)a_{n+2} + 3(n+1)a_n\} x^n = 0$$

Setting the coefficients of x^n equal to zero gives the recursion formula

$$a_{n+2} = -\frac{3}{n+2} a_n \qquad n = 0, 1, 2, \ldots$$

Starting with $n = 0$ and working with even values of n gives

$$a_2 = -\frac{3}{2} a_0 \qquad a_4 = -\frac{3}{4} a_2 = \frac{3^2}{4 \cdot 2} a_0$$

$$a_6 = -\frac{3}{6} a_4 = -\frac{3^3}{6 \cdot 4 \cdot 2} a_0 = -\frac{3^3}{2^3 3!} a_0$$

By continuing to determine more a_{2n}, the general term is found to be

$$a_{2n} = \frac{(-1)^n 3^n}{2^n n!} a_0$$

The odd-subscripted a's are

$$a_3 = -a_1 \qquad a_5 = -\frac{3}{5} a_3 = \frac{3}{5} a_1$$

$$a_7 = -\frac{3}{7} a_5 = -\frac{3^2}{7 \cdot 5} a_1 = -\frac{3^3}{7 \cdot 5 \cdot 3} a_0$$

The general term is

$$a_{2n+1} = \frac{(-1)^n 3^n}{(2n+1) \ldots 5 \cdot 3 \cdot 1} a_1$$

Substituting these results for the a_n into the power series gives

$$y(x) = a_0 \sum_{n=0}^{\infty} \frac{(-1)^n 3^n}{2^n n!} x^{2n} + a_1 \sum_{n=0}^{\infty} \frac{(-1)^n 3^n}{(2n+1)!!} x^{2n+1}$$

where we have introduced the standard notation $(2n+1)!! = (2n+1) \cdots 5 \cdot 3 \cdot 1$. Problem 7–8 has you show that each power series converges for all values of x. Can you identify the first series here as $e^{-3x^2/2}$?

7.2 Series Solutions of Legendre's Equation

Consider the equation

$$(1 - x^2) y''(x) - 2xy'(x) + \alpha(\alpha + 1) y(x) = 0 \qquad (7.11)$$

where $\alpha(\alpha + 1)$ is a constant, which with foresight we write in this form. Equation 7.11, which is called *Legendre's equation*, is one of the most famous and important differential equations in science and engineering. We shall learn in Chapter 16 that Equation 7.11 arises naturally in problems involving spherical coordinates, where $x = \cos\theta$, but it also occurs in a variety of other instances as well.

If we substitute

$$y(x) = \sum_{n=0}^{\infty} a_n x^n$$

into Equation 7.11, we find that (Problem 7–11)

$$a_{n+2} = -\frac{(\alpha - n)(\alpha + n + 1)}{(n+1)(n+2)} a_n \qquad n \geq 0 \qquad (7.12)$$

Starting with $n = 0$ and using even values of n, we find that

$$
\begin{aligned}
a_2 &= -\frac{\alpha(\alpha + 1)}{1 \cdot 2} a_0 \\
a_4 &= -\frac{(\alpha - 2)(\alpha + 3)}{3 \cdot 4} a_2 = \frac{\alpha(\alpha - 2)(\alpha + 1)(\alpha + 3)}{4!} a_0 \\
a_6 &= -\frac{(\alpha - 4)(\alpha + 5)}{5 \cdot 6} a_4 = -\frac{\alpha(\alpha - 2)(\alpha - 4)(\alpha + 1)(\alpha + 3)(\alpha + 5)}{6!} a_0
\end{aligned}
\tag{7.13}
$$

or generally

$$
a_{2n} = (-1)^n \frac{\alpha(\alpha - 2) \cdots (\alpha - 2n + 2)(\alpha + 1)(\alpha + 3) \cdots (\alpha + 2n - 1)}{(2n)!} a_0 \quad n \geq 1
\tag{7.14}
$$

Similarly, we find that

$$
\begin{aligned}
a_3 &= -\frac{(\alpha - 1)(\alpha + 2)}{3!} a_1 \\
a_5 &= \frac{(\alpha - 1)(\alpha - 3)(\alpha + 2)(\alpha + 4)}{5!} a_1
\end{aligned}
\tag{7.15}
$$

or generally

$$
a_{2n+1} = (-1)^n \frac{(\alpha - 1)(\alpha - 3) \cdots (\alpha - 2n + 1)(\alpha + 2)(\alpha + 4) \cdots (\alpha + 2n)}{(2n + 1)!} a_1 \quad n \geq 1
\tag{7.16}
$$

The two linearly independent solutions of Equation 7.11 then are

$$
y_1(x) = \sum_{n=0}^{\infty} a_{2n} x^{2n} \quad \text{and} \quad y_2(x) = \sum_{n=0}^{\infty} a_{2n+1} x^{2n+1}
\tag{7.17}
$$

For arbitrary values of α, both series of Equation 7.17 diverge at $x = \pm 1$. Yet in many applications, $x = \cos\theta$, where θ is the polar angle in spherical coordinates (see Chapter 14). Consequently, we often require a solution that is finite at the points $x = \pm 1$, which correspond to $\theta = 0$ and $\theta = \pi$. It turns out that this is actually easy to accomplish. If we set α equal to zero or a positive integer, then one of the two series in Equation 7.17 will truncate, resulting in a polynomial. We then eliminate the other (divergent) series by setting its coefficients equal to zero. Thus, we will always have a polynomial solution to Equation 7.11 for integer values of α.

To see that this is so, first let $\alpha = 0$. This gives $a_2 = a_4 = \cdots = 0$ in Equation 7.14, or $a_{2n} = 0$ for $n \geq 1$. Thus, we have $y_1(x) = a_0$, which clearly converges. Although $y_2(x)$ in Equation 7.17 does not converge for $\alpha = 0$, we can effectively eliminate it by setting $a_1 = 0$ in Equation 7.16. Then all the odd-subscripted coefficients are equal to zero, which makes $y_2(x)$ equal to zero. Therefore, the complete solution is $y(x) = a_0$ for $\alpha = 0$.

We now let $\alpha = 1$. In this case, $a_3 = a_5 = \cdots = 0$ in Equation 7.16, or $a_{2n+1} = 0$ for $n \geq 1$, which gives $y_2(x) = a_1 x$, which clearly converges. Although $y_1(x)$ in Equation 7.17 does not converge for $\alpha = 1$, if we set $a_0 = 0$, then all the even-subscripted coefficients are equal to zero, which makes $y_1(x) = 0$. Therefore, the complete solution is $y(x) = a_1 x$ for $\alpha = 1$.

We now let $\alpha = 2$. In this case, $a_4 = a_6 = \cdots = 0$, or $a_{2n} = 0$ for $n \geq 2$. This gives $y_1(x) = a_0 + a_2 x^2 = a_0(1 - 3x^2)$, where we used the fact that $a_2 = -\alpha(\alpha + 1)a_0/2 = -3a_0$ from Equation 7.13. If we set $a_1 = 0$, then all the odd-subscripted coefficients are equal to zero, which makes $y_2(x) = 0$. Therefore, the complete solution is $y(x) = a_0(1 - 3x^2)$ for $\alpha = 2$.

If we continue in this manner, we generate a polynomial solution to Equation 7.11 for $\alpha = 0, 1, 2, \ldots$ (Problem 7–12). If we denote these solutions by $f_n(x)$, where n is the value of α, then we have (setting the arbitrary constants a_0 and a_1 equal to unity)

$$f_0(x) = 1 \qquad f_1(x) = x$$

$$f_2(x) = 1 - 3x^2 \qquad f_3(x) = x - \frac{5x^3}{3} \tag{7.18}$$

and so on. Each of these polynomials is a solution of Equation 7.11 that is finite at $x = 1$.

EXAMPLE 7–2
Show that $f_3(x) = x - 5x^3/3$ is a solution to Equation 7.11 (with $\alpha = 3$).

SOLUTION:

$$f_3'(x) = 1 - 5x^2; \qquad f_3''(x) = -10x$$

Substitute these and $f_3(x)$ into Equation 7.11 with $\alpha = 3$ to get

$$(1 - x^2)f_3''(x) - 2xf_3'(x) + 12f_3(x) = -10x + 10x^3 - 2x + 10x^3 + 12x - 20x^3 = 0$$

Any multiple of the polynomials given in Equations 7.18 is also a solution to Equation 7.11, so we are free to choose any multiplicative constants. It's conventional to choose the constants such that $P_n(x) = c_n f_n(x) = 1$ when $x = 1$. Thus, we define

$$P_0(x) = 1 \quad P_1(x) = x \quad P_2(x) = \frac{1}{2}(3x^2 - 1) \quad P_3(x) = \frac{1}{2}(5x^3 - 3x) \quad (7.19)$$

and so on. The polynomials defined by Equations 7.19 are called *Legendre polynomials*. Note that the $P_n(x)$ are even or odd, depending upon the value of n. The first few Legendre polynomials are plotted in Figure 7.1.

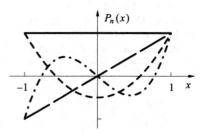

Figure 7.1. The first few Legendre polynomials plotted against x: $P_0(x)$ (solid), $P_1(x)$ (long dashed), $P_2(x)$ (short dashed), and $P_3(x)$ (long-short dashed). Note that $P_n(x)$ has n zeros between -1 and 1.

EXAMPLE 7–3
A set of functions $\{\phi_n(x)\}$ such that

$$\int_a^b \phi_n(x)\phi_m(x)\,dx = 0 \qquad \text{if } m \neq n$$

is said to be *orthogonal* over the interval (a, b). Show that the first few Legendre polynomials are orthogonal over the interval $(-1, 1)$.

SOLUTION: Using the above definition of orthogonality, we have

$$\int_{-1}^{1} P_0(x)P_1(x)\,dx = \int_{-1}^{1} P_0(x)P_3(x)\,dx = \int_{-1}^{1} P_1(x)P_2(x)\,dx = 0$$

because of the parity (in other words, the even-odd character) of the $P_n(x)$. In addition,

$$\int_{-1}^{1} P_0(x)P_2(x)\,dx = \frac{1}{2}\int_{0}^{1}(3x^2 - 1)\,dx = 0$$

$$\int_{-1}^{1} P_1(x)P_3(x)\,dx = \frac{1}{2}\int_{0}^{1}(5x^4 - 3x^2)\,dx = 0$$

You should keep in mind that when $\alpha = 0, 1, 2, \ldots$ in Equations 7.13 through 7.16, we obtain a polynomial solution from one of Equations 7.14 or 7.16 *plus* an infinite series that diverges at $x = \pm 1$ from the other. For example, if $\alpha = 0$, Equations 7.14 lead to $P_0(x) = 1$, while Equations 7.16 give

$$y_2(x) = a_1\left(x + \frac{x^3}{3} + \frac{x^5}{5} + \frac{x^7}{7} + \cdots\right) \tag{7.20}$$

This infinite series is equal to $(1/2)\ln[(1 + x)/(1 - x)]$ (Problem 7–16). It is customary to denote this second solution by $Q_0(x)$, so we write

$$Q_0(x) = \frac{1}{2}\ln\frac{1 + x}{1 - x} = x + \frac{x^3}{3} + \frac{x^5}{5} + \cdots \tag{7.21}$$

Thus, the complete solution to Equation 7.11 with $\alpha = 0$ is a linear combination of $P_0(x)$ and $Q_0(x)$ (Figure 7.2):

$$y(x) = c_1 P_0(x) + c_2 Q_0(x) \tag{7.22}$$

The general solution to Legendre's equation when α is zero or a positive integer is

$$y(x) = c_1 P_n(x) + c_2 Q_n(x) \qquad n = 0, 1, 2, \ldots \tag{7.23}$$

where $P_n(x)$ is the nth-degree polynomial that is finite for all values of x and $Q_n(x)$ is a logarithmic function that diverges at $x = \pm 1$. Because we want $y(x)$ to be finite at $x = \pm 1$ in the vast majority of applications, we choose $c_2 = 0$ and work with the Legendre polynomials. This happens so frequently that you tend to forget there are indeed nonpolynomial solutions to Legendre's equation even when n is an integer.

Before we leave this chapter, let's consider the equation

$$4x^2 y''(x) + (3x + 1)y(x) = 0 \tag{7.24}$$

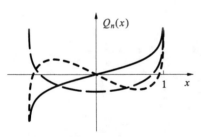

Figure 7.2. The first few second solutions, $Q_0(x)$ (solid), $Q_1(x)$ (long dash), and $Q_2(x)$ (short dash), to Legendre's equation plotted against x. Note that they all diverge at $x = \pm 1$.

When we substitute Equation 7.2 into this equation, we find that all the a_n's come out to be zero (Problem 7–21). There is no power series solution to Equation 7.24. Equations 7.1 and 7.11 yield two power series solutions and Equation 7.24 yields none. The only difference between these equations are the coefficients of the terms in $y(x)$ and its derivatives, so clearly there is some property of these coefficients that dictates when a power series solution can or cannot be obtained. There is actually a beautiful theory that describes the nature of the solutions to differential equations in terms of their coefficients. Instead of a simple power series as in Equation 7.2, many differential equations require a solution of the form

$$y(x) = x^r \sum_{n=0}^{\infty} a_n x^n$$

where r need not even be an integer. This theory is usually found under the method of Frobenius in differential equations books. We won't discuss this method because the purpose of this chapter is just to give a brief introduction to series solutions of differential equations (see, however, Problem 7–22).

Problems

7–1. Rewrite the following summations so that the first nonzero term starts at $n = 0$:

(a) $\sum_{n=1}^{\infty} n a_n x^{n-1}$ (b) $\sum_{n=2}^{\infty} n(n-1) a_n x^{n-2}$ (c) $\sum_{n=2}^{\infty} (n-2) c_{n-2} x^n$

7–2. Determine a general expression for a_n in terms of a_0 from the following recursion formulas:

(a) $a_{n+1} = -\dfrac{2a_n}{n+1}$ (b) $a_{n+1} = \dfrac{n+2}{2(n+1)} a_n$ (c) $a_{n+1} = -\dfrac{a_n}{(n+1)^2}$

Take $n \geq 0$ in each case.

7–3. Determine general expressions for a_{2n} in terms of a_0 and a_{2n+1} in terms of a_1 for $a_{n+2} = -a_n/(n+1)(n+2)$.

7–4. Solve the equation $y'(x) + y(x) = 0$ using the power series of the form of Equation 7.2.

7–5. Solve the equation $(1-x^2)y''(x) - 2xy'(x) + 2y(x) = 0$ using the power series of the form of Equation 7.2.

7–6. Solve the equation $(1-x^2)y''(x) - 6xy'(x) - 4y(x) = 0$ using the power series of the form of Equation 7.2.

7–7. Show that each series in the previous problem can be summed to

$$y(x) = \frac{a_0}{(1-x^2)^2} + \frac{a_1(3x - x^3)}{3(1-x^2)^2}$$

7–8. Show that each of the series solutions in Example 7–1 converges for all values of x.

7–9.   Determine the values of x for which each of the following series converge:

(a) $\displaystyle\sum_{n=1}^{\infty} \frac{(-1)^{n+1}x^n}{n}$ (b) $\displaystyle\sum_{n=0}^{\infty} \frac{(-1)^n x^{2n}}{(2n)!}$ (c) $\displaystyle\sum_{n=0}^{\infty} \frac{x^n}{2^n}$

7–10. The solutions to the equation $y''(x) + y(x) = 0$ are $\sin x$ and $\cos x$, an odd function and an even function, respectively. There must be some property of the differential equation that gives this result. Show that we can choose a function $y_1(x)$ such that if $y_1(x)$ is a solution, then so is $y_1(-x)$, and that $y_1(-x)$ is a constant multiple of $y_1(x)$. Show that this constant is ± 1, or that $y_1(x)$ must be either an even or an odd function of x.

7–11. Derive Equation 7.12.

7–12. Extend the polynomial solutions given in Equation 7.18 up to $f_5(x)$.

7–13. Show that the polynomials $f_4(x)$ and $f_5(x)$ that you generated in the previous problem satisfy Legendre's equation.

7–14. Use the polynomials $f_4(x)$ and $f_5(x)$ that you generated in Problem 7–12 to derive expressions for $P_4(x)$ and $P_5(x)$.

7–15. Show that the Maclaurin series of $\ln[(1+x)/(1-x)]/2$ is given by Equation 7.20.

7–16. Show that $\ln[(1+x)/(1-x)]$ is a solution to Legendre's equation for $\alpha = 0$.

7–17. Show that Legendre's equation can be written as $[(1-x^2)y'(x)]' + \alpha(\alpha+1)y(x) = 0$.

7–18. In this problem, we'll show in general that the Legendre polynomials are orthogonal over the interval $(-1, 1)$. Start with Legendre's equation in the previous problem for $\alpha = n$ and $\alpha = m$. Multiply the first by $P_m(x)$ and the second by $P_n(x)$ and integrate both results from -1 to $+1$ by parts. Now equate the results and show the orthogonality condition.

7–19. Argue that the solutions to Legendre's equation can be chosen to be either even or odd functions of x (see Problem 7–10). Are they?

7–20. Often, in physical problems, x in Equation 7.11 is equal to $\cos\theta$, where θ is the polar angle in spherical coordinates ($0 \le \theta \le \pi$). Express Equation 7.11 in terms of θ.

7–21. Show that all the $a_n = 0$ when Equation 7.2 is substituted into Equation 7.24.

7–22. In this problem, we shall develop a little of the theory behind power series solutions to differential equations. Consider the linear differential equation

$$A(x)y''(x) + P(x)y'(x) + Q(x)y(x) = 0 \qquad (1)$$

where $A(x)$, $P(x)$, and $Q(x)$ are polynomials containing no common factors. Suppose we want to solve equation 1 in some interval containing the point x_0. If $A(x_0) \neq 0$, then the point x_0 is called an *ordinary point*. In this case, we can divide equation 1 by $A(x)$ to obtain

$$y''(x) + p(x)y'(x) + q(x)y(x) = 0 \qquad (2)$$

where $p(x) = P(x)/A(x)$ and $q(x) = Q(x)/A(x)$. The functions $p(x)$ and $q(x)$ in equation 2 will usually be ratios of polynomials if $A(x)$, $P(x)$, and $Q(x)$ are polynomials, in which case they will have Taylor series expansions about the point x_0. We're now ready to state an important theorem concerning ordinary points:

If x_0 is an ordinary point of

$$y''(x) + p(x)y'(x) + q(x)y(x) = 0 \tag{3}$$

then the general solution of equation 3 consists of two linearly independent power series

$$y(x) = c_1 \sum_{n=0}^{\infty} a_n(x - x_0)^n + c_2 \sum_{n=0}^{\infty} b_n(x - x_0)^n$$

where c_1 and c_2 are arbitrary constants and the a's and b's are determined as in Section 7.1. Furthermore, the two power series have convergent Taylor series at $x = x_0$, and the interval of convergence of each one is at least as large as the smaller of the intervals of convergence of the series expansions of $p(x)$ and $q(x)$.

Let's consider the solution of

$$(1 - x^2)y''(x) - 6xy'(x) - 4y(x) = 0 \tag{4}$$

in the neighborhood of the point $x = 0$. This point is an ordinary point because $1 - x^2 \neq 0$ at $x = 0$. Show that

$$p(x) = -6 \sum_{n=0}^{\infty} x^{2n+1} \quad \text{and} \quad q(x) = -4 \sum_{n=0}^{\infty} x^{2n}$$

Show that the interval of convergence of each of these series is $-1 < x < 1$. It turns out that the power series solutions to equation 4 are (Problem 7–6)

$$y(x) = a_0 \sum_{n=0}^{\infty} (n + 1)x^{2n} + a_1 \sum_{n=0}^{\infty} \frac{2n + 3}{3} x^{2n+1} \tag{5}$$

Use the ratio test to show that the two series in equation 5 converge for $|x| < 1$.

7–23. Use the method of the previous problem to predict the interval of convergence of the power series solutions to $(1 + 4x^2)y''(x) - 8y(x) = 0$ about the point $x = 0$.

7–24. Given that $y(x) = e^{-3x^2/2}$ is one solution to the differential equation in Example 7–1, find the other solution using the method of reduction of order. Leave your result as an indefinite integral. Use a CAS to show that the power series of your result is the same as the one in Example 7–1.

CHAPTER 8

Orthogonal Polynomials

In Chapter 7, we learned that certain special functions such as Legendre polynomials can be defined as solutions to differential equations. These functions have become standard functions because they occur in a wide variety of physical problems. Legendre's differential equation and its solutions arise naturally in problems having spherical symmetry and occur in many other applications as well.

There are a number of other special functions that have become part of the repertoire of applied mathematics, and many of them form sets of orthogonal polynomials. In Section 8.1, we shall revisit Legendre polynomials and learn more about their properties. We'll see that they can be defined in a number of ways besides the solution of their differential equation (Legendre's equation). One of the most important properties of Legendre polynomials is that they can be used for the expansion of a fairly arbitrary function in the interval $-1 \leq x \leq 1$.

In Section 8.2, we'll develop a general theory of orthogonal polynomials that encompasses the Legendre polynomials as a special case. This will lead us to Laguerre polynomials, Hermite polynomials, Chebyshev polynomials, and others. The similar properties of all these special polynomials lie ultimately in their defining differential equations.

8.1 Legendre Polynomials

We saw in Chapter 7 that the Legendre polynomials are solutions to the differential equation

$$(1 - x^2)y''(x) - 2xy'(x) + n(n + 1)y(x) = 0 \qquad (8.1)$$

where n is an integer. The first few Legendre polynomials are

$$P_0(x) = 1 \qquad P_1(x) = x \qquad P_2(x) = \frac{1}{2}(3x^2 - 1)$$

$$P_3(x) = \frac{1}{2}(5x^3 - 3x) \qquad P_4(x) = \frac{1}{8}(35x^4 - 30x^2 + 3) \tag{8.2}$$

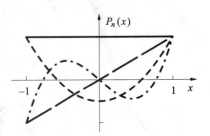

Figure 8.1. The Legendre polynomials, $P_0(x)$ (solid), $P_1(x)$ (long dashed), $P_2(x)$ (short dashed), and $P_3(x)$ (dash-dot).

and are plotted in Figure 8.1. Note that $P_n(x)$ has exactly n distinct zeros for $-1 < x < 1$. Note also that $P_n(x)$ is an even function if n is even and an odd function if n is odd.

It shouldn't be apparent at this point, but the Legendre polynomials form an orthogonal set of functions over the interval $(-1, 1)$. (However, see Problem 7–18.) For example,

$$\int_{-1}^{1} P_1(x) P_3(x) \, dx = \frac{1}{2} \int_{-1}^{1} x(5x^3 - 3x) \, dx = 0$$

Generally,

$$\int_{-1}^{1} P_n(x) P_m(x) \, dx = 0 \qquad \text{if } m \neq n \tag{8.3}$$

The orthogonality property of the Legendre polynomials follows from the differential equation that defines them. Let's go back to Equation 8.1. We first write it in the form (Problem 8–1)

$$-[(1 - x^2) P_n'(x)]' = n(n + 1) P_n(x) \tag{8.4}$$

with a similar equation for $P_m(x)$

$$-[(1 - x^2) P_m'(x)]' = m(m + 1) P_m(x) \tag{8.5}$$

Multiply Equation 8.4 by $P_m(x)$ and integrate from -1 to $+1$:

$$-\int_{-1}^{1} [(1 - x^2) P_n'(x)]' P_m(x) \, dx = n(n + 1) \int_{-1}^{1} P_n(x) P_m(x) \, dx$$

Integrate the left side by parts to obtain

$$-\left[(1 - x^2) P_n'(x) P_m(x)\right]_{-1}^{1} + \int_{-1}^{1} (1 - x^2) P_n'(x) P_m'(x) \, dx =$$

$$\int_{-1}^{1} (1 - x^2) P_n'(x) P_m'(x) \, dx = n(n + 1) \int_{-1}^{1} P_n(x) P_m(x) \, dx \tag{8.6}$$

Now multiply Equation 8.5 by $P_n(x)$, integrate both sides from -1 to $+1$, and then integrate the left side by parts to get

$$\int_{-1}^{1} (1 - x^2) P_n'(x) P_m'(x) \, dx = m(m + 1) \int_{-1}^{1} P_n(x) P_m(x) \, dx \tag{8.7}$$

Subtract Equation 8.7 from Equation 8.6 to get

$$[n(n + 1) - m(m + 1)] \int_{-1}^{1} P_n(x) P_m(x) \, dx = 0$$

If $n \neq m$, then we get Equation 8.3, the orthogonality condition of the Legendre polynomials.

The Legendre polynomials also satisfy a number of recursion formulas that can be used to express one Legendre polynomial in terms of others. For example,

$$(n+1)P_{n+1}(x) - (2n+1)x\,P_n(x) + n\,P_{n-1}(x) = 0 \tag{8.8}$$

EXAMPLE 8–1

Use Equation 8.8 to derive expressions for $P_2(x)$ and $P_3(x)$ from $P_0(x) = 1$ and $P_1(x) = x$.

SOLUTION: Let $n = 1$ in Equation 8.8:

$$2P_2(x) = 3x\,P_1(x) - P_0(x)$$

or

$$P_2(x) = \frac{1}{2}(3x^2 - 1)$$

For $n = 2$:

$$3P_3(x) = 5x\,P_2(x) - 2P_1(x)$$

or

$$P_3(x) = \frac{1}{2}(5x^3 - 3x)$$

The function

$$G(x, t) = \frac{1}{(1 - 2xt + t^2)^{1/2}} \tag{8.9}$$

is called a *generating function* for Legendre polynomials in the sense that

$$G(x, t) = \sum_{n=0}^{\infty} P_n(x)t^n \qquad |t| < 1 \tag{8.10}$$

The coefficient of t^n is the nth Legendre polynomial. Thus, $G(x, t)$ is said to generate the Legendre polynomials from its Maclaurin expansion in t.

EXAMPLE 8–2

Use Equation 8.9 to generate the first three Legendre polynomials.

SOLUTION: Use the expansion

$$(1 - z)^{-1/2} = 1 + \frac{z}{2} + \frac{3}{8}z^2 + \frac{5}{16}z^3 + \cdots$$

with $z = 2xt - t^2$.

$$G(x, t) = 1 + xt - \frac{t^2}{2} + \frac{3}{8}[4x^2t^2 - 4xt^3 + O(t^4)] + \frac{5}{16}[8x^3t^3 + O(t^4)]$$

$$= 1 + xt + \frac{3x^2 - 1}{2}t^2 + \frac{5x^3 - 3x}{2}t^3 + O(t^4)$$

$$= P_0(x) + P_1(x)t + P_2(x)t^2 + P_3(x)t^3 + O(t^4)$$

As you can see from Example 8–2, Equation 8.9 is awkward to use to generate Legendre polynomials, but it is useful for developing general properties of Legendre polynomials. For example, Problems 8–5 and 8–6 have you use Equation 8.9 to show that

$$\int_{-1}^{1} P_n^2(x)\, dx = \frac{2}{2n + 1} \tag{8.11}$$

Combining this result with Equation 8.3 gives

$$\int_{-1}^{1} P_n(x)P_m(x)\, dx = \frac{2}{2n + 1}\delta_{nm} \tag{8.12}$$

where δ_{nm} is a special symbol with the meaning

$$\delta_{nm} = \begin{cases} 0 & m \neq n \\ 1 & m = n \end{cases} \tag{8.13}$$

This symbol is called the *Kronecker delta* and occurs frequently in applied mathematics. We shall use the Kronecker delta often, and Problem 8–8 gives you some practice with manipulations involving it.

EXAMPLE 8–3
Show that the first few Legendre polynomials obey Equation 8.11.

SOLUTION:

$$\int_{-1}^{1} P_0^2(x)\, dx = \int_{-1}^{1} dx = 2 \qquad \int_{-1}^{1} P_1^2(x)\, dx = \int_{-1}^{1} x^2 dx = \frac{2}{3}$$

$$\int_{-1}^{1} P_2^2(x)\, dx = \frac{1}{4} \int_{-1}^{1} (3x^2 - 1)^2 dx = \frac{2}{5}$$

A useful property of Legendre polynomials, as well as other orthogonal polynomials that we shall encounter in the next section, is that it is possible to expand a suitably behaved function $f(x)$ as an infinite series of Legendre polynomials,

$$f(x) = \sum_{n=0}^{\infty} a_n P_n(x) \tag{8.14}$$

Such an expansion is called a *Fourier–Legendre series*. We can determine the a_n by multiplying both sides of Equation 8.14 by $P_m(x)$, integrating over x, and using Equation 8.12 to obtain (see Problem 8–8)

$$\int_{-1}^{1} f(x)P_m(x)\,dx = \sum_{n=0}^{\infty} a_n \int_{-1}^{1} P_n(x)P_m(x)\,dx = \sum_{n=0}^{\infty} \frac{2a_n}{2n+1}\delta_{nm}$$

$$= \frac{2a_m}{2m+1}$$

or

$$a_n = \frac{2n+1}{2}\int_{-1}^{1} f(x)P_n(x)\,dx \tag{8.15}$$

where we used $\sum_m c_m\delta_{nm} = c_n$ in going from the first line to the second line (see Problem 8–8).

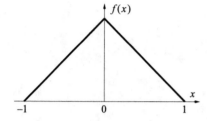

Figure 8.2. The function

$$f(x) = \begin{cases} 1+x & -1 \le x < 0 \\ 1-x & 0 \le x \le 1 \end{cases}$$

used in Example 8–4 plotted against x.

EXAMPLE 8–4
Expand the function (see Figure 8.2)

$$f(x) = \begin{cases} 1+x & -1 \le x < 0 \\ 1-x & 0 \le x \le 1 \end{cases}$$

in terms of Legendre polynomials.

SOLUTION:

$$f(x) = \sum_{n=0}^{\infty} a_n P_n(x)$$

where

$$a_n = \frac{2n+1}{2}\int_{-1}^{1} f(x)P_n(x)\,dx$$

Thus,

$$a_0 = \frac{1}{2}\left[\int_{-1}^{0}(1+x)\,dx + \int_{0}^{1}(1-x)\,dx\right] = \frac{1}{2}$$

Because $f(x)$ is an even function of x,

$$a_1 = \frac{3}{2}\left[\int_{-1}^{0} x(1+x)\,dx + \int_{0}^{1} x(1-x)\,dx\right] = 0 = a_3 = a_5 = \cdots$$

$$a_2 = \frac{5}{2}\left[\int_{-1}^{0}(1+x)P_2(x)\,dx + \int_{0}^{1}(1-x)P_2(x)\,dx\right]$$

$$= 5\int_{0}^{1}(1-x)P_2(x)\,dx = -\frac{5}{8}$$

$$a_4 = 9\int_{0}^{1}(1-x)P_4(x)\,dx = -\frac{9}{48}$$

You can use any CAS to calculate numerous a_n, and Figure 8.3 shows a few partial sums of the series compared to $f(x)$.

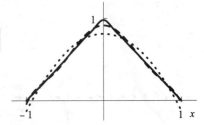

When we write an equation like Equation 8.14, there are several questions to consider. Does the series converge, and if so, for what values of x does it converge? Even if the right side does converge, does it equal $f(x)$? Here is a theorem that addresses these questions:

Figure 8.3. Partial sums of the series in Example 8–4 using 2 (dotted), 4 (dashed), and 16 (solid) nonzero terms. (See Figure 8.2 for a plot of $f(x)$.)

> If $f(x)$ and its first derivative are piecewise continuous in the interval $-1 < x < 1$, then Equation 8.14, with the a_n given by Equation 8.15, converges to $[\,f(x+) + f(x-)\,]/2$ for all x in the interval $-1 < x < 1$.

Note that $[\,f(x+) + f(x-)\,]/2$ is the average of $f(x)$ at the discontinuity if $f(x)$ has a jump discontinuity at x and is simply equal to $f(x)$ if $f(x)$ is continuous at x.

This theorem says that Equation 8.14 is valid even if $f(x)$ is discontinuous. It is remarkable that a Fourier–Legendre series, which is a series of continuous functions, can represent functions with discontinuities. A Maclaurin series, for example, requires not only that $f(x)$ be continuous, but that all its derivatives be continuous. We shall see in the next chapter that Fourier series can also represent discontinuous functions. As an example, consider the Heaviside step function

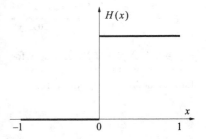

$$H(x) = \begin{cases} 0 & -1 \le x < 0 \\ 1 & 0 \le x \le 1 \end{cases} \qquad (8.16)$$

Note that $H(x)$ has a jump discontinuity at $x = 0$; $H(x-) = 0$, $H(x+) = 1$, and $[\,H(x+) + H(x-)\,]/2 = 1/2$ (see Figure 8.4). The coefficients a_n for the Fourier–Legendre series expansion of $H(x)$ are given by

Figure 8.4. The piecewise continuous Heaviside step function given by Equation 8.16.

$$a_n = \frac{2n+1}{2} \int_{-1}^{1} H(x) P_n(x)\,dx = \frac{2n+1}{2} \int_{0}^{1} dx\, P_n(x)$$

Using any computer algebra system, we can evaluate the a_n and plot partial sums of the expansion. Figure 8.5 shows partial sums for 5 and 20 terms. It turns out that $a_0 = 1/2$ and all the a_{2n} for $n \ge 1$ are equal to zero in this expansion. Do you see why (Problem 8–26)? Note that all three curves pass through the point $f(x) = [\,f(x+) + f(x-)\,]/2 = 1/2$ at $x = 0$.

Before we leave this section, we shall discuss one final topic. Suppose we approximate some function $f(x)$ by a finite number of terms in Equation 8.14. How can we assess the accuracy of the approximation? Consider the integral

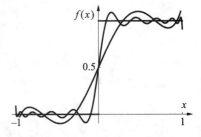

$$D_N^2 = \int_{-1}^{1} \left[f(x) - \sum_{n=0}^{N} \alpha_n P_n(x) \right]^2 dx \qquad (8.17)$$

Figure 8.5. The 5-term and 20-term partial sums of the Fourier–Legendre expansion of the Heaviside step function defined by Equation 8.16. Note that all three curves pass through the point $f(x) = [\,f(x+) + f(x-)\,]/2 = 1/2$ at $x = 0$.

The quantity D_N^2 is called the *mean square error* between $f(x)$ and the sum of the N terms in Equation 8.14. The magnitude of the mean square error is a measure

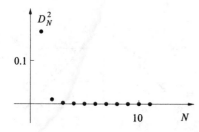

Figure 8.6. A plot of D_N^2 against N for the expansion given in Example 8–4.

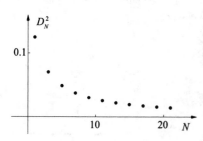

Figure 8.7. A plot of D_N^2 against N for the expansion of the Heaviside step function defined in Equation 8.16.

of the accuracy of the approximation of expressing $f(x)$ in terms of a sum of N Legendre polynomials. Problem 8–11 has you show that D_N^2 will be a minimum with respect to the α_n if they are equal to the a_n given by Equation 8.15. This result tells us that the "best approximation" of $f(x)$ by a finite sum of Legendre polynomials is given by

$$f(x) \approx \sum_{n=0}^{N} a_n P_n(x)$$

with the a_n given by Equation 8.15 if we define "best approximation" as the one that minimizes the mean square error between $f(x)$ and the finite sum. Figures 8.6 and 8.7 show D_N^2 plotted against N for the expansion in Example 8–4 and the expansion of $f(x)$ given by Equation 8.16. Note that the convergence of the continuous function in Example 8–4 is much faster than that of the discontinuous function given by Equation 8.16.

8.2 Orthogonal Polynomials

As we implied in the previous section, the Legendre polynomials are just one of a number of "name" polynomials that arise in applied mathematics. In this section, we shall present a general theory of orthogonal polynomials that encompasses any one of them as a special case.

Consider a set of functions $\phi_0(x)$, $\phi_1(x)$, ... This set is said to be orthogonal over an interval $a \leq x \leq b$ with a weight function $r(x) \geq 0$ if

$$\int_a^b r(x)\phi_i(x)\phi_j(x)\,dx = 0 \qquad i \neq j \tag{8.18}$$

Note the presence of the weight function $r(x)$, which is equal to one for the Legendre polynomials. If, in addition to Equation 8.18, we have

$$\int_a^b r(x)\phi_i^2(x)\,dx = 1 \tag{8.19}$$

the set $\{\phi_n(x)\}$ is said to be *orthonormal* with respect to the weight function $r(x)$ over the interval $a \leq x \leq b$. We can combine Equations 8.18 and 8.19 into one equation:

$$\int_a^b r(x)\phi_i(x)\phi_j(x)\,dx = \delta_{ij} \tag{8.20}$$

Equation 8.20 is the condition for the orthonormality of the $\{\phi_i(x)\}$.

The interval and the weight function specify the polynomials. We can see this by construction. Take $\phi_0(x) = 1$. Now let $\phi_1(x) = c_1 x + c_2$ and use the fact that

$$\int_a^b r(x)\phi_0(x)\phi_1(x)\,dx = 0 = c_1 \int_a^b r(x)x\,dx + c_2 \int_a^b r(x)\,dx \tag{8.21}$$

This equation gives us c_1 in terms of c_2, or $\phi_1(x)$ to within a multiplicative constant. We can determine this constant by requiring that $\phi_1(1) = 1$, as we did in the case of Legendre polynomials, or by requiring that $\phi_1(x)$ be normalized, or by any other convenient convention. To find $\phi_2(x)$, we let it be equal to $c_3x^2 + c_4x + c_5$ and require that it be orthogonal to $\phi_0(x)$ and $\phi_1(x)$. This gives $\phi_2(x)$ to within a multiplicative constant. Continuing this procedure generates a set of orthogonal polynomials. This procedure is called *Gram–Schmidt orthogonalization*.

EXAMPLE 8–5

Use Gram–Schmidt orthogonalization to generate the first few polynomials that are orthogonal over the interval $0 \leq x < \infty$ with weight function e^{-x}. Fix the multiplicative constant by requiring that $\phi_n(0) = n!$. (This is just a convention.)

SOLUTION: Let $\phi_0(x) = 1$ and $\phi_1(x) = c_1x + c_2$. Equation 8.21 gives

$$c_1 \int_0^\infty e^{-x} x \, dx + c_2 \int_0^\infty e^{-x} dx = c_1 + c_2 = 0$$

or $\phi_1(x) = c_1(1 - x)$, or $\phi_1(x) = 1 - x$. Requiring that $\phi_2(x) = c_3x^2 + c_4x + c_5$ be orthogonal to $\phi_0(x) = 1$ and $\phi_1(x) = 1 - x$ gives

$$c_3 \int_0^\infty e^{-x} x^2 dx + c_4 \int_0^\infty e^{-x} x \, dx + c_5 \int_0^\infty e^{-x} dx = 0$$

and

$$c_3 \int_0^\infty e^{-x}(1 - x)x^2 dx + c_4 \int_0^\infty e^{-x}(1 - x)x \, dx + c_5 \int_0^\infty e^{-x}(1 - x) dx = 0$$

These two equations give $2c_3 + c_4 + c_5 = 0$ and $-4c_3 - c_4 = 0$, and so $\phi_2(x) = c_3(x^2 - 4x + 2)$. The condition that $\phi_2(0) = 2!$ gives $c_3 = 1$, and so $\phi_2(x) = x^2 - 4x + 2$. Continuing, we find that

$$\phi_3(x) = -x^3 + 9x^2 - 18x + 6$$

These polynomials are called *Laguerre polynomials*, $L_n(x)$. Laguerre polynomials occur in the quantum-mechanical treatment of a hydrogen atom.

The defining intervals and weight functions of some other commonly occurring sets of orthogonal polynomials are given in Table 8.1. We devoted the entire first section of this chapter to the Legendre polynomials. The associated Laguerre polynomials reduce to the Laguerre polynomials (Example 8–5) when $\alpha = 0$ and occur in the quantum-mechanical treatment of a hydrogen atom. The Hermite polynomials, $H_n(x)$, occur in the quantum-mechanical treatment of a harmonic

Table 8.1. Some commonly occurring orthogonal polynomials.

Name	Symbol	Interval	Weight function
Legendre	$P_n(x)$	$-1 \leq x \leq 1$	1
Chebyshev(Tchebychef)	$T_n(x)$	$-1 \leq x \leq 1$	$(1-x^2)^{-1/2}$
Laguerre	$L_n(x)$	$0 \leq x < \infty$	e^{-x}
Associated Laguerre	$L_n^\alpha(x)$	$0 \leq x < \infty$	$x^\alpha e^{-x}$
Hermite	$H_n(x)$	$-\infty < x < \infty$	e^{-x^2}

oscillator, and the Chebyshev polynomials are used in numerical analysis. The first few Hermite polynomials, $H_n(x)$, are (Problem 8–14)

$$H_0(x) = 1 \qquad H_1(x) = 2x \qquad H_2(x) = 4x^2 - 2$$
$$H_3(x) = 8x^3 - 12x \qquad H_4(x) = 16x^4 - 48x^2 + 12 \tag{8.22}$$

Note that $H_n(-x) = (-1)^n H_n(x)$ and that the coefficient of x^n is 2^n (by convention).

EXAMPLE 8–6

Show that $H_0(x)$ is orthogonal to $H_1(x)$ and $H_2(x)$ with respect to the weight function e^{-x^2} over the interval $(-\infty, \infty)$.

SOLUTION:

$$\int_{-\infty}^{\infty} e^{-x^2} H_0(x)H_1(x)\,dx = 2\int_{-\infty}^{\infty} xe^{-x^2}\,dx = 0$$

because the integrand is an odd function of x.

$$\int_{-\infty}^{\infty} e^{-x^2} H_0(x)H_2(x)\,dx = \int_{-\infty}^{\infty} (4x^2 - 2)e^{-x^2}\,dx$$
$$= 8\int_0^{\infty} x^2 e^{-x^2}\,dx - 4\int_0^{\infty} e^{-x^2}\,dx$$
$$= 8\frac{\sqrt{\pi}}{4} - 4\frac{\sqrt{\pi}}{2} = 0$$

All orthogonal polynomials satisfy a recursion formula of a form similar to Equation 8.8. Table 8.2 lists the recursion formulas for the orthogonal polynomials in Table 8.1.

Table 8.2. The recursion formulas of the orthogonal polynomials listed in Table 8.1.

Recursion formula

$P_n(x)$	$(n+1)P_{n+1}(x) - (2n+1)x P_n(x) + n P_{n-1}(x) = 0 \qquad n \geq 1$
$T_n(x)$	$T_{n+1}(x) - 2x T_n(x) + T_{n-1}(x) = 0 \qquad n \geq 1$
$L_n(x)$	$L_{n+1}(x) + (x - 1 - 2n)L_n(x) + n^2 L_{n-1}(x) = 0 \qquad n \geq 1$
$L_n^\alpha(x)$	$(n+1-\alpha)L_{n+1}^\alpha + (n+1)(x + \alpha - 2n - 1)L_n^\alpha(x) + (n+1)n^2 L_{n-1}^\alpha(x) = 0 \qquad n \geq \alpha + 1$
$H_n(x)$	$H_{n+1}(x) - 2x H_n(x) + 2n H_{n-1}(x) = 0 \qquad n \geq 1$

EXAMPLE 8–7
Use the recursion formula in Table 8.2 and the fact that $T_0(x) = 1$, $T_1(x) = x$, and $T_2(x) = 2x^2 - 1$ to derive an expression for $T_3(x)$.

SOLUTION: From Table 8.2, with $n = 2$,

$$T_3(x) = 2x T_2(x) - T_1(x) = 2x(2x^2 - 1) - x$$
$$= 4x^3 - 3x$$

Given the recursion formulas, it is possible to derive generating functions. Table 8.3 lists the generating functions associated with the recursion formulas in Table 8.2.

Table 8.3. The generating functions of the orthogonal polynomials listed in Table 8.1.

Generating function

$P_n(x)$	$(1 - 2xt + t^2)^{-1/2} = \displaystyle\sum_{n=0}^{\infty} P_n(x)t^n \qquad |t| < 1$
$T_n(x)$	$\dfrac{1 - xt}{1 - 2xt + t^2} = \displaystyle\sum_{n=0}^{\infty} T_n(x)t^n \qquad |t| < 1$
$L_n(x)$	$\dfrac{e^{-xt/(1-t)}}{1-t} = \displaystyle\sum_{n=0}^{\infty} \dfrac{L_n(x)}{n!} t^n \qquad |t| < 1$
$L_n^\alpha(x)$	$(-t)^\alpha \dfrac{e^{-xt/(1-t)}}{(1-t)^{\alpha+1}} = \displaystyle\sum_{n=\alpha}^{\infty} \dfrac{L_n^\alpha(x)}{n!} t^n \qquad |t| < 1$
$H_n(x)$	$e^{2xt - t^2} = \displaystyle\sum_{n=0}^{\infty} \dfrac{H_n(x)}{n!} t^n$

EXAMPLE 8–8

Use the generating function in Table 8.3 to generate the first three Hermite polynomials.

SOLUTION:

$$e^{2xt-t^2} = 1 + (2xt - t^2) + \frac{(2xt - t^2)^2}{2!} + O(t^3)$$

$$= 1 + (2x)t + (2x^2 - 1)t^2 + O(t^3)$$

$$= H_0(x) + H_1(x)t + \frac{H_2(x)}{2!}t^2 + O(t^3)$$

Equating similar powers of t gives $H_0(x) = 1$, $H_1(x) = 2x$, and $H_2(x) = 4x^2 - 2$, in agreement with Equations 8.22.

Generating orthogonal polynomials using generating functions can become pretty laborious by hand as n increases, but any CAS can generate them with ease. As we pointed out in the previous section, generating functions can be used to determine values of integral conditions such as

$$\int_a^b r(x)\phi_n(x)\phi_m(x)\,dx = h_n\delta_{nm} \tag{8.23}$$

Problem 8–16 walks you through this for Hermite polynomials. Table 8.4 lists the integral conditions (Equation 8.23) for the orthogonal polynomials given in Table 8.1.

Table 8.4. The integral conditions (Equation 8.23) for the orthogonal polynomials listed in Table 8.1.

	Integral condition
$P_n(x)$	$\displaystyle\int_{-1}^{1} P_n(x)P_m(x)\,dx = \frac{2}{2n+1}\delta_{nm}$
$L_n^\alpha(x)$	$\displaystyle\int_0^\infty e^{-x}x^\alpha L_n^\alpha(x)L_m^\alpha(x)\,dx = \frac{(n!)^3}{(n-\alpha)!}\delta_{nm}$
$T_n(x)$	$\displaystyle\int_{-1}^{1} \frac{T_n(x)T_m(x)}{(1-x^2)^{1/2}}\,dx = \begin{cases} \dfrac{\pi}{2}\delta_{nm} & n \neq 0 \\ \pi\delta_{nm} & n = 0 \end{cases}$
$H_n(x)$	$\displaystyle\int_0^\infty e^{-x^2} H_n(x)H_m(x)\,dx = 2^n\sqrt{\pi}\,n!\delta_{nm}$

We can also use generating functions to derive the differential equations that the orthogonal polynomials satisfy. Table 8.5 lists the differential equation associated with each set of polynomials.

The final topic that we shall discuss in this section is the approximation of functions in terms of a sum of orthogonal polynomials. As we did in the previous

Table 8.5. The differential equations for the orthogonal
polynomials listed in Table 8.1.

Differential equation

$P_n(x)$	$(1 - x^2)y''(x) - 2xy'(x) + n(n + 1)y(x) = 0$
$T_n(x)$	$(1 - x^2)y''(x) - xy'(x) + n^2 y(x) = 0$
$L_n^\alpha(x)$	$xy''(x) + (\alpha + 1 - x)y'(x) + (n - \alpha)y(x) = 0$
$H_n(x)$	$y''(x) - 2xy'(x) + 2ny(x) = 0$

section for the Legendre polynomials, we seek the coefficients α_n such that

$$D_N^2 = \int_a^b r(x) \left[f(x) - \sum_{n=0}^{N} \alpha_n \phi_n(x) \right]^2 dx \qquad (8.24)$$

is minimized. If we differentiate D_N^2 with respect to the α_n and set the results equal
to zero, we find that (Problem 8–23)

$$\alpha_n = a_n = \frac{\int_a^b r(x)f(x)\phi_n(x)}{\int_a^b r(x)\phi_n^2(x)dx} \qquad (8.25)$$

Furthermore, if the integral of $r(x)f^2(x)$ is finite over $a \le x \le b$, then

$$\lim_{N\to\infty} D_N^2 = \lim_{N\to\infty} \int_a^b r(x) \left[f(x) - \sum_{n=0}^{N} a_n \phi_n(x) \right]^2 dx = 0 \qquad (8.26)$$

This type of convergence is called *convergence in the mean*.

As a final word, the definitions of the various orthogonal polynomials that we
have discussed in this chapter are not unique. Unfortunately, various authors use
various definitions. For example, some authors, particularly in the statistics literature, use a weight function of $e^{-x^2/2}$ instead of e^{-x^2} to define Hermite polynomials. Also, there are several definitions of the Laguerre polynomials and associated
Laguerre polynomials, which differ by whether or not the $n!$ is included in the
generating function definition (see Table 8.3) or whether $L_n(0)$ is chosen to be
equal to 1 or to $n!$. We have used the definitions found in most quantum chemistry books, which go back to Pauling and Wilson (see References at the end of
the book). All these definitions are just conventions and never affect any physical
results, but you've got to be careful when you jump from one source to another.

Problems

8–1. Show that Equations 8.1 and 8.4 are equivalent.

8–2. Show explicitly that $P_1(x)$ is orthogonal to $P_2(x)$ and to $P_4(x)$.

8–3. Use Equation 8.8 to derive an expression for $P_4(x)$ from $P_2(x)$ and $P_3(x)$.

8–4. Consider the two summations $S_1 = \sum_{n=1}^{3} a_n$ and $S_2 = \sum_{n=1}^{2} b_n$. It is important to
realize that in writing the product of S_1 and S_2 as a double summation, it is

necessary to use two different summation indices and to write the product as

$$S_1 S_2 = \sum_{n=1}^{3} \sum_{m=1}^{2} a_n b_m$$

Expand this double sum and show that it equals $S_1 S_2 = (a_1 + a_2 + a_3)(b_1 + b_2)$. Show that

$$S_1 S_2 \neq \sum_{n=1}^{3} \sum_{n=1}^{2} a_n b_n$$

Does this notation even make sense?

8–5. In this problem and the next problem, we're going to use the generating function of the Legendre polynomials (Equation 8.9) to show that they are orthogonal. Start with

$$G(x, t)G(x, u) = \sum_{n=0}^{\infty} \sum_{m=0}^{\infty} P_n(x)P_m(x)t^n u^m$$

$$= (1 - 2xt + t^2)^{-1/2}(1 - 2xu + u^2)^{-1/2}$$

Notice that we write the product $G(x, t)G(x, u)$ as a double summation with *different* summation indices. It would be wrong to write $G(x, t)G(x, u)$ as a summation of the form $\sum_{n=0}^{\infty} \sum_{n=0}^{\infty}$ (see Problem 8–4). Now argue that the Legendre polynomials are orthogonal if $\int_{-1}^{1} G(x, t)G(x, u)\,dx =$ function of the product of t and u and not upon t and u individually.

8–6. Carry out the integration in the previous problem and show explicitly that the result is a function of tu only. Use this result to verify Equation 8.11. The algebra here is a little involved; the final result is $\dfrac{1}{z} \ln \dfrac{1+z}{1-z}$ where $z = (ut)^{1/2}$.

8–7. Use the generating function in Equation 8.9 to show that $P_n(1) = 1$ and that $P_n(-1) = (-1)^n$.

8–8. Show that

$$\sum_{n=1}^{\infty} c_n \delta_{nm} = c_m \tag{1}$$

and that

$$\sum_{n} \sum_{m} a_n b_m \delta_{nm} = \sum_{n} a_n b_n = \sum_{m} a_m b_m \tag{2}$$

Equation 1 shows that the Kronecker delta is a discrete version of the Dirac delta function (Section 4.4).

8–9. The integral $I = \int_{-1}^{1} x P_n(x) P_m(x)\,dx$ occurs in atomic spectroscopy. Show that

$$I = \frac{2(n+1)}{(2n+1)(2n+3)} \delta_{m,n+1} + \frac{2n}{(2n+1)(2n-1)} \delta_{m,n-1}$$

8–10. The electrostatic potential at an arbitrary point in space due to a point charge at $x = 0$, $y = 0$, $z = 1$ is given by $1/R$, where R is the distance from the charge to the arbitrary point. Show that

$$\frac{1}{R} = \frac{1}{(1 - 2r\cos\theta + r^2)^{1/2}}$$

where r and θ are shown in Figure 8.8. Now show that

$$\frac{1}{R} = \sum_{n=0}^{\infty} \frac{P_n(\cos\theta)}{r^{n+1}} \qquad \text{if } r > 1$$

This result is important in electrostatics.

8–11. Show that D_N^2 given by Equation 8.17 is a minimum with respect to α_n if they are equal to the a_n given by Equation 8.15.

8–12. Expand the function $f(x) = \sin\pi x$ in a series of Legendre polynomials. Plot the first few partial sums. You should consider using a CAS for this problem.

8–13. Starting with Equation 8.17, show that

$$\int_{-1}^{1} f^2(x)\,dx \geq \sum_{n=0}^{N} \frac{2}{2n+1}\alpha_n^2$$

This inequality is called *Bessel's inequality* for Legendre polynomials.

8–14. Use Gram–Schmidt orthogonalization to generate the first few polynomials that are orthogonal and normalized with respect to e^{-x^2} over the interval $(-\infty, \infty)$ and use the convention that the coefficient of x^n is 2^n.

8–15. Derive an expression for $H_4(x)$ from $H_2(x)$ and $H_3(x)$ using the recursion formula in Table 8.2.

8–16. In this problem, we're going to use the generating function for Hermite polynomials given in Table 8.3 to determine the integral condition given in Table 8.4. First write out the product of $G(x, t)$ and $G(x, u)$, multiply by e^{-x^2}, and then integrate over x from $-\infty$ to ∞ to obtain

$$\int_{-\infty}^{\infty} e^{-x^2} e^{2xt-t^2} e^{2xu-u^2}\,dx = \sum_{n=0}^{\infty}\sum_{m=0}^{\infty} \frac{t^n}{n!}\frac{t^m}{m!} \int_{-\infty}^{\infty} e^{-x^2} H_n(x)H_m(x)\,dx \quad (1)$$

Notice that we write the product $G(x, t)G(x, u)$ as a double summation, with *different* summation indices. It would be wrong to write $G^2(x, t)$ as a summation of the form $\sum_{n=0}^{\infty}\sum_{n=0}^{\infty}$ (see Problem 8–4). You can evaluate the integral on the left by completing the square. Show that

$$\int_{-\infty}^{\infty} e^{-x^2} e^{2xt-t^2} e^{2xu-u^2}\,dx = e^{-2ut} \int_{-\infty}^{\infty} e^{-[x-(t+u)]^2}\,dx$$

$$= e^{-2ut} \int_{-\infty}^{\infty} e^{-z^2}\,dz = \sqrt{\pi}\,e^{-2ut}$$

Therefore, the left side of equation 1 is a function of the product ut only. What does this say about the integrals on the right side of equation 1? Now use the notation in Equation 8.23 to write equation 1 as

$$\sum_{n=0}^{\infty} \frac{(ut)^n}{(n!)^2} h_n = \sqrt{\pi}\,e^{-2ut}$$

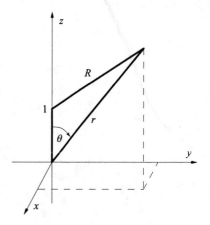

Figure 8.8. The geometry used for the derivation of the result in Problem 8–10.

Expand the function on the left in a Maclaurin series and finally show that

$$h_n = \int_{-\infty}^{\infty} e^{-x^2} H_n^2(x)\, dx = \sqrt{\pi}\, 2^n n!$$

in agreement with Table 8.4.

8–17. The integral $I = \int_{-\infty}^{\infty} e^{-x^2} H_n(x) x H_m(x)\, dx$ occurs in a discussion of the vibrational spectrum of a diatomic molecule when it is modeled as a harmonic oscillator. Show that this integral is equal to zero unless $m = n \pm 1$.

8–18. The average potential energy of a quantum-mechanical harmonic oscillator is directly related to the integral $I = \int_{-\infty}^{\infty} e^{-x^2} H_n(x) x^2 H_n(x)\, dx$. Show that $I = (n + \frac{1}{2})\sqrt{\pi}\, 2^n n!$.

8–19. Use the recursion formula for the Laguerre polynomials to verify the formula for $L_3(x)$ from the expressions in Example 8–5.

8–20. Use the generating function for the Laguerre polynomials in Table 8.3 to show that

$$\int_0^{\infty} e^{-x} L_n(x) L_m(x)\, dx = (n!)^2 \delta_{nm}$$

(See Problems 8–5, 8–6, and 8–16.)

8–21. The associated Laguerre polynomials can be generated by means of the formula

$$L_n^\alpha(x) = \frac{d^\alpha}{dx^\alpha} L_n(x)$$

Use this formula to generate $L_1^1(x)$, $L_2^1(x)$, $L_3^1(x)$, $L_2^2(x)$, $L_3^2(x)$, and $L_3^3(x)$ from $L_1(x)$, $L_2(x)$, and $L_3(x)$ given in Example 8–5.

8–22. Use the associated Laguerre polynomials that you generated in the previous problem to verify the integral condition in Table 8.4.

8–23. Derive Equation 8.25.

8–24. All the orthogonal polynomials in this chapter can be generated by a differentiation formula called a *Rodrigues formula*. For example, the Rodrigues formula for the Hermite polynomials is

$$H_n(x) = (-1)^n e^{x^2} \frac{d^n}{dx^n} e^{-x^2}$$

Use this formula to generate the first few Hermite polynomials.

8–25. A Rodrigues formula (see the previous problem) for the Laguerre polynomials is

$$L_n(x) = e^x \frac{d^n}{dx^n} (x^n e^{-x})$$

Use this formula to generate the four Laguerre polynomials in Example 8–5.

8–26. It turns out that $a_0 = 1/2$ and that all the coefficients a_{2n} with $n \geq 1$ in the Legendre expansion of the function defined in Equation 8.16 are equal to zero, although the function is not quite an odd function of x. Can you explain why this is so? *Hint:* Consider the function $f(x) - a_0$.

8–27. Use any CAS to reproduce Figures 8.3 and 8.6.

8–28. Use any CAS to reproduce Figures 8.5 and 8.7.

CHAPTER 9

Fourier Series

This chapter is devoted to a single topic, Fourier series, one of the most useful and important tools of applied mathematics. At the turn of the 19th century, the French mathematician and physicist Joseph Fourier analyzed the flow and the distribution of energy as heat in solid bodies. In the course of this work, Fourier found it necessary to express temperature distributions as infinite series of sines and cosines of the form

$$f(x) = \frac{a_0}{2} + \sum_{n=1}^{\infty} \left(a_n \cos \frac{n\pi x}{l} + b_n \sin \frac{n\pi x}{l} \right)$$

where the a_n and b_n are constants whose values depend upon $f(x)$, which is defined over the interval $(-l, l)$. This type of series is now called a *Fourier series*. It might not be unexpected that a Fourier series would converge to a function $f(x)$ if $f(x)$ is continuous over the interval, $-l < x < l$, but the amazing thing about a Fourier series is that $f(x)$ does not even have to be continuous. We shall see in this chapter that a Fourier series will converge to $f(x)$ even if $f(x)$ is discontinuous, such as might occur initially across the boundary of a hot solid quenched in a cold liquid. At the time, it was incredible that a series of continuous functions could converge to a discontinuous function, and Fourier's work was severely criticized. Nevertheless, Fourier's work not only survived almost two centuries of mathematical scrutiny, but has fostered several areas of modern mathematical research. A good website for Fourier series is *http://en.wikipedia.org/wiki/Fourier_series*.

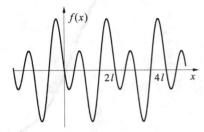

Figure 9.1. A periodic function with period $2l$.

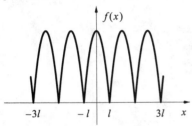

Figure 9.2. The periodic function defined by $f(x) = l^2 - x^2$ in the interval $(-l, l)$ and $f(x) = f(x + 2l)$ outside this interval.

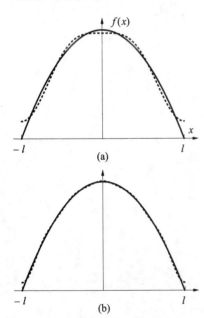

(a)

(b)

Figure 9.3. The function $f(x) = l^2 - x^2$ (solid) for $(-l, l)$ together with (a) the two-term partial sum (dotted) and (b) the five-term partial sum (dotted) of the Fourier series in Example 9–1.

9.1 Fourier Series As an Expansion in Orthogonal Functions

As we stated in the introduction to this chapter, Fourier made great use of expansions of the form

$$f(x) = \frac{a_0}{2} + \sum_{n=1}^{\infty} \left(a_n \cos \frac{n\pi x}{l} + b_n \sin \frac{n\pi x}{l} \right) \tag{9.1}$$

All the terms in Equation 9.1 are periodic functions of period $2l$ (Figure 9.1). (Recall that a function has period $2l$ if $f(x + 2l) = f(x)$ for all values of x.) Therefore, the series in Equation 9.1 is particularly suited for the expansion of functions of period $2l$. In fact, if $f(x)$ is defined on the interval $(-l, l)$ and has period $2l$, then the series in Equation 9.1 is called the *Fourier series* of $f(x)$ if the coefficients are given by

$$a_n = \frac{1}{l} \int_{-l}^{l} f(x) \cos \frac{n\pi x}{l} \, dx \qquad n = 0, 1, 2, \ldots \tag{9.2}$$

$$b_n = \frac{1}{l} \int_{-l}^{l} f(x) \sin \frac{n\pi x}{l} \, dx \qquad n = 1, 2, \ldots \tag{9.3}$$

These coefficients, called *Fourier coefficients*, are obtained using the orthogonality conditions

$$\int_{-l}^{l} \sin \frac{n\pi x}{l} \sin \frac{m\pi x}{l} \, dx = \int_{-l}^{l} \cos \frac{n\pi x}{l} \cos \frac{m\pi x}{l} \, dx = l \, \delta_{nm}$$

$$\int_{-l}^{l} \sin \frac{n\pi x}{l} \cos \frac{m\pi x}{l} \, dx = 0 \tag{9.4}$$

much like we did in the previous chapter. The inclusion of the "2" in the denominator of the a_0 term in Equation 9.1 allows us to use Equation 9.2 to calculate a_0. Otherwise, the integral for a_0 would have to be listed separately.

EXAMPLE 9–1
Determine the Fourier series of $f(x) = l^2 - x^2$ for $(-l, l)$ and $f(x) = f(x + 2l)$ outside this interval (Figure 9.2).

SOLUTION:

$$a_n = \frac{1}{l} \int_{-l}^{l} (l^2 - x^2) \cos \frac{n\pi x}{l} \, dx$$

$$= \frac{1}{n^3 \pi^3} \left[-2ln\pi x \cos \frac{n\pi x}{l} + (2l^2 + l^2 n^2 \pi^2 - n^2 \pi^2 x^2) \sin \frac{n\pi x}{l} \right]_{-l}^{l}$$

$$= \frac{4l^2}{\pi^2} \frac{(-1)^{n+1}}{n^2} \qquad n \neq 0$$

$$a_0 = \frac{1}{l} \int_{-l}^{l} (l^2 - x^2) \, dx = \frac{4l^2}{3}$$

The b_n are equal to zero because $(l^2 - x^2)\sin(n\pi x/l)$ is an odd function of x. Thus,

$$f(x) = \frac{2l^2}{3} + \frac{4l^2}{\pi^2} \sum_{n=1}^{\infty} \frac{(-1)^{n+1}}{n^2} \cos \frac{n\pi x}{l}$$

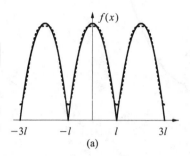
(a)

Figure 9.3 shows $f(x) = l^2 - x^2$ and the two- and five-term partial sums of the Fourier series representation of $f(x)$ over the interval $(-l, l)$. Realize that the Fourier series represents not only $f(x)$ in the interval $(-l, l)$ but its periodic extension as well (Figure 9.4).

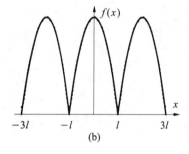
(b)

Figure 9.4. The periodic extension of Figure 9.3.

EXAMPLE 9–2
Determine the Fourier series of $f(x) = x$ in the interval $(-l, l)$ and $f(x + 2l) = f(x)$ outside this interval (Figure 9.5).

SOLUTION: The Fourier coefficients are

$$a_n = \frac{1}{l} \int_{-l}^{l} f(x)\cos \frac{n\pi x}{l}\, dx = \frac{1}{l} \int_{-l}^{l} x \cos \frac{n\pi x}{l}\, dx = 0$$

$$b_n = \frac{1}{l} \int_{-l}^{l} x \sin \frac{n\pi x}{l}\, dx = \frac{2}{l} \left[\frac{\sin(n\pi x/l)}{n^2\pi^2/l^2} - \frac{x\cos(n\pi x/l)}{n\pi/l} \right]_0^l$$

$$= \frac{(-1)^{n+1}2l}{n\pi}$$

The $a_n = 0$ because the integrand is an odd function of x in the interval $(-l, l)$. The Fourier series of $f(x)$ is

$$f(x) = \frac{2l}{\pi} \sum_{n=1}^{\infty} \frac{(-1)^{n+1}}{n} \sin \frac{n\pi x}{l}$$

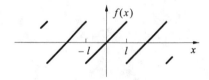

Figure 9.5. The periodic function defined by $f(x) = x$ in the interval $(-l, l)$ and by $f(x) = f(x + 2l)$ outside this interval.

Figure 9.6 shows $f(x)$ and two partial sums of its Fourier series representation plotted over the interval $(-l, l)$. Once again, realize that the Fourier series of $f(x)$ represents $f(x)$ not only over the interval $(-l, l)$ but over its periodic extensions as well. (See Figure 9.7.)

Notice that the rate of convergence in Example 9–2 is much slower than that in Example 9–1. The coefficients in Example 9–1 decay as $1/n^2$, whereas those in Example 9–2 decay as $1/n$. The reason for this is that $f(x)$ is a discontinuous function in Example 9–2 and $f(x)$ is a continuous function in Example 9–1. In a sense, it is easier for the continuous functions $\sin n\pi x/l$ and $\cos n\pi x/l$ to reproduce a continuous function than it is to reproduce a discontinuous function. In fact, it is remarkable that a Fourier series can represent a function with discontinuities.

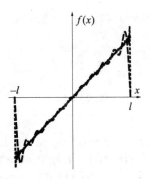

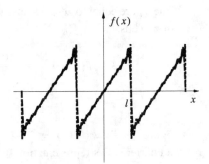

Figure 9.6. The function $f(x) = x$ (solid) in the interval $(-l, l)$ together with a 10-term (dashed), and a 100-term (dotted) partial sum of the Fourier series in Example 9–2 plotted over the interval $(-l, l)$.

Figure 9.7. The periodic extension of Figure 9.6.

A Maclaurin series, for example, requires not only that $f(x)$ be continuous, but that all its derivatives be continuous.

The Fourier series in Example 9–1 consisted of only cosines and the one in Example 9–2 consisted of only sines. The reason is that $f(x)$ in Example 9–1 is an even function of x and $f(x)$ in Example 9–2 is an odd function of x in the interval $(-l, l)$ (Problem 9–9). Generally, Fourier series consist of both sines and cosines, as the next Example shows.

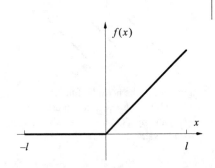

Figure 9.8. The function $f(x)$ from Example 9–3.

EXAMPLE 9–3

Determine the Fourier series for

$$f(x) = \begin{cases} 0 & -l \leq x < 0 \\ x & 0 \leq x < l \end{cases}$$

(See Figure 9.8.)

SOLUTION: The Fourier coefficients are given by

$$a_n = \frac{1}{l} \int_0^l x \cos \frac{n\pi x}{l}\, dx = \frac{l}{n^2 \pi^2}[(-1)^n - 1] \qquad n \neq 0$$

$$= \begin{cases} 0 & n \text{ even} \quad n \neq 0 \\ -\dfrac{2l}{n^2 \pi^2} & n \text{ odd} \quad n \neq 0 \end{cases}$$

$$a_0 = \frac{l}{2} \qquad n = 0$$

$$b_n = \frac{1}{l} \int_0^l x \sin \frac{n\pi x}{l}\, dx = \frac{(-1)^{n+1} l}{n\pi}$$

The Fourier series is

$$f(x) = \frac{l}{4} - \frac{2l}{\pi^2} \sum_{n=1}^{\infty} \frac{1}{(2n-1)^2} \cos \frac{(2n-1)\pi x}{l} + \frac{l}{\pi} \sum_{n=1}^{\infty} \frac{(-1)^{n+1}}{n} \sin \frac{n\pi x}{l}$$

The 10-term and 100-term partial sums of $f(x)$ over the interval $(-l, l)$ are shown in Figure 9.9. Note that the Fourier coefficients decay as $1/n$ because the periodic extension of this function is discontinuous at $x = \pm nl$, where n is an odd integer.

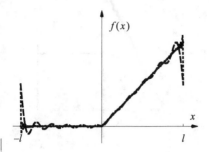

Figure 9.9. The function $f(x)$ (solid) from Example 9–3 together with a 10-term (dashed), and a 100-term (dotted) partial sum of the Fourier series in Example 9–3 plotted over the interval $(-l, l)$.

Frequently in physical problems, the independent variable represents time and the function to be expanded as a Fourier series is a periodic signal. Recall that $\sin t$ goes through one cycle as t goes from 0 to 2π (or from any point t_0 to $t_0 + 2\pi$). Therefore, $\sin 2\pi t$ goes through one cycle as t goes from 0 to 1, and $\sin 2\pi \nu t$ goes through ν cycles as t goes from 0 to 1. The function $\sin 2\pi \nu t$ represents a sinusoidal signal with a frequency of ν cycles per second, or ν *Hertz* (Hz). Because there are 2π radians in one cycle (one circle), $\omega = 2\pi \nu$ is the angular frequency in units of radians per second. If a signal has a frequency of ν cycles per second, then the time between successive maxima or minima is $1/\nu$ seconds per cycle, which means that the period, τ, of the signal is $\tau = 1/\nu$, or $\tau = 2\pi/\omega$.

Let's consider the square wave shown in Figure 9.10. We can express $f(t)$ mathematically by

$$f(t) = \begin{cases} -1 & -t_0 \leq t < 0 \\ 1 & 0 \leq t < t_0 \end{cases}$$

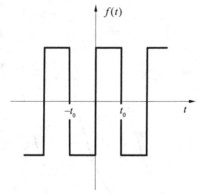

Figure 9.10. A square wave of period $2t_0$.

with $f(t) = f(t + \tau)$. The period of $f(t)$ is $\tau = 2t_0$ and its (angular) frequency is $\omega = 2\pi/\tau = \pi/t_0$. If we write $f(t)$ as

$$f(t) = \frac{a_0}{2} + \sum_{n=1}^{\infty} \left(a_n \cos \frac{n\pi t}{t_0} + b_n \sin \frac{n\pi t}{t_0} \right)$$

then

$$a_n = \frac{1}{t_0} \int_{-t_0}^{t_0} dt \, f(t) \cos \frac{n\pi t}{t_0} = 0$$

because the integrand is an odd function of t and

$$b_n = \frac{1}{t_0} \int_{-t_0}^{t_0} dt \, f(t) \sin \frac{n\pi t}{t_0} = \frac{2}{n\pi} [1 - (-1)^n]$$

Thus,

$$f(t) = \frac{2}{\pi} \sum_{n=1}^{\infty} \frac{[1 - (-1)^n]}{n} \sin \frac{n\pi t}{t_0} = \frac{4}{\pi} \sum_{n=1}^{\infty} \frac{1}{2n-1} \sin(2n-1)\omega t$$

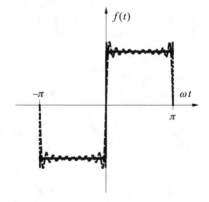

Figure 9.11. The square wave in Figure 9.10 and the 10-term partial sum of its Fourier series plotted against ωt.

where $\omega = \pi/t_0$. Figure 9.11 shows the 10-term partial sum of this series. Note the slow convergence due to the denominator being of order n. Figure 9.12 shows

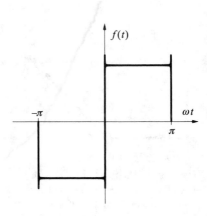

the partial sum consisting of 500 terms, showing that it is possible to represent a square wave by a Fourier series, provided enough terms are taken. (By the way, this is easily done using any CAS.)

All our examples so far have been on a symmetric interval $(-l, l)$. Suppose we want to expand $f(x) = x^2$ in $(0, 2l)$ with $f(x + 2l) = f(x)$ (see Figure 9.13). Problem 9–14 has you show that since $f(x)$ is periodic with period $2l$, the integrals for the Fourier coefficients (Equations 9.2 and 9.3) can be written as

$$a_n = \frac{1}{l} \int_{-l}^{l} f(x) \cos \frac{n\pi x}{l} \, dx = \frac{1}{l} \int_{-l+c}^{l+c} f(x) \cos \frac{n\pi x}{l} \, dx \qquad (9.5)$$

and

$$b_n = \frac{1}{l} \int_{-l}^{l} f(x) \sin \frac{n\pi x}{l} \, dx = \frac{1}{l} \int_{-l+c}^{l+c} f(x) \sin \frac{n\pi x}{l} \, dx \qquad (9.6)$$

where c is an arbitrary constant.

Figure 9.12. The partial sum consisting of 500 terms of the Fourier series of the square wave in Figure 9.10 plotted against ωt.

Therefore, the Fourier coefficents of the expansion of $f(x) = x^2$ in $(0, 2l)$ and $f(x) = f(x + 2l)$ are given by Equations 9.5 and 9.6 with $c = l$, or by

$$a_n = \frac{1}{l} \int_{0}^{2l} f(x) \cos \frac{n\pi x}{l} \, dx = \frac{1}{l} \int_{0}^{2l} x^2 \cos \frac{n\pi x}{l} \, dx = \frac{4l^2}{n^2 \pi^2} \qquad n \neq 0$$

$$a_0 = \frac{1}{l} \int_{0}^{2l} x^2 \, dx = \frac{8l^2}{3}$$

and

$$b_n = \frac{1}{l} \int_{0}^{2l} dx \, x^2 \sin \frac{n\pi x}{l} = -\frac{4l^2}{\pi n}$$

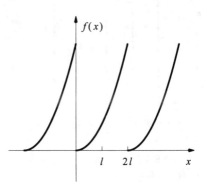

Thus,

$$f(x) = \frac{4l^2}{3} + \frac{4l^2}{\pi} \sum_{n=1}^{\infty} \left(\frac{1}{\pi n^2} \cos \frac{n\pi x}{l} - \frac{1}{n} \sin \frac{n\pi x}{l} \right)$$

Figure 9.13. The function defined on the interval $(0, 2l)$ by $f(x) = x^2$ and $f(x + 2l) = f(x)$.

The 5-term and 50-term partial sums of $f(x)$ are shown in Figure 9.14 for the interval $(0, 2l)$ and in Figure 9.15 for the periodic extension of $f(x)$. The $1/n$ terms in the Fourier series representation of $f(x)$ are due to the discontinuities of the periodic extension of $f(x)$ at $x = \pm 2nl$ with $n = 0, 1, 2, \ldots$.

EXAMPLE 9–4
Expand the function

$$f(x) = \begin{cases} 1 & -1/2 \leq x < 1/2 \\ 0 & 1/2 \leq x < 3/2 \end{cases}$$

with $f(x) = f(x + 2)$. (See Figure 9.16.)

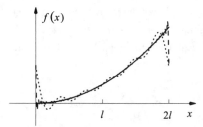

Figure 9.14. The function defined on the interval $(0, 2l)$ by $f(x) = x^2$ (solid) and the 5-term (dotted) and 50-term (dashed) partial sums of the Fourier series representation of $f(x)$.

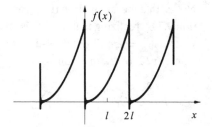

Figure 9.15. The 50-term Fourier series representation of $f(x)$ in Figure 9.14, showing that the Fourier series representation is a periodic function of period $2l$.

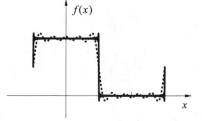

Figure 9.16. The function $f(x)$ from Example 9–4.

SOLUTION: We use Equations 9.5 and 9.6 with $l = 1$ and $c = 1/2$:

$$a_n = \int_{-1/2}^{3/2} f(x) \cos n\pi x \, dx = \int_{-1/2}^{1/2} \cos n\pi x \, dx = \frac{2}{n\pi} \sin \frac{n\pi}{2}$$

$$= \begin{cases} 0 & n \text{ even} \\ \dfrac{2(-1)^{n+1}}{n\pi} & n \text{ odd} \end{cases}$$

$$a_0 = 1$$

$$b_n = \int_{-1/2}^{1/2} \sin n\pi x \, dx = 0$$

Thus,

$$f(x) = \frac{1}{2} + \frac{2}{\pi} \sum_{n=1}^{\infty} \frac{(-1)^{n+1}}{2n - 1} \cos(2n - 1)\pi x$$

The 5-term and 50-term partial sums of $f(x)$ over the interval $(-1/2, 3/2)$ are plotted in Figure 9.17.

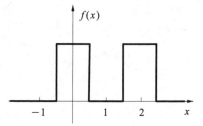

Figure 9.17. The partial sums of the Fourier series representation of $f(x)$ given in Example 9–4 consisting of 5 (dotted) and 50 (dashed) terms.

For simplicity only, most of the examples of Fourier series that we will discuss in this chapter will involve functions defined over a symmetric interval $(-l, l)$, but the above Example shows that it is not at all necessary that $f(x)$ be defined over a symmetric interval.

9.2 Complex Fourier Series

Another form of a Fourier series is the complex Fourier series

$$f(t) = \sum_{n=-\infty}^{\infty} c_n e^{in\omega_0 t} \qquad (9.7)$$

Notice that the summation runs from $-\infty$ to ∞. We can determine the c_n in Equation 9.7 by realizing that $\{e^{in\omega_0 t}\}$ is an orthogonal set over the interval $(-\tau/2, \tau/2)$ where $\omega_0 \tau = 2\pi$ (Problem 9–15). Multiply Equation 9.7 by $e^{-ik\omega_0 t}$ and integrate from $-\tau/2$ to $\tau/2$ to obtain

$$\int_{-\tau/2}^{\tau/2} f(t)e^{-ik\omega_0 t}\, dt = \sum_{n=-\infty}^{\infty} c_n \int_{-\tau/2}^{\tau/2} e^{i(n-k)\omega_0 t}\, dt = \sum_{n=-\infty}^{\infty} c_n \tau \delta_{nk} = c_k \tau$$

so

$$c_k = \frac{1}{\tau} \int_{-\tau/2}^{\tau/2} f(t)e^{-ik\omega_0 t}\, dt \tag{9.8}$$

Let's determine the complex Fourier series representation of $f(t) = t$ in the interval $(-\tau/2, \tau/2)$ with $f(t + \tau) = f(t)$. In this case,

$$c_k = \frac{1}{\tau} \int_{-\tau/2}^{\tau/2} te^{-ik\omega_0 t}\, dt = \frac{1}{\tau} \left[\frac{e^{-ik\omega_0 t}}{-k^2\omega_0^2}(-ik\omega_0 t - 1) \right]_{-\tau/2}^{\tau/2}$$

$$= \frac{i}{k\omega_0} \cos k\pi = \frac{(-1)^k i}{k\omega_0} \qquad k \neq 0$$

$$c_0 = \frac{1}{\tau} \int_{-\tau/2}^{\tau/2} t\, dt = 0$$

(Problem 9–16). Therefore,

$$f(t) = \frac{i}{\omega_0} \sum_{\substack{n=-\infty \\ \neq 0}}^{\infty} \frac{(-1)^n}{n} e^{in\omega_0 t} \tag{9.9}$$

9.3 Convergence of Fourier Series

The conditions under which a Fourier series converges to $f(x)$ are similar to those for a Fourier–Legendre series that we presented in the previous chapter. One set of conditions is

> If $f(x)$ and its first derivative are piecewise continuous on the interval $(-l, l)$ and have period $2l$, then the Fourier series of $f(x)$ converges to $f(x)$ at all points where it is continuous and to $[f(x+) + f(x-)]/2$ at points where it is discontinuous.

It turns out that these conditions are sufficient but not necessary.

Because $f(x) = [f(x+) + f(x-)]/2$ at points where it is continuous, we can express the above theorem mathematically by writing

$$\frac{1}{2}[f(x+) + f(x-)] = \frac{a_0}{2} + \sum_{n=1}^{\infty} \left(a_n \cos \frac{n\pi x}{l} + b_n \sin \frac{n\pi x}{l} \right) \tag{9.10}$$

where

$$a_n = \frac{1}{l} \int_{-l}^{l} f(x) \cos \frac{n\pi x}{l}\, dx \qquad b_n = \frac{1}{l} \int_{-l}^{l} f(x) \sin \frac{n\pi x}{l}\, dx \qquad (9.11)$$

if $f(x)$ satisfies the above conditions. Certainly, almost all functions that we encounter in physical problems will satisfy our conditions for convergence.

EXAMPLE 9–5

Use the Fourier series in Example 9–2 to derive

$$\frac{\pi}{4} = 1 - \frac{1}{3} + \frac{1}{5} - \frac{1}{7} + \frac{1}{9} + \cdots$$

SOLUTION: Let $x = l/2$ in the series to get

$$\frac{l}{2} = \frac{2l}{\pi} \sum_{n=1}^{\infty} \frac{(-1)^{n+1}}{n} \sin \frac{n\pi}{2}$$

$$= \frac{2l}{\pi}\left(1 - \frac{1}{3} + \frac{1}{5} - \frac{1}{7} + \frac{1}{9} + \cdots\right)$$

or

$$\frac{\pi}{4} = 1 - \frac{1}{3} + \frac{1}{5} - \frac{1}{7} + \frac{1}{9} + \cdots$$

This was the first series involving π ever discovered. It converges much too slowly to be useful for calculating the value of π, however, but it was a wonderfully curious result when it was first discovered.

We've observed several times that the rate of decay of Fourier coefficients depends upon whether or not the periodic extension of $f(x)$ is continuous. We found that if the periodic extension of $f(x)$ is discontinuous, then its Fourier coefficients decay as $1/n$, and that they decay at least as fast as $1/n^2$ if the periodic extension of $f(x)$ is continuous. We can formalize these observations in the following theorem:

If $f(x)$ and its first k derivatives are piecewise continuous on the interval $(-l, l)$ and have period $2l$ and if the periodic extensions of $f(x)$, $f'(x)$, $\ldots$, $f^{(k-1)}(x)$ are all piecewise continuous, then the Fourier coefficients of $f(x)$ decay at least as rapidly as $1/n^{k+1}$.

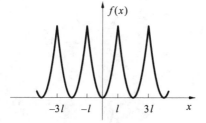

Figure 9.18. The function $f(x) = x^2$, $-l \le x < l$, and its periodic extension defined by $f(x + 2l) = f(x)$. Note that $f(x)$ is continuous for all values of x but that its first derivative is discontinuous at $x = \pm l$, $\pm 3l$, $\ldots$.

EXAMPLE 9–6

Determine the order of the Fourier coefficients of $f(x) = x^2$ on the interval $(-l, l)$.

SOLUTION: Figure 9.18 shows $f(x) = x^2$ in the interval $(-l, l)$ and its periodic extension. The function $f(x)$ is continuous, but its first derivative is discontinuous at odd multiples of $\pm l$. Therefore, $k = 1$ in the above theorem, and so we expect the coefficients in the Fourier series for $f(x)$ to decay as $1/n^2$ (at least) (Problem 9–22).

Problems

The first eight problems are a refresher of the evaluation of integrals that occur frequently in this chapter.

9–1. Use Euler's formula, $e^{ix} = \cos x + i \sin x$, to show that $\sin ax \ \sin bx = \frac{1}{2} \cos(a - b)x - \frac{1}{2} \cos(a + b)x$ and that $\cos ax \ \cos bx = \frac{1}{2} \cos(a - b)x + \frac{1}{2} \cos(a + b)x$.

9–2. Use Euler's formula, $e^{ix} = \cos x + i \sin x$, to show that $\sin^2 x = \frac{1}{2}(1 - \cos 2x)$ and that $\cos^2 x = \frac{1}{2}(1 + \cos 2x)$.

9–3. Use the relations in Problem 9–1 to show that

$$\int \sin ax \ \sin bx \ dx = \frac{\sin(a - b)x}{2(a - b)} - \frac{\sin(a + b)x}{2(a + b)} + c \text{ and that}$$

$$\int \cos ax \ \cos bx \ dx = \frac{\sin(a - b)x}{2(a - b)} + \frac{\sin(a + b)x}{2(a + b)} + c.$$

9–4. Use the relations in Problem 9–2 to show that $\int \sin^2 ax \ dx = \frac{x}{2} - \frac{1}{4a} \sin 2ax + c$ and that $\int \cos^2 ax \ dx = \frac{x}{2} + \frac{1}{4a} \sin 2ax + c$.

9–5. Use Euler's formula, $e^{ix} = \cos x + i \sin x$, to show that $\cos ax \ \sin bx = \frac{1}{2} \sin(a + b)x + \frac{1}{2} \sin(a - b)x$ and then show that $\int \sin ax \ \cos ax \ dx = \frac{1}{2a} \sin^2 ax + c_1 = -\frac{\cos^2 ax}{2a} + c_2$, where $c_2 = \frac{1}{2a} + c_1$, and

$$\int \sin ax \ \cos bx \ dx = -\frac{\cos(a - b)x}{2(a - b)} - \frac{\cos(a + b)x}{2(a + b)} + c.$$

9–6. Use the results of Problems 9–1 through 9–5 to show that $\int_{-\pi}^{\pi} \sin nx \ \sin mx \ dx = \int_{-\pi}^{\pi} \cos nx \ \cos mx \ dx = \int_{-\pi}^{\pi} \sin nx \ \cos mx \ dx = 0$, when m and n are integers and $m \neq n$, and $\int_{-\pi}^{\pi} \sin^2 nx \ dx = \int_{-\pi}^{\pi} \cos^2 nx \ dx = \pi$, when $n = m$.

9–7. Generalize the result of Problem 9–6 to Equation 9.4.

9–8. Use integration by parts to show that $\int u \cos au \ du = \frac{u \sin au}{a} + \frac{\cos au}{a^2} + c$ and that $\int u \sin au \ du = \frac{\sin au}{a^2} - \frac{u \cos au}{a} + c$.

9–9. Show that the $a_n = 0$ in Equation 9.2 if $f(x)$ is an odd function and that the $b_n = 0$ in Equation 9.3 if $f(x)$ is an even function.

9–10. Find the Fourier series of $f(x) = \begin{cases} 0 & -\pi \leq x < 0 \\ 1 & 0 \leq x < \pi \end{cases}$.

9–11. Find the Fourier series of $f(x) = x^2$ for $-\pi \leq x < \pi$.

9–12. Find the Fourier series of $f(x) = \begin{cases} \dfrac{x}{l} & 0 \le x < l \\ \dfrac{2l - x}{l} & l \le x < 2l \end{cases}$.

9–13. Find the Fourier series of $f(x) = \begin{cases} 0 & -2 \le x < 0 \\ 2 & 0 \le x < 2 \end{cases}$.

9–14. Verify Equations 9.5 and 9.6.

9–15. Show that $\{e^{in\omega_0 t}\}$ is an orthogonal set over the interval $(-\tau/2, \tau/2)$ where $\omega_0 \tau = 2\pi$ and n is an integer that goes from $-\infty$ to ∞.

9–16. Verify Equation 9.9.

9–17. Use a CAS to plot the partial sums of the Fourier series representation in Problem 9–10.

9–18. Use a CAS to plot the partial sums of the Fourier series representation in Problem 9–11.

9–19. Use a CAS to plot the partial sums of the Fourier series representation in Problem 9–12.

9–20. Evaluate the series in Example 9–1 at $x = 0$ and $x = l$. Do these answers make sense? *Hint*: Use the fact that $\sum_{n=1}^{\infty} 1/n^2 = \pi^2/6$ and that $\sum_{n=1}^{\infty} (-1)^{n+1}/n^2 = \pi^2/12$.

9–21. The Fourier series of $f(x) = x^2$ in the interval $(0, 2\pi)$ with $f(x + 2\pi) = f(x)$ is

$$f(x) = \frac{4\pi^2}{3} + \sum_{n=1}^{\infty} \left(\frac{4}{n^2} \cos nx - \frac{4\pi}{n} \sin nx \right)$$

What does $f(2\pi)$ equal? Does this value make sense? What does $f(0)$ equal? Does this value make sense? *Hint*: You need to use $\sum_{n=1}^{\infty} 1/n^2 = \pi^2/6$.

9–22. Verify that the Fourier coefficients of $f(x) = x^2$ in the interval $(-l, l)$ decay as $1/n^2$.

9–23. What do you predict for the n dependence of the Fourier coefficients of

(a) $|\cos x|$ over $(-\pi, \pi)$ (b) $f(x) = \begin{cases} -1 & -l \le x < 0 \\ 1 & 0 \le x < l \end{cases}$

(c) $x^3 - x$ over $(-1, 1)$ (d) $(x - 1)^2$ over $(-1, 1)$

9–24. Find a polynomial over the interval $(0, 1)$ whose Fourier coefficients decay at least as rapidly as $1/n^4$.

CHAPTER 10

Fourier Transforms

Until the 1970s, Fourier transforms were not part of the vocabulary of very many chemists. With the advent of Fourier transform spectroscopy, however, they are now used by all chemists, not just by a few theorists and physical chemists. In this chapter, we shall see just what a Fourier transform is and why they have permeated so much of chemistry.

10.1 Fourier's Integral Theorem

A Fourier transform arises in a natural way by inquiring about the behavior of a Fourier series as the period of the function being expanded goes to infinity. In other words, what form does a Fourier series take on if the function being expanded is not periodic? To address this question, let's start with the complex form of a Fourier series (Equation 9.7):

$$f(t) = \sum_{n=-\infty}^{\infty} c_n e^{in\omega_0 t} \tag{10.1}$$

where $f(t)$ is a periodic function of period $\tau = 2\pi/\omega_0$ and (Equation 9.8)

$$c_n = \frac{1}{\tau} \int_{-\tau/2}^{\tau/2} f(t) e^{-in\omega_0 t} \, dt \tag{10.2}$$

Now replace the multiplicative factor $1/\tau$ by $\omega_0/2\pi$ and substitute Equation 10.2 into Equation 10.1 to yield

$$f(t) = \sum_{n=-\infty}^{\infty} \left[\frac{\omega_0}{2\pi} \int_{-\tau/2}^{\tau/2} f(u) e^{-in\omega_0 u} \, du \right] e^{in\omega_0 t} \tag{10.3}$$

We have used u as the dummy integration variable in the expression for c_n so that we don't mix it up with the t in Equation 10.1. If we let τ become very large, then ω_0 becomes very small. Denote ω_0 by $\Delta\omega$ and write Equation 10.3 as

$$f(t) = \frac{1}{2\pi} \sum_{n=-\infty}^{\infty} F(n\Delta\omega)\Delta\omega \tag{10.4}$$

where

$$F(n\Delta\omega) = \int_{-\tau/2}^{\tau/2} f(u)e^{in\Delta\omega(t-u)}\, du \tag{10.5}$$

The limit of Equation 10.4 as $\Delta\omega \to 0$ looks just like the Riemann sum definition of an integral (Equation 2.1), so we denote $n\Delta\omega$ by ω and write the limit of Equations 10.4 and 10.5 as $\Delta\omega \to 0$ ($\tau \to \infty$) as

$$f(t) = \frac{1}{2\pi} \int_{-\infty}^{\infty}\int_{-\infty}^{\infty} f(u)e^{i\omega(t-u)}\, du\, d\omega \tag{10.6}$$

Equation 10.6 is a central formula of Fourier transforms and is called *Fourier's integral theorem*. It leads directly to a Fourier transform pair. If we let

$$\frac{1}{(2\pi)^{1/2}} \int_{-\infty}^{\infty} f(u)e^{-i\omega u}\, du = \hat{F}(\omega) \tag{10.7}$$

then Equation 10.6 becomes

$$f(t) = \frac{1}{(2\pi)^{1/2}} \int_{-\infty}^{\infty} \hat{F}(\omega)e^{i\omega t}\, d\omega \tag{10.8}$$

Equations 10.7 and 10.8 constitute a *Fourier transform pair*. We say that $\hat{F}(\omega)$ is the Fourier transform of $f(t)$.

10.2 Some Fourier Transform Pairs

For example, if

$$f(t) = e^{-\alpha|t|} \qquad -\infty < t < \infty \tag{10.9}$$

then

$$\hat{F}(\omega) = \frac{1}{(2\pi)^{1/2}} \int_{-\infty}^{\infty} e^{-\alpha|t|}e^{-i\omega t}\, dt$$

$$= \frac{1}{(2\pi)^{1/2}} \left\{ \int_{-\infty}^{\infty} e^{-\alpha|t|}\cos \omega t\, dt - i \int_{-\infty}^{\infty} e^{-\alpha|t|}\sin \omega t\, dt \right\}$$

Because $e^{-\alpha|t|}$ is an even function of t, only the first term survives and we have

$$\hat{F}(\omega) = \frac{2}{(2\pi)^{1/2}} \int_{0}^{\infty} e^{-\alpha t}\cos \omega t\, dt = \left(\frac{2}{\pi}\right)^{1/2} \frac{\alpha}{\omega^2 + \alpha^2} \tag{10.10}$$

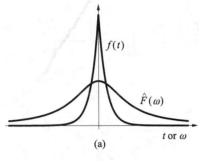

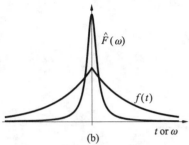

(a)

(b)

Figure 10.1. The Fourier transform pair, $f(t) = e^{-\alpha|t|}$ and $\hat{F}(\omega) = \alpha(2/\pi)^{1/2}/(\omega^2 + \alpha^2)$ plotted for (a) $\alpha = 2$ and (b) $\alpha = 1/2$. Note the reciprocal nature of the widths of the plots of $f(t)$ and $\hat{F}(\omega)$.

Equation 10.10 is often called a *cosine transform*. We can recover $f(t)$ from $\hat{F}(\omega)$ by using Equation 10.8:

$$f(t) = \frac{1}{(2\pi)^{1/2}} \int_{-\infty}^{\infty} \left(\frac{2}{\pi}\right)^{1/2} \frac{\alpha}{\omega^2 + \alpha^2} e^{i\omega t} \, d\omega$$

$$= \frac{2\alpha}{\pi} \int_0^{\infty} \frac{\cos \omega t}{\omega^2 + \alpha^2} \, d\omega = e^{-\alpha|t|}$$

where we have recognized that $\alpha/(\omega^2 + \alpha^2)$ is an even function of ω. Thus, we see that $e^{-\alpha|t|}$ and $(2/\pi)^{1/2}\alpha/(\omega^2 + \alpha^2)$ constitute a Fourier transform pair. If you plot Equations 10.9 and 10.10 for various values of α, you'll see that there is a reciprocal relation between the widths of the plots (Problem 10–4). In other words, as one curve gets wider (narrower), the other curve gets narrower (wider). We can see this reciprocal nature in Figure 10.1, where $f(t) = e^{-\alpha|t|}$ and $\hat{F}(\omega) = \alpha(2/\pi)^{1/2}/(\omega^2 + \alpha^2)$ are plotted together for $\alpha = 2$ (in part a) and $\alpha = 1/2$ (in part b). Table 10.1 lists a few other Fourier transform pairs.

Starting from Equation 10.6, we could have defined $\hat{F}(\omega)$ with a factor of $1/2\pi$ in front instead of $1/(2\pi)^{1/2}$, in which case Equation 10.8 would not have any factor in front. We chose to split the factor of $1/2\pi$ in Equation 10.6 between $f(t)$ and $\hat{F}(\omega)$ in Equations 10.7 and 10.8 in order to make these two equations more symmetric. Some authors use the less symmetric definitions, so you must always be aware just what the definition of $\hat{F}(\omega)$ is before you go from source to source.

Table 10.1. Some Fourier transform pairs. In all cases, $a > 0$.

$f(t)$	$\hat{F}(\omega)$	$f(t)$	$\hat{F}(\omega)$
$e^{i\omega_0 t}$	$(2\pi)^{1/2}\delta(\omega - \omega_0)$	$e^{-a^2 t^2}$	$\dfrac{1}{(2a^2)^{1/2}}e^{-\omega^2/4a^2}$
$e^{-a|t|}$	$\left(\dfrac{2}{\pi}\right)^{1/2}\dfrac{a}{\omega^2 + a^2}$	$\dfrac{1}{t^2 + a^2}$	$\left(\dfrac{\pi}{2a^2}\right)^{1/2}e^{-a|\omega|}$

EXAMPLE 10–1
Find the Fourier transform of the Gaussian distribution (Equation 4.19)

$$f(t) = \frac{1}{(2\pi\sigma^2)^{1/2}}e^{-t^2/2\sigma^2}$$

and verify your result.

SOLUTION: Using ω as the transform variable, we have

$$\hat{F}(\omega) = \frac{1}{2\pi\sigma} \int_{-\infty}^{\infty} e^{-i\omega t} e^{-t^2/2\sigma^2} \, dt$$

$$= \frac{1}{2\pi\sigma} \int_{-\infty}^{\infty} (e^{-t^2/2\sigma^2} \cos \omega t + i e^{-t^2/2\sigma^2} \sin \omega t) \, dt$$

The second term here vanishes because the integrand is an odd function of t, so using the fact that the first integrand is an even function of t gives

$$\hat{F}(\omega) = \frac{1}{\pi\sigma} \int_0^\infty e^{-t^2/2\sigma^2} \cos\omega t \, dt = \frac{1}{(2\pi)^{1/2}} e^{-\sigma^2\omega^2/2}$$

The inverse of $\hat{F}(\omega)$ is given by

$$f(t) = \frac{1}{(2\pi)^{1/2}} \int_{-\infty}^\infty e^{i\omega t} \hat{F}(\omega) \, d\omega$$

$$= \frac{1}{2\pi} \int_{-\infty}^\infty (e^{-\sigma^2\omega^2/2} \cos\omega t + i e^{-\sigma^2\omega^2/2} \sin\omega t) \, d\omega$$

Once again, the second term vanishes because the integrand is an odd function of ω, so using the fact that the first integrand is an even function of ω, we have

$$f(t) = \frac{1}{\pi} \int_0^\infty e^{-\sigma^2\omega^2/2} \cos\omega t \, d\omega$$

$$= \frac{1}{(2\pi\sigma^2)^{1/2}} e^{-t^2/2\sigma^2}$$

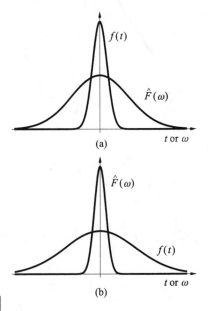

(a)

(b)

Figure 10.2. The Fourier transform pair $f(t) = e^{-t^2/2\sigma^2}/(2\pi\sigma^2)^{1/2}$ and $\hat{F}(\omega) = e^{-\sigma^2\omega^2/2}/(2\pi)^{1/2}$ plotted together for (a) $\sigma = 2$ and (b) $\sigma = 1/2$. Note the reciprocal nature of the widths of the plots of $f(t)$ and $\hat{F}(\omega)$.

Example 10–1 tells us that the Fourier transform of a Gaussian function is another Gaussian function. Furthermore, there is a reciprocal relation between the widths of these Gaussian functions. Figure 10.2 shows this reciprocal relation where $f(t) = e^{-t^2/2\sigma^2}/(2\pi\sigma^2)^{1/2}$ and $\hat{F}(\omega) = e^{-\sigma^2\omega^2/2}/(2\pi)^{1/2}$ are plotted together for $\sigma = 2$ (in part a) and $\sigma = 1/2$ (in part b).

The following Example illustrates another case of the reciprocal relation between the relative widths of Fourier transform pairs.

EXAMPLE 10–2

Find the Fourier transform of the finite wave train (Figure 10.3):

$$f(t) = \begin{cases} \cos\omega_0 t & -\dfrac{N\pi}{\omega_0} < t < \dfrac{N\pi}{\omega_0} \\ 0 & \text{otherwise} \end{cases}$$

A short pulse of the form of $f(t)$ is used in FT–NMR to generate a band of frequencies used to excite the nuclei in a sample. Discuss the relationship between the widths of $f(t)$ and its Fourier transform $\hat{F}(\omega)$.

SOLUTION: The wave train is an even function of t, so

$$\hat{F}(\omega) = \left(\frac{2}{\pi}\right)^{1/2} \int_0^{N\pi/\omega_0} \cos\omega_0 t \cos\omega t \, dt$$

$$= \frac{1}{(2\pi)^{1/2}} \left[\frac{\sin(\omega - \omega_0)N\pi/\omega_0}{\omega - \omega_0} + \frac{\sin(\omega + \omega_0)N\pi/\omega_0}{\omega + \omega_0} \right]$$

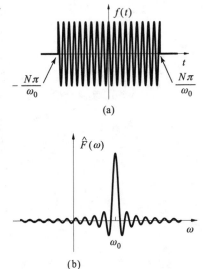

(a)

(b)

Figure 10.3. The Fourier transform pair (a) $f(t)$ and (b) $\hat{F}(\omega)$ given in Example 10–2.

Let's restrict our discussion to positive values of ω. For large values of ω_0, only the first term is important (Problem 10–21) and is shown in Figure 10.3. The zeros of $\hat{F}(\omega)$ occur at the points $(\omega - \omega_0)N\pi/\omega_0 = n\pi$ for $n = \pm 1, \pm 2, \ldots$, or at

$$\frac{\omega}{\omega_0} = 1 \pm \frac{1}{N}, 1 \pm \frac{2}{N}, \ldots$$

Because the contributions outside the central peak of $\hat{F}(\omega)$ are small (see Figure 10.3), we may take $\Delta\omega = 2\omega_0/N$ as a measure of the width of $\hat{F}(\omega)$. The width of the wave train is $\Delta t = 2N\pi/\omega_0$, and so we have the reciprocal relation

$$\Delta\omega\Delta t = 4\pi$$

This result has a close relationship with the Heisenberg uncertainty principle in energy and time. The energy in quantum mechanics is related to angular frequency by $E = \hbar\omega$, where $\hbar$ is the Planck constant divided by 2π. Therefore, $\Delta\omega\Delta t = 4\pi$ can be written as $\Delta E\Delta t = 4\pi\hbar$. The Heisenberg uncertainty principle says that $\Delta E\Delta t \geq \hbar/2$, and so we see that our finite wave train does indeed satisfy the uncertainty principle.

We have used the notation $f(t)$ and $\hat{F}(\omega)$ because we have in mind time signals and their corresponding spectra. The time domain and the frequency domain are related through Fourier transforms. There is a similar relationship between the position of a particle and its momentum in quantum mechanics. If $\psi(x)$ is the wave function of a particle, in the sense that $\psi^*(x)\psi(x)\,dx$ is the probability of finding the particle between x and $x + dx$, then the Fourier transform of $\psi(x)$,

$$\phi(p) = \frac{1}{(2\pi\hbar)^{1/2}} \int_{-\infty}^{\infty} \psi(x)e^{-ipx/\hbar}\,dx \qquad (10.11)$$

has the physical interpretation that $\phi^*(p)\phi(p)\,dp$ is the probability that the particle has a momentum between p and $p+dp$. (Chapter 21 discusses probabilities.) The inverse of Equation 10.11 is

$$\psi(x) = \frac{1}{(2\pi\hbar)^{1/2}} \int_{-\infty}^{\infty} \phi(p)e^{ipx/\hbar}\,dp \qquad (10.12)$$

Suppose that the (coordinate space) wave function of a particle is given by

$$\psi(x) = \left(\frac{1}{\pi a^2}\right)^{1/4} e^{-x^2/2a^2} \qquad -\infty < x < \infty \qquad (10.13)$$

We say that this function is normalized, in the sense that $\int_{-\infty}^{\infty} \psi^2(x)\,dx = 1$, meaning physically that the particle must be found somewhere. The corresponding wave function in momentum space is given by Equation 10.11:

$$\phi(p) = \frac{1}{(2\pi\hbar)^{1/2}} \int_{-\infty}^{\infty} \psi(x) e^{-ipx/\hbar}\, dx = \left(\frac{a^2}{\pi\hbar^2}\right)^{1/4} e^{-a^2 p^2/2\hbar^2} \qquad -\infty < x < \infty$$
(10.14)

It is easy to show that $\phi(p)$ is normalized, in the sense that (Problem 10–17) $\int_{-\infty}^{\infty} \phi^2(p)\, dp = 1$.

Just as the results in Examples 10–1 and 10–2 have a reciprocal relation, Equations 10.13 and 10.14 also have a reciprocal relation (Figure 10.4). Figure 10.4 shows that $\psi^2(x)$ given by Equation 10.13 and $\phi^2(p)$ given by Equation 10.14 have a reciprocal relation between the widths of their plots. We illustrate this relation quantitatively in the following Example.

EXAMPLE 10–3

We shall learn in Chapter 21 that a quantitative measure of the widths of curves like those in Figure 10.4 are the integrals

$$(\Delta x)^2 = \int_{-\infty}^{\infty} x^2 \psi^2(x)\, dx \qquad \text{and} \qquad (\Delta p)^2 = \int_{-\infty}^{\infty} p^2 \phi^2(p)\, dp$$

where Δx and Δp are our measures of the widths. Show that $\Delta x \Delta p = \hbar/2$.

SOLUTION:

$$(\Delta x)^2 = \left(\frac{1}{\pi a^2}\right)^{1/2} \int_{-\infty}^{\infty} x^2 e^{-x^2/a^2}\, dx = 2\left(\frac{1}{\pi a^2}\right)^{1/2} \int_{0}^{\infty} x^2 e^{-x^2/a^2}\, dx = \frac{a^2}{2}$$

and

$$(\Delta p)^2 = \left(\frac{a^2}{\pi\hbar^2}\right)^{1/2} \int_{-\infty}^{\infty} p^2 e^{-a^2 p^2/\hbar^2}\, dp = 2\left(\frac{a^2}{\pi\hbar^2}\right)^{1/2} \int_{0}^{\infty} p^2 e^{-a^2 p^2/\hbar^2}\, dp = \frac{\hbar^2}{2a^2}$$

and so the product $\Delta x \Delta p = \hbar/2$. This relation is a statement of the Heisenberg uncertainty principle for the position and momentum of a particle. As you can see, the Heisenberg uncertainty principle and Fourier transforms are intrinsically bound together.

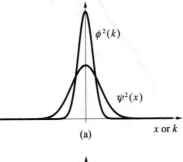
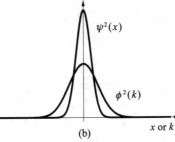

Figure 10.4. The Fourier transform pair $\psi^2(x) = (1/\pi a^2)^{1/2} e^{-x^2/a^2}$ and $\phi^2(p) = (a^2/2\pi\hbar^2)^{1/2} e^{-\sigma^2 p^2/2\hbar}$ plotted for (a) $a = 2$ and (b) $a = 1$. Note the reciprocal nature of the widths of the plots. We have set $\hbar = 1$ for convenience.

We can use Equations 10.7 and 10.8 to formally derive a useful equation for the Dirac delta function (Section 4.4). Let $f(u) = \delta(u - u_0)$ in Equation 10.7 to obtain

$$\hat{F}(\omega) = \frac{1}{(2\pi)^{1/2}} \int_{-\infty}^{\infty} \delta(u - u_0) e^{-i\omega u}\, du = \frac{e^{-i\omega u_0}}{(2\pi)^{1/2}}$$

Now substitute this result into Equation 10.8 to get

$$\delta(u - u_0) = \frac{1}{(2\pi)^{1/2}} \int_{-\infty}^{\infty} \hat{F}(\omega)e^{i\omega u}\, d\omega = \frac{1}{2\pi} \int_{-\infty}^{\infty} e^{i\omega(u - u_0)}\, d\omega \qquad (10.15)$$

We shall use this result in the final section.

10.3 Fourier Transforms and Spectroscopy

Fourier transform infrared spectra and NMR spectra (FT–IR and FT–NMR) are obtained by observing the time behavior of a system and then taking its cosine transform. Under certain conditions, the time behavior is given by

$$f(t) = e^{-\alpha t} \cos \omega_0 t \qquad t \geq 0 \qquad (10.16)$$

where ω_0 is some "natural" frequency of the system (such as the resonance frequency of a proton spin flip). The cosine transform of $f(t)$ is (Problem 10–10)

$$\hat{F}(\omega) = \left(\frac{2}{\pi}\right)^{1/2} \frac{\alpha}{\alpha^2 + (\omega - \omega_0)^2} + \left(\frac{2}{\pi}\right)^{1/2} \frac{\alpha}{\alpha^2 + (\omega + \omega_0)^2} \qquad (10.17)$$

When this function is plotted against ω, the first term peaks at $\omega = \omega_0$, and the second term is essentially always negligible for spectroscopically important values of ω_0 and ω (Problem 10–22). Thus, we need focus on only the first term:

$$\hat{F}(\omega) \approx \left(\frac{2}{\pi}\right)^{1/2} \frac{\alpha}{\alpha^2 + (\omega - \omega_0)^2} \qquad (10.18)$$

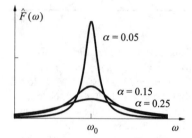

Figure 10.5. The Lorentzian function, Equation 10.18, plotted against ω for $\omega_0 = 2$ and for various values of α. Notice that the curves become narrower and more peaked as α decreases.

for the discussion that follows. Equation 10.18, which is plotted in Figure 10.5, is called a *Lorentzian function*. Figure 10.5 shows that the width of $\hat{F}(\omega)$ is controlled by α; the smaller α is, the narrower will be $\hat{F}(\omega)$. Thus, in practice, α is a measure of the width of a spectral line (Problem 10–11). Experimentally, the shape of a spectral line is often well represented by a Lorentzian function.

We'll now show how a Fourier transform can be used to extract frequency information from time signals. This is one of the most important uses of Fourier transforms in chemistry. Figure 10.6a plots $e^{-\alpha t} \cos \omega_0 t$ against time for $\alpha = 0.075$

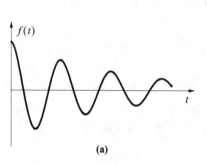

(a)

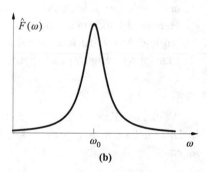

(b)

Figure 10.6. (a) The function $e^{-at} \cos \omega_0 t$ against time for $a = 0.075$ and $\omega_0 = 1$. (b) The cosine transform of $f(t)$, $F(\omega) = (2/\pi)^{1/2} a / [a^2 + (\omega - \omega_0)^2]$ plotted against ω.

and $\omega_0 = 1$. It is pretty clear from the figure that the plotted function consists of only one frequency. Figure 10.6b plots the cosine transform, which clearly shows that just one frequency is involved. What about the function

$$f(t) = 0.70\, e^{-0.010t} \cos(1.2\, t) + 1.25\, e^{-0.025t} \cos(2.0\, t) + 0.75\, e^{-0.075t} \cos(2.6\, t)$$

$$(10.19)$$

plotted in Figure 10.7a, however? We cannot possibly determine visually how many and what frequencies are involved. If we take the cosine transform of the function in Figure 10.7a, however, we obtain

$$\hat{F}(\omega) = \left(\frac{2}{\pi}\right)^{1/2} \left[0.70 \frac{0.010}{(0.010)^2 + (\omega - 1.2)^2} + 1.25 \frac{0.025}{(0.025)^2 + (\omega - 2.0)^2} \right.$$

$$\left. + 0.75 \frac{0.075}{(0.075)^2 + (\omega - 2.6)^2} \right] \qquad (10.20)$$

which is plotted in Figure 10.7b. This plot clearly shows that three frequencies are involved in the time record. Thus, a cosine transform extracts the frequencies involved in the time behavior of a system. Experimentally, the time behavior is obtained numerically (rather than analytically). There is an extremely efficient numerical algorithm called Fast Fourier Transform (FFT) for the numerical evaluation of Fourier transforms. Prior to the development of FFT, the numerical evaluation of Fourier transforms was rather cumbersome. The ready availability of FFT is one of the reasons why Fourier transform spectroscopy has become so widely used.

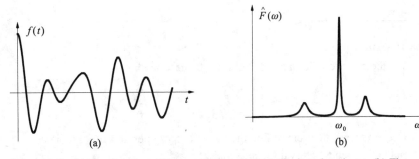

Figure 10.7. (a) The function $f(t)$ given by Equation 10.19 plotted against t. (b) The cosine transform of $f(t)$, Equation 10.20, plotted against ω.

10.4 Parseval's Theorem

As a final topic in this chapter, we shall derive a result known as *Parseval's theorem* for Fourier transforms. We start with

$$\int_{-\infty}^{\infty} |f(t)|^2\, dt = \int_{-\infty}^{\infty} f^*(t) f(t)\, dt$$

where we have allowed for the possibility that $f(t)$ is complex for generality. We first express $|f(t)|^2$ as a double integral using Equation 10.8 for $f(t)$.

$$|f(t)|^2 = \frac{1}{(2\pi)^{1/2}} \int_{-\infty}^{\infty} \hat{F}^*(\omega)e^{-i\omega t}\,d\omega \times \frac{1}{(2\pi)^{1/2}} \int_{-\infty}^{\infty} \hat{F}(\omega')e^{i\omega' t}\,d\omega$$

$$= \frac{1}{2\pi} \int_{-\infty}^{\infty} \int_{-\infty}^{\infty} \hat{F}^*(\omega)e^{-i\omega t}\hat{F}(\omega')\,e^{i\omega' t}\,d\omega\,d\omega'$$

Note that we have used two different dummy variables of integration here. The final result would make no sense if we used ω in both integrals. Now integrate $|f(t)|^2$ over t to obtain

$$\int_{-\infty}^{\infty} |f(t)|^2\,dt = \frac{1}{2\pi} \int_{-\infty}^{\infty} \int_{-\infty}^{\infty} \int_{-\infty}^{\infty} \hat{F}^*(\omega)\hat{F}(\omega')e^{i(\omega'-\omega)t}\,d\omega\,d\omega'dt$$

Using Equation 10.11 for the integration over t, we get

$$\int_{-\infty}^{\infty} |f(t)|^2\,dt = \int_{-\infty}^{\infty} \int_{-\infty}^{\infty} \hat{F}^*(\omega)\hat{F}(\omega')\,\delta(\omega' - \omega)\,d\omega\,d\omega'$$

If we now integrate over ω', we get

$$\int_{-\infty}^{\infty} |f(t)|^2\,dt = \int_{-\infty}^{\infty} |\hat{F}(\omega)|^2\,d\omega \qquad (10.21)$$

Equation 10.21 is known as Parseval's theorem for Fourier transforms. This derivation is a good illustration of the manipulative utility of the delta function.

Parseval's theorem guarantees that if a quantum-mechanical wave function is normalized (see Equation 10.13), then its Fourier transform will also be normalized (see Equation 10.14). Thus, a Fourier transform preserves the normalization of quantum-mechanical wave functions.

EXAMPLE 10–4

Let the electric field in a radiated wave be described by

$$E(t) = \begin{cases} 0 & t < 0 \\ e^{-t/\tau} \sin \omega_0 t & t > 0 \end{cases}$$

Use Parseval's theorem to find the power (energy per unit time) radiated in the frequency interval $(\omega, \omega + d\omega)$.

SOLUTION: The Fourier transform of $E(t)$ is

$$\hat{E}(\omega) = \frac{1}{(2\pi)^{1/2}} \int_{0}^{\infty} e^{-t/\tau}e^{-i\omega t} \sin \omega_0 t\,dt$$

We'll evaluate this integral by writing $\sin \omega_0 t$ as $(e^{i\omega_0 t} - e^{-i\omega_0 t})/2i$.

$$\hat{E}(\omega) = \frac{1}{2(2\pi)^{1/2}} \left(\frac{1}{\omega - \omega_0 - \dfrac{i}{\tau}} - \frac{1}{\omega + \omega_0 - \dfrac{i}{\tau}} \right)$$

The total power radiated is given by

$$\int_{-\infty}^{\infty} |E(t)|^2 dt = \int_{-\infty}^{\infty} |\hat{E}(\omega)|^2 d\omega$$

Using the fact that the first term in $|\hat{E}(\omega)|$ will dominate for $\omega > 0$, we have that the power radiated in the frequency interval $(\omega, \omega + d\omega)$ is given by

$$|\hat{E}(\omega)|^2 d\omega \approx \frac{1}{8\pi} \frac{d\omega}{(\omega - \omega_0)^2 + \frac{1}{\tau^2}}$$

This frequency spectrum is shown in Figure 10.8. This is a Lorentzian curve like Equation 10.18. When $\omega = \omega_0 \pm 1/\tau$, the height of the spectrum falls off by a factor of 2, as shown in the figure. Therefore, the width at half power is given by $2/\tau$ (Problem 10–11). In this case, we have that the width of a frequency spectrum varies inversely as the duration of the signal, similar to the result for the finite wave train discussed in Example 10–2.

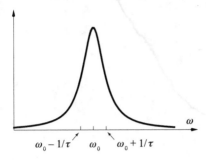

Figure 10.8. The frequency spectrum associated with the radiated wave described in Example 10–4 for $\omega_0 = 25$ and $\tau = 1/4$.

Problems

10–1. Find the Fourier transform of $f(t) = 1/(t^2 + a^2)$.

10–2. Find the Fourier transform of $f(t) = \begin{cases} 1 & -a \leq t \leq a \\ 0 & \text{otherwise} \end{cases}$

Discuss the reciprocal relation between the widths of $f(t)$ and $\hat{F}(\omega)$.

10–3. Find the Fourier transform of $e^{-|t|/\tau} \cos \omega_0 t$.

10–4. Plot Equations 10.9 and 10.10 for various values of α and show that there is a reciprocal relation between the widths of the plots.

10–5. Show that $\hat{F}(-\omega) = \hat{F}^*(\omega)$ if $f(t)$ is real.

10–6. Use Equation 10.11 to show that the two sides of Equation 10.6 are equal.

10–7. Show that if $f(u)$ is an even function of u, then Fourier's integral theorem, Equation 10.6, reduces to

$$f(t) = \frac{2}{\pi} \int_0^\infty \int_0^\infty f(u) \cos \omega t \cos \omega u \, du \, d\omega \qquad (1)$$

Now show that if we let

$$\hat{F}_C(\omega) = \left(\frac{2}{\pi}\right)^{1/2} \int_0^\infty f(u) \cos \omega u \, du = \left(\frac{2}{\pi}\right)^{1/2} \int_0^\infty f(t) \cos \omega t \, dt \qquad (2)$$

then it follows from equation 1 that

$$f(t) = \left(\frac{2}{\pi}\right)^{1/2} \int_0^\infty \hat{F}_C(\omega) \cos \omega t \, d\omega \qquad (3)$$

Equations 2 and 3 constitute a cosine transform pair.

10–8.   Show that the cosine transform of $e^{-\alpha t}$, $\alpha \geq 0$, is (see the previous problem)

$$\hat{F}_C(\omega) = \left(\frac{2}{\pi}\right)^{1/2} \frac{\alpha}{\alpha^2 + \omega^2}$$

Use this result to show that

$$\int_0^\infty \frac{\cos ax}{x^2 + b^2} dx = \frac{\pi}{2b} e^{-ab}$$

10–9.   Use the result of the previous problem to show that

$$\int_0^\infty \frac{x \sin ax}{x^2 + b^2} = \frac{\pi}{2} e^{-ab}$$

10–10. Show that the cosine transform of Equation 10.16 is given by Equation 10.17.

10–11. Show that the Lorentzian function given by Equation 10.18 has a maximum at $\omega = \omega_0$ and is half maximum at $\omega - \omega_0 = \pm \alpha$. Show that the width at half maximum is 2α. This is the standard measure of the width of a Lorentzian function.

10–12. Using the result of Problem 10–2, show that

$$\int_{-\infty}^\infty \frac{\sin az \cos xz}{z} dz = \begin{cases} \pi & |x| < a \\ 0 & |x| > a \end{cases}$$

10–13. Use the result of the previous problem to show that $\displaystyle\int_0^\infty \frac{\sin x}{x} dx = \frac{\pi}{2}$.

10–14. Plot $e^{-\alpha t} \cos \omega_0 t$ and its cosine transform for $\alpha = 0.10$ and $\omega_0 = 2$. Compare your result to Figure 10.6.

10–15. Plot $f(t) = e^{-0.050t} \cos t + 2e^{-0.025t} \cos 2t$ and its cosine transform. Can you see two frequencies in your plot of $f(t)$? How about for

$$g(t) = e^{-0.010t} \cos 1.2t + 1.75\, e^{-0.025t} \cos 2t + 0.50\, e^{-0.075t} \cos 2.6t$$

Plot its cosine transform.

10–16. Show that the wave function $\psi(x)$ given by Equation 10.13 is normalized. (Be sure to square $\psi(x)$.)

10–17. Show that the wave function $\phi(p)$ given by Equation 10.14 is normalized. (Be sure to square $\phi(p)$.)

10–18. Use $f(x) = e^{-|x|}$, $-\infty < x < \infty$, and Parseval's theorem to show that $\displaystyle\int_0^\infty \frac{du}{(1 + u^2)^2} = \frac{\pi}{4}$.

10–19. Use the result of the previous problem to evaluate $\displaystyle\int_0^\infty \frac{du}{(\alpha^2 + u^2)^2}$.

10–20. In this problem we'll determine the momentum distribution of a particle in a box. The (coordinate) wave function is

$$\psi_n(x) = \begin{cases} \left(\dfrac{2}{a}\right)^{1/2} \sin \dfrac{n\pi x}{a} & 0 \leq x \leq a \\ 0 & \text{otherwise} \end{cases}$$

with $n = 1, 2, \ldots$. Use Equation 10.14 to show that

$$\phi_n(p) = \left(\frac{\pi \hbar^3}{a^3}\right)^{1/2} \frac{n}{\left(\frac{n\pi\hbar}{a}\right)^2 - p^2}[1 - (-1)^n e^{-ipa/\hbar}]$$

or that

$$\phi_n(p) = \begin{cases} i\left(\frac{4\pi\hbar^3}{a^3}\right)^{1/2} \dfrac{ne^{-ipa/2\hbar}\sin\frac{pa}{2\hbar}}{\left(\frac{n\pi\hbar}{a}\right)^2 - p^2} & n \text{ even} \\[4ex] \left(\frac{4\pi\hbar^3}{a^3}\right)^{1/2} \dfrac{ne^{-ipa/2\hbar}\cos\frac{pa}{2\hbar}}{\left(\frac{n\pi\hbar}{a}\right)^2 - p^2} & n \text{ odd} \end{cases}$$

where $-\infty < p < \infty$. Plot $\phi_n^*(p)\phi_n(p)$ for $n = 1$ through 4 and discuss the result. You should use a CAS to produce the plots. Using a CAS, show that $\phi_n(p)$ is normalized.

10–21. Use any CAS to plot only the first term and then both terms of $\hat{F}(\omega)$ in Example 10–2 for various values of ω_0 (1/2, 1, and 2, for example) and compare your results.

10–22. Use any CAS to plot only the first term and then both terms of $\hat{F}(\omega)$ in Equation 10.17 for various values of ω_0 and α and compare your results.

10–23. Use any CAS to make up a problem similar to Problem 10–15.

CHAPTER 11

Operators

The concept of an operator plays a central role in quantum mechanics. Physical quantities, such as momentum, kinetic energy, position, and angular momentum, are represented by operators in quantum mechanics. The most famous operator in quantum mechanics is the Hamiltonian operator, which corresponds to the energy. In fact, the Schrödinger equation can be expressed as an operator equation of the form

$$\hat{H}\psi = E\psi$$

where $\hat{H}$ is the Hamiltonian operator, E is the energy, and ψ is a wave function. In this chapter, we shall discuss some important general properties of operators, and then discuss Hermitian operators in particular.

11.1 Linear Operators

An *operator* is a symbol that tells you to do something to whatever follows the symbol. For example, we can consider dy/dx to be the d/dx operator operating on the function $y(x)$. We can express this concept in a formal operator notation by letting $\hat{D} = d/dx$ and writing $dy/dx = \hat{D}y(x)$. Thus, we have $\hat{D}x^4 = 4x^3$, for example. Some other examples of operators are SQR (square what follows), $\int_0^1 dx\ \square$ (integrate what follows from 0 to 1), 3 (multiply by 3), and so on. We usually denote an operator by a capital letter with a carat over it (e.g., $\hat{A}$). Thus, we write

$$\hat{A}f(x) = g(x)$$

to indicate that the operator $\hat{A}$ operates on $f(x)$ to give a new function $g(x)$.

EXAMPLE 11–1
Perform the following operations:

(a) $\hat{A}x^2$, $\hat{A} = \int_0^2 dx \, \square$

(b) $\hat{A}x^2$, $\hat{A} = \dfrac{d^2}{dx^2} + 2\dfrac{d}{dx} + 3$

(c) $\hat{A}\sin ax$, $\hat{A} = -\dfrac{d^2}{dx^2}$

SOLUTION:

(a) $\hat{A}x^2 = \int_0^2 dx \, x^2 = \dfrac{8}{3}$

(b) $\hat{A}x^2 = \dfrac{d^2}{dx^2}x^2 + 2\dfrac{d}{dx}x^2 + 3x^2 = 2 + 4x + 3x^2$

(c) $\hat{A}\sin ax = -\dfrac{d^2}{dx^2}\sin ax = a^2 \sin ax$

In quantum mechanics, we deal only with *linear operators*. An operator is said to be linear if

$$\hat{A}\left[c_1 f_1(x) + c_2 f_2(x)\right] = c_1 \hat{A} f_1(x) + c_2 \hat{A} f_2(x) \tag{11.1}$$

where c_1 and c_2 are (possibly complex) constants. Clearly, the "differentiate" and "integrate" operators are linear because

$$\frac{d}{dx}\left[c_1 f_1(x) + c_2 f_2(x)\right] = c_1 \frac{df_1}{dx} + c_2 \frac{df_2}{dx}$$

and

$$\int dx \left[c_1 f_1(x) + c_2 f_2(x)\right] = c_1 \int dx \, f_1(x) + c_2 \int dx \, f_2(x)$$

The "square" operator, SQR, on the other hand, is nonlinear because

$$\text{SQR}\left[c_1 f_1(x) + c_2 f_2(x)\right] = c_1^2 f_1^2(x) + c_2^2 f_2^2(x) + 2c_1 c_2 f_1(x) f_2(x)$$
$$\neq c_1 f_1^2(x) + c_2 f_2^2(x)$$

and, therefore, does not satisfy the definition given by Equation 11.1.

EXAMPLE 11–2
Determine whether the following operators are linear or nonlinear:

(a) $\hat{A} f(x) = \text{SQRT } f(x)$ (take the square root)

(b) $\hat{A} f(x) = x^2 f(x)$ (multiply by x^2)

SOLUTION:

(a) $\hat{A}\,[c_1 f_1(x) + c_2 f_2(x)] = \text{SQRT}\,[c_1 f_1(x) + c_2 f_2(x)]$

$\qquad = [c_1 f_1(x) + c_2 f_2(x)]^{1/2} \neq c_1 f_1^{1/2}(x) + c_2 f_2^{1/2}(x)$

and so SQRT is a nonlinear operator.

(b) $\hat{A}\,[c_1 f_1(x) + c_2 f_2(x)] = x^2\,[c_1 f_1(x) + c_2 f_2(x)]$

$\qquad = c_1 x^2 f_1(x) + c_2 x^2 f_2(x) = c_1 \hat{A} f_1(x) + c_1 \hat{A} f_1(x)$

and so x^2 is a linear operator.

A problem that we frequently encounter in physical chemistry is the following: Given an operator $\hat{A}$, find a function $\psi(x)$ and a constant a such that

$$\hat{A}\psi(x) = a\psi(x) \qquad\qquad (11.2)$$

Note that the result of operating on the function $\psi(x)$ by $\hat{A}$ is simply to give $\psi(x)$ back again, only multiplied by a constant factor. Clearly, $\hat{A}$ and $\psi(x)$ have a very special relationship to each other. The function $\psi(x)$ is called an *eigenfunction* of the operator $\hat{A}$, and a is called an *eigenvalue*. The problem of determining $\psi(x)$ and a for a given $\hat{A}$ is called an *eigenvalue problem*.

It's commonly the case that an operator $\hat{A}$ has a whole set of eigenfunctions $\psi_n(x)$ and corresponding eigenvalues a_n, in which case we write Equation 11.2 as

$$\hat{A}\psi_n(x) = a_n \psi_n(x) \qquad n = 1,\ 2,\ 3,\ldots \qquad (11.3)$$

An example is the operator $\hat{A} = -d^2/dx^2$ and the eigenfunctions $\psi_n(x) = \sin n\pi x$ for $n = 1,\ 2,\ldots$. Equation 11.3 reads

$$\hat{A}\psi_n(x) = -\frac{d^2}{dx^2}\sin n\pi x = n^2\pi^2 \sin n\pi x \qquad n = 1,\ 2,\ldots$$

and so we see that the eigenvalues of $\hat{A}$ are given by $a_n = n^2\pi^2$. (You might recognize these equations as those of a quantum-mechanical particle in a box in simplified notation.)

Operators can be imaginary or complex quantities. For example, the x component of the linear momentum is represented in quantum mechanics by the operator

$$\hat{P}_x = -i\hbar\frac{d}{dx} \qquad\qquad (11.4)$$

EXAMPLE 11–3

Show that e^{ikx} is an eigenfunction of $\hat{P}_x$. What is the eigenvalue?

SOLUTION: We apply $\hat{P}_x$ to e^{ikx} and find

$$\hat{P}_x e^{ikx} = -i\hbar\frac{d}{dx}e^{ikx} = \hbar k e^{ikx}$$

and so we see that e^{ikx} is an eigenfunction and $\hbar k$ is the eigenvalue of the operator $\hat{P}_x$. Actually, the value of k is unrestricted here, so there is a continuous range of eigenfunctions and eigenvalues in this eigenvalue problem.

Linear operators have the following important property. Suppose we have the two eigenvalue equations

$$\hat{A}\psi_1 = a\psi_1 \qquad \text{and} \qquad \hat{A}\psi_2 = a\psi_2$$

Both ψ_1 and ψ_2 have the same eigenvalue a. We say that the eigenvalues of $\hat{A}$ are two-fold degenerate. If this is the case, then any linear combination of ψ_1 and ψ_2, say $c_1\psi_1 + c_2\psi_2$, is also an eigenfunction of $\hat{A}$ with the eigenvalue a. The proof relies on the linear property of $\hat{A}$:

$$\hat{A}(c_1\psi_1 + c_2\psi_2) = c_1\hat{A}\psi_1 + c_2\hat{A}\psi_2$$
$$= c_1a\psi_1 + c_2a\psi_2 = a(c_1\psi_1 + c_2\psi_2)$$

EXAMPLE 11–4

Consider the eigenvalue equation

$$\frac{d^2\Phi(\phi)}{d\phi^2} = -m^2\Phi(\phi)$$

where m is a real (not imaginary or complex) number. The two eigen-functions of $\hat{A} = d^2/d\phi^2$ are

$$\Phi_m(\phi) = e^{im\phi} \qquad \text{and} \qquad \Phi_{-m}(\phi) = e^{-im\phi}$$

We can easily show that each of these eigenfunctions has the eigenvalue $-m^2$. Show that any linear combination of $\Phi_m(\phi)$ and $\Phi_{-m}(\phi)$ is also an eigenfunction of $\hat{A} = d^2/d\phi^2$.

SOLUTION:

$$\frac{d^2}{d\phi^2}(c_1 e^{im\phi} + c_2 e^{-im\phi}) = c_1\frac{d^2 e^{im\phi}}{d\phi^2} + c_2\frac{d^2 e^{-im\phi}}{d\phi^2}$$
$$= -c_1 m^2 e^{im\phi} - c_2 m^2 e^{-im\phi}$$
$$= -m^2(c_1 e^{im\phi} + c_2 e^{-im\phi})$$

Example 11–4 shows that this result is directly due to the linear property of $\hat{A}$. Although we have considered only a two-fold degeneracy, the result is general.

11.2 Commutators of Operators

If we denote kinetic energy by T and momentum by p, then the relation between kinetic energy and momentum is $T = p^2/2m$. We can express this result for the x direction in operator notation by writing

$$\hat{T}_x = \frac{1}{2m}\hat{P}_x^2 \tag{11.5}$$

We can interpret the operator $\hat{P}_x^2$ by considering the case of two operators acting sequentially, as in $\hat{A}\hat{B}f(x)$. In cases such as this, we apply each operator in turn, working from right to left. Thus,

$$\hat{A}\hat{B}f(x) = \hat{A}[\hat{B}f(x)] = \hat{A}h(x)$$

where $h(x) = \hat{B}f(x)$. If $\hat{A} = \hat{B}$, we have $\hat{A}\hat{A}f(x)$ and denote this term as $\hat{A}^2 f(x)$. Note that $\hat{A}^2 f(x) \neq [\hat{A}f(x)]^2$ for arbitrary $f(x)$.

Using the fact that $\hat{P}_x^2$ means two successive applications of $\hat{P}_x$, we see that we can write the kinetic energy operator as

$$\hat{T}_x f(x) = \frac{1}{2m}\hat{P}_x^2 f(x) = \frac{1}{2m}\left(-i\hbar\frac{d}{dx}\right)\left(-i\hbar\frac{d}{dx}\right)f(x)$$

$$= -\frac{\hbar^2}{2m}\frac{d^2}{dx^2}f(x) \tag{11.6}$$

where $f(x)$ is an arbitrary, suitably behaved function of x. When evaluating operator expressions, it is good practice to allow the operator to act upon a function $f(x)$ as we have done here. We'll see in Example 11–6 that you can get an incorrect result if you do not include $f(x)$.

EXAMPLE 11–5
Given $\hat{A} = d/dx$ and $\hat{B} = x^2$ (multiply by x^2), show (a) that $\hat{A}^2 f(x) \neq [\hat{A}f(x)]^2$ and (b) that $\hat{A}\hat{B}f(x) \neq \hat{B}\hat{A}f(x)$ for arbitrary $f(x)$.

SOLUTION:

(a) $\hat{A}^2 f(x) = \dfrac{d}{dx}\left(\dfrac{df}{dx}\right) = \dfrac{d^2f}{dx^2}$

$[\hat{A}f(x)]^2 = \left(\dfrac{df}{dx}\right)^2 \neq \dfrac{d^2f}{dx^2}$

for arbitrary $f(x)$.

(b) $\hat{A}\hat{B}f(x) = \dfrac{d}{dx}[x^2 f(x)] = 2xf(x) + x^2\dfrac{df}{dx}$

$\hat{B}\hat{A}f(x) = x^2\dfrac{df}{dx} \neq \hat{A}\hat{B}f(x)$

for arbitrary $f(x)$. Thus, we see that the order of the application of operators must be specified. If $\hat{A}$ and $\hat{B}$ are such that

$$\hat{A}\hat{B}f(x) = \hat{B}\hat{A}f(x)$$

for any suitably behaved $f(x)$, then the two operators are said to *commute*. The two operators in this Example, however, do not commute.

As we have just seen in Example 11–5, an important difference between operators and ordinary algebraic quantities is that operators do not necessarily commute. If

$$\hat{A}\hat{B}f(x) = \hat{B}\hat{A}f(x) \qquad \text{(commutative)} \tag{11.7}$$

for arbitrary $f(x)$, then we say that $\hat{A}$ and $\hat{B}$ *commute*. If

$$\hat{A}\hat{B}f(x) \neq \hat{B}\hat{A}f(x) \qquad \text{(noncommutative)} \tag{11.8}$$

for arbitrary $f(x)$, then we say that $\hat{A}$ and $\hat{B}$ do not commute. For example, if $\hat{A} = d/dx$ and $\hat{B} = x$ (multiply by x), then

$$\hat{A}\hat{B}f(x) = \frac{d}{dx}[xf(x)] = f(x) + x\frac{df}{dx}$$

and

$$\hat{B}\hat{A}f(x) = x\frac{d}{dx}f(x) = x\frac{df}{dx}$$

Therefore, $\hat{A}\hat{B}f(x) \neq \hat{B}\hat{A}f(x)$, and $\hat{A}$ and $\hat{B}$ do not commute. In this particular case, we have

$$\hat{A}\hat{B}f(x) - \hat{B}\hat{A}f(x) = f(x)$$

or

$$(\hat{A}\hat{B} - \hat{B}\hat{A})f(x) = \hat{I}f(x) \tag{11.9}$$

where we have introduced the identity operator $\hat{I}$, which simply multiplies $f(x)$ by unity. Because $f(x)$ is arbitrary, we can write Equation 11.9 as an operator equation by suppressing $f(x)$ on both sides of the equation to give

$$\hat{A}\hat{B} - \hat{B}\hat{A} = \hat{I} \tag{11.10}$$

Realize that an operator equality like this is valid only if it is true for all suitably behaved $f(x)$. The combination of $\hat{A}$ and $\hat{B}$ appearing in Equation 11.10 occurs often and is called the *commutator*, $[\hat{A}, \hat{B}]$, of $\hat{A}$ and $\hat{B}$:

$$[\hat{A}, \hat{B}] = \hat{A}\hat{B} - \hat{B}\hat{A} \tag{11.11}$$

If $[\hat{A}, \hat{B}]f(x) = 0$ for all (suitably behaved) $f(x)$ on which the commutator acts, then we write that $[\hat{A}, \hat{B}] = 0$ and we say that $\hat{A}$ and $\hat{B}$ commute.

EXAMPLE 11–6

Let $\hat{A} = d/dx$ and $\hat{B} = x^2$. Evaluate the commutator $[\hat{A}, \hat{B}]$.

SOLUTION: We let $\hat{A}$ and $\hat{B}$ act upon an arbitrary function $f(x)$:

$$\hat{A}\hat{B}f(x) = \frac{d}{dx}[x^2 f(x)] = 2xf(x) + x^2\frac{df}{dx}$$

$$\hat{B}\hat{A}f(x) = x^2\frac{d}{dx}f(x) = x^2\frac{df}{dx}$$

By subtracting these two results, we obtain

$$(\hat{A}\hat{B} - \hat{B}\hat{A})f(x) = 2xf(x)$$

Because, and only because, $f(x)$ is arbitrary, we write

$$[\hat{A}, \hat{B}] = 2x\hat{I}$$

As we said earlier, when evaluating a commutator, it is essential to include a function $f(x)$ as we have done; otherwise, we can obtain a spurious result. To this end, note well that

$$[\hat{A}, \hat{B}] = \hat{A}\hat{B} - \hat{B}\hat{A} = \frac{d}{dx}x^2 - x^2\frac{d}{dx}$$

$$\neq 2x - x^2\frac{d}{dx}$$

a result that is obtained by forgetting to include the function $f(x)$. Such errors will not occur if an arbitrary function $f(x)$ is included from the outset.

EXAMPLE 11–7

The position operator in quantum mechanics is given by $\hat{X} = x$ (multiply by x) and the x component of the momentum operator is given by $\hat{P}_x = -i\hbar d/dx$. Evaluate $[\hat{P}_x, \hat{X}]$.

SOLUTION: We let $[\hat{P}_x, \hat{X}]$ act upon an arbitrary function $f(x)$:

$$[\hat{P}_x, \hat{X}]f(x) = \hat{P}_x\hat{X}f(x) - \hat{X}\hat{P}_x f(x)$$

$$= -i\hbar\frac{d}{dx}[xf(x)] + xi\hbar\frac{d}{dx}f(x)$$

$$= -i\hbar f(x) - i\hbar x\frac{df}{dx} + i\hbar x\frac{df}{dx}$$

$$= -i\hbar f(x)$$

Because $f(x)$ is arbitrary, we write this result as

$$[\hat{P}_x, \hat{X}] = -i\hbar\hat{I}$$

where $\hat{I}$ is the unit operator (the multiply-by-one operator).

11.3 Hermitian Operators

It turns out that quantum-mechanical operators must be linear, but they also must satisfy a more subtle requirement. One of the tenets of quantum mechanics is that if you undertake a measurement of the physical quantity corresponding to a given operator, $\hat{A}$, then the *only* values of that quantity that you will ever observe are eigenvalues of $\hat{A}$. We have seen, however, that operators can be complex quantities (see Equation 11.4), but certainly the eigenvalues must be real quantities if they are to correspond to the result of experimental measurement. In an equation, we have

$$\hat{A}\psi = a\psi \tag{11.12}$$

where $\hat{A}$ and even ψ may be complex but a must be real. We shall insist, then, that quantum-mechanical operators have only real eigenvalues. Clearly, this places a restriction on the operator $\hat{A}$.

To see what this restriction is, we multiply Equation 11.12 from the left by ψ^*, the complex conjugate of ψ, and integrate to obtain

$$\int \psi^*\hat{A}\psi\,dx = a\int \psi^*\psi\,dx \tag{11.13}$$

Now take the complex conjugate of Equation 11.12,

$$(\hat{A}\psi)^* = a^*\psi^* = a\psi^* \tag{11.14}$$

where the equality $a^* = a$ recognizes that a must be real. Multiply Equation 11.14 from the left by ψ and integrate:

$$\int \psi(\hat{A}\psi)^*dx = a\int \psi\psi^*dx = a\int \psi^*\psi\,dx \tag{11.15}$$

Equating the left sides of Equations 11.13 and 11.15 gives

$$\int \psi^*\hat{A}\psi\,dx = \int \psi(\hat{A}\psi)^*dx \tag{11.16}$$

The operator $\hat{A}$ must satisfy Equation 11.16 to assure that its eigenvalues are real. An operator that satisfies Equation 11.16 for any well-behaved function is called a *Hermitian operator*. Thus, we can write the definition of a Hermitian operator as an operator that satisfies the relation

$$\int_{-\infty}^{\infty} f^*\hat{A}f\,dx = \int_{-\infty}^{\infty} f(\hat{A}f)^*dx \quad \text{(Hermitian)} \tag{11.17}$$

where $f(x)$ is any well-behaved function. Hermitian operators have real eigenvalues. Quantum-mechanical operators corresponding to physically observable properties are Hermitian.

How do you determine if an operator is Hermitian? Consider the operator $\hat{A} = d/dx$. Does $\hat{A}$ satisfy Equation 11.17? Let's substitute $\hat{A} = d/dx$ into Equation 11.17 and integrate by parts:

$$\int_{-\infty}^{\infty} f^* \frac{d}{dx} f \, dx = \int_{-\infty}^{\infty} f^* \frac{df}{dx} \, dx = \left[f^* f \right]_{-\infty}^{\infty} - \int_{-\infty}^{\infty} f \frac{df^*}{dx} \, dx$$

For a function to be acceptable as a wave function in quantum mechanics, it must vanish at infinity, and so the first term on the right side here is zero. Therefore, we have

$$\int_{-\infty}^{\infty} f^* \frac{d}{dx} f \, dx = - \int_{-\infty}^{\infty} f \frac{d}{dx} f^* \, dx$$

For an arbitrary function $f(x)$, d/dx does *not* satisfy Equation 11.17 and so is *not* Hermitian.

Let's consider the x component of the momentum operator $\hat{P}_x = -i\hbar d/dx$. Substitution of $\hat{P}_x$ into the right side of Equation 11.17 gives

$$\int_{-\infty}^{\infty} f(\hat{P}_x f)^* \, dx = \int_{-\infty}^{\infty} f \left(-i\hbar \frac{df}{dx} \right)^* dx = i\hbar \int_{-\infty}^{\infty} f \frac{df^*}{dx} \, dx$$

Similarly, substitution of $\hat{P}_x$ into the left side of Equation 11.17 and integration by parts gives

$$\int_{-\infty}^{\infty} f^* \hat{P}_x f \, dx = \int_{-\infty}^{\infty} f^* \left(-i\hbar \frac{d}{dx} \right) f \, dx = -i\hbar \int_{-\infty}^{\infty} f^* \frac{df}{dx} \, dx$$

$$= -i\hbar \left[f^* f \right]_{-\infty}^{\infty} + i\hbar \int_{-\infty}^{\infty} f \frac{df^*}{dx} \, dx = i\hbar \int_{-\infty}^{\infty} f \frac{df^*}{dx} \, dx$$

where, as usual, we assume that f vanishes at infinity. Thus, we see that $\hat{P}_x$ does, indeed, satisfy Equation 11.17. Therefore, the momentum operator is a Hermitian operator.

EXAMPLE 11–8
Prove that the (one-dimensional) kinetic energy operator

$$\hat{T} = -\frac{\hbar^2}{2m} \frac{d^2}{dx^2}$$

is Hermitian.

SOLUTION: Integrating by parts twice, we obtain

$$\int_{-\infty}^{\infty} f^* \hat{T} f \, dx = -\frac{\hbar^2}{2m} \int_{-\infty}^{\infty} f^* \frac{d^2 f}{dx^2} \, dx = -\frac{\hbar^2}{2m} \left[f^* \frac{df}{dx} \right]_{-\infty}^{\infty} + \frac{\hbar^2}{2m} \int_{-\infty}^{\infty} \frac{df^*}{dx} \frac{df}{dx} dx$$

$$= \frac{\hbar^2}{2m} \left[\frac{df^*}{dx} f \right]_{-\infty}^{\infty} - \frac{\hbar^2}{2m} \int_{-\infty}^{\infty} \frac{d^2 f^*}{dx^2} f \, dx = -\frac{\hbar^2}{2m} \int_{-\infty}^{\infty} f \frac{d^2 f^*}{dx^2} \, dx$$

where, as always, we shall assume that f vanishes at infinity. Therefore, we have

$$\int_{-\infty}^{\infty} f^* \hat{T} f \, dx = \int_{-\infty}^{\infty} f \left(-\frac{\hbar^2}{2m} \frac{d^2}{dx^2} \right) f^* \, dx$$

$$= \int_{-\infty}^{\infty} f \left(-\frac{\hbar^2}{2m} \frac{d^2 f}{dx^2} \right)^* dx = \int_{-\infty}^{\infty} f(\hat{T} f)^* \, dx$$

Thus, Equation 11.17 is satisfied, and the kinetic energy operator is Hermitian.

The definition of a Hermitian operator that is given by Equation 11.17 is not the most general definition. A more general definition of a Hermitian operator is given by

$$\int_{-\infty}^{\infty} f^*(x)\hat{A}g(x)\,dx = \int_{-\infty}^{\infty} f(x)[\hat{A}g(x)]^*\,dx \qquad \text{(Hermitian)} \qquad (11.18)$$

where $f(x)$ and $g(x)$ are any two well-behaved functions. It so happens that it is possible to prove that Equation 11.18 follows from Equation 11.17, and so the definition given by Equation 11.17 suffices if you know this. Problem 11–13 leads you through the proof.

We have been led naturally to the definition and use of Hermitian operators by requiring that quantum-mechanical operators have real eigenvalues. Not only are the eigenvalues of Hermitian operators real, but their eigenfunctions satisfy a rather special condition as well. Consider the two eigenvalue equations

$$\hat{A}\psi_n = a_n\psi_n \qquad\qquad \hat{A}\psi_m = a_m\psi_m \qquad (11.19)$$

We multiply the first of Equations 11.19 by ψ_m^* and integrate to obtain

$$\int_{-\infty}^{\infty} \psi_m^*\hat{A}\psi_n\,dx = a_n \int_{-\infty}^{\infty} \psi_m^*\psi_n\,dx \qquad (11.20)$$

Now take the complex conjugate of the second of Equations 11.19, multiply by ψ_n, and integrate to obtain

$$\int_{-\infty}^{\infty} \psi_n(\hat{A}\psi_m)^*\,dx = a_m^* \int_{-\infty}^{\infty} \psi_n\psi_m^*\,dx = a_m^* \int_{-\infty}^{\infty} \psi_m^*\psi_n\,dx \qquad (11.21)$$

By subtracting Equations 11.20 and 11.21, we obtain

$$\int_{-\infty}^{\infty} \psi_m^*\hat{A}\psi_n\,dx - \int_{-\infty}^{\infty} \psi_n(\hat{A}\psi_m)^*\,dx = (a_n - a_m^*) \int_{-\infty}^{\infty} \psi_m^*\psi_n\,dx \qquad (11.22)$$

Because $\hat{A}$ is Hermitian, Equation 11.18 shows that the left side of Equation 11.22 is zero, and so we have

$$(a_n - a_m^*) \int_{-\infty}^{\infty} \psi_m^*\psi_n\,dx = 0 \qquad (11.23)$$

There are two possibilities to consider in Equation 11.23, $n = m$ and $n \neq m$. When $n = m$, the integral is positive because $\psi_n^*(x)\psi_n(x) \geq 0$ for all values of x, and so we have

$$a_n = a_n^* \qquad (11.24)$$

which is just another proof that the eigenvalues are real.

When $n \neq m$, we have

$$(a_n - a_m) \int_{-\infty}^{\infty} \psi_m^*\psi_n\,dx = 0 \qquad m \neq n \qquad (11.25)$$

Now if the system is nondegenerate, $a_n \neq a_m$, and

$$\int_{-\infty}^{\infty} \psi_m^* \psi_n \, dx = 0 \qquad n \neq m \tag{11.26}$$

Thus, we see that the $\psi_n(x)$ are orthogonal. We have just proved that the eigenfunctions of a Hermitian operator are orthogonal, at least for a nondegenerate system. The quantum-mechanical system of a particle in a box is nondegenerate, and its wave functions are given by (see Example 6–4)

$$\psi_n(x) = \begin{cases} \left(\dfrac{2}{a}\right)^{1/2} \sin \dfrac{n\pi x}{a} & 0 < x < a \\ 0 & \text{otherwise} \end{cases} \tag{11.27}$$

for $n = 1, \ 2, \ 3, \dots$. The factor of $(2/a)^{1/2}$ assures that the $\psi_n(x)$ are normalized, in the sense that

$$\int_o^a \psi_n^2(x) \, dx = \frac{2}{a} \int_0^a \sin^2 \frac{n\pi x}{a} \, dx = 1$$

A set of functions that are both normalized and orthogonal to each other is called an *orthonormal* set. We can express the condition of orthonormality by writing

$$\int_{-\infty}^{\infty} \psi_m^* \psi_n \, dx = \delta_{nm} \tag{11.28}$$

where

$$\delta_{nm} = \begin{cases} 1 & m = n \\ 0 & m \neq n \end{cases} \tag{11.29}$$

where δ_{nm} is the Kronecker delta. (See Equation 8.13 and Problem 8–8.)

EXAMPLE 11–9

Show that the functions

$$\psi_m(\theta) = (2\pi)^{-1/2} e^{im\theta} \qquad m = 0, \ \pm 1, \ \pm 2, \dots$$

form an orthonormal set over the interval $(0, 2\pi)$.

SOLUTION: To prove that a set of functions forms an orthonormal set, we must show that they satisfy Equation 11.28. To see if they do, we have

$$\int_0^{2\pi} \psi_m^*(\theta) \psi_n(\theta) \, d\theta = \frac{1}{2\pi} \int_0^{2\pi} e^{-im\theta} e^{in\theta} \, d\theta$$

$$= \frac{1}{2\pi} \int_0^{2\pi} e^{i(n-m)\theta} \, d\theta$$

$$= \frac{1}{2\pi} \int_0^{2\pi} \cos(n-m)\theta \, d\theta + \frac{i}{2\pi} \int_0^{2\pi} \sin(n-m)\theta \, d\theta$$

For $n \neq m$, the final two integrals vanish because they are over complete cycles of the cosine and sine. For $n = m$, the last integral vanishes because $\sin 0 = 0$ and the next to last gives 2π because $\cos 0 = 1$. Thus,

$$\int_0^{2\pi} \psi_m^*(\theta)\psi_n(\theta)\,d\theta = \delta_{mn}$$

and the $\psi_m(\theta)$ form an orthonormal set.

When we proved that the eigenfunctions of a Hermitian operator are orthogonal, we assumed that the system was nondegenerate; that is, that $a_n - a_m \neq 0$ if $n \neq m$ in Equation 11.25. Even if the system is degenerate, however, we can use the Gram–Schmidt orthogonalization procedure described in Section 8.2 to form a set of orthogonal eigenfunctions as linear combinations of the degenerate set. Thus, we can say that the eigenfunctions of a Hermitian operator are orthogonal, or can be made orthogonal.

Problems

11–1. Evaluate $g = \hat{A}f$, where $\hat{A}$ and f are given below:

$\hat{A}$	f	$\hat{A}$	f
(a) SQRT	x^4	(c) $\int_0^1 dx\ \square$	$x^3 - 2x + 3$
(b) $\dfrac{d^3}{dx^3} + x^3$	e^{-ax}	(d) $\dfrac{\partial^2}{\partial x^2} + \dfrac{\partial^2}{\partial y^2} + \dfrac{\partial^2}{\partial z^2}$	$x^3 y^2 z^4$

11–2. Determine whether the following operators are linear or nonlinear:
(a) $\hat{A}f(x) = \text{SQR}\,f(x)$ [square $f(x)$]
(b) $\hat{A}f(x) = f^*(x)$ [form the complex conjugate of $f(x)$]
(c) $\hat{A}f(x) = [f(x)]^{-1}$ [take the reciprocal of $f(x)$]
(d) $\hat{A}f(x) = \ln f(x)$ [take the logarithm of $f(x)$]

11–3. In each case, show that f is an eigenfunction of the operator given. Find the eigenvalue.

$\hat{A}$	f	$\hat{A}$	f
(a) $\dfrac{d^2}{dx^2}$	$\cos \omega x$	(c) $\dfrac{d^2}{dx^2} + 2\dfrac{d}{dx} + 3$	$e^{\alpha x}$
(b) $\dfrac{d}{dt}$	$e^{i\omega t}$	(d) $\dfrac{\partial}{\partial y}$	$x^2 e^{6y}$

11–4. Show that $(\cos ax)(\cos by)(\cos cz)$ is an eigenfunction of the operator,

$$\nabla^2 = \frac{\partial^2}{\partial x^2} + \frac{\partial^2}{\partial y^2} + \frac{\partial^2}{\partial z^2}$$

which is called the Laplacian operator. What is the eigenvalue?

11-5.   Write out the operator $\hat{A}^2$ for $\hat{A} =$

(a) $\dfrac{d^2}{dx^2}$ (b) $\dfrac{d}{dx} + x$ (c) $\dfrac{d^2}{dx^2} - 2x\dfrac{d}{dx} + 1$

Hint: Be sure to include $f(x)$ before carrying out the operations.

11-6.   Determine whether or not the following pairs of operators commute.

	$\hat{A}$	$\hat{B}$		$\hat{A}$	$\hat{B}$
(a)	$\dfrac{d}{dx}$	$\dfrac{d^2}{dx^2} + 2\dfrac{d}{dx}$	(c)	SQR	SQRT
(b)	x	$\dfrac{d}{dx}$	(d)	$\dfrac{\partial}{\partial x}$	$\dfrac{\partial}{\partial y}$

11-7.   In ordinary algebra, $(P + Q)(P - Q) = P^2 - Q^2$. Expand $(\hat{P} + \hat{Q})(\hat{P} - \hat{Q})$. Under what conditions do we find the same result as in the case of ordinary algebra?

11-8.   Show that

$$\int_0^a e^{\pm i 2\pi nx/a}\, dx = 0 \qquad n \neq 0$$

for integral values of n.

11-9.   Show that the set of functions $\psi_n(x) = (2a)^{-1/2} e^{i\pi nx/a}$, where $n = 0,\ \pm 1,\ \pm 2,\ \ldots$, is orthonormal over the interval $-a \leq x \leq a$.

11-10.   Evaluate the commutator $[\hat{A},\ \hat{B}]$, where $\hat{A}$ and $\hat{B}$ are given below.

	$\hat{A}$	$\hat{B}$		$\hat{A}$	$\hat{B}$
(a)	$\dfrac{d^2}{dx^2}$	x	(c)	$\displaystyle\int_0^x du\ \square$	$\dfrac{d}{dx}$
(b)	$\dfrac{d}{dx} - x$	$\dfrac{d}{dx} + x$	(d)	$\dfrac{d^2}{dx^2} - x$	$\dfrac{d}{dx} + x^2$

11-11.   Which of the following operators is Hermitian: d/dx, $i\,d/dx$, d^2/dx^2, $i\,d^2/dx^2$, $x\,d/dx$, and x? Assume that the functions on which these operators operate are appropriately well behaved at infinity.

11-12.   Show that if $\hat{A}$ is Hermitian, then $\hat{A} - a$, where a is a constant, is Hermitian. Show that the sum of two Hermitian operators is Hermitian.

11-13.   To prove that Equation 11.18 follows from Equation 11.17, first write Equation 11.17 with f and with g:

$$\int f^* \hat{A} f\, dx = \int f\,(\hat{A}f)^*\, dx \qquad \text{and} \qquad \int g^* \hat{A} g\, dx = \int g\,(\hat{A}g)^*\, dx$$

Now let c_1 and c_2 be arbitrary complex constants and write Equation 11.17 as

$$\int (c_1 f + c_2 g)^* \hat{A}\,(c_1 f + c_2 g)\, dx = \int (c_1 f + c_2 g)[\hat{A}\,(c_1 f + c_2 g)]^*\, dx$$

If we expand both sides and use the first two equations, we find that

$$c_1^* c_2 \int f^* \hat{A} g\, dx + c_2^* c_1 \int g^* \hat{A} f\, dx = c_1 c_2^* \int f(\hat{A}g)^*\, dx + c_1^* c_2 \int g(\hat{A}f)^*\, dx$$

Rearrange this into

$$c_1^* c_2 \int [f^* \hat{A} g - g(\hat{A} f)^*] \, dx = c_1 c_2^* \int [f(\hat{A} g)^* - g^* \hat{A} f] \, dx$$

Notice that the two sides of this equation are complex conjugates of each other. If $z = x + iy$ and $z = z^*$, then show that this implies that z is real. Thus, both sides of this equation are real. But because c_1 and c_2 are arbitrary complex constants, the only way for both sides to be real is for both integrals to equal zero. Show that this implies Equation 11.18.

11–14. Show that if $\hat{A}$ and $\hat{B}$ are Hermitian, then

$$\int \psi_n (\hat{B} \hat{A} \psi_m)^* \, dx = \int \psi_m^* \hat{A} \hat{B} \psi_n \, dx$$

Hint: Use Equation 11.18.

11–15. Use the result of the previous problem to show that if $\hat{A}$ and $\hat{B}$ are Hermitian, then $\hat{A} \hat{B}$ is Hermitian only if $\hat{A}$ and $\hat{B}$ commute.

11–16. Let $f(\hat{A})$ be a polynomial function of $\hat{A}$:

$$f(\hat{A}) = a_0 + a_1 \hat{A} + a_2 \hat{A}^2 + \cdots + a_N \hat{A}^N$$

Show that if ψ is an eigenfunction of $\hat{A}$ with eigenvalue β, then

$$f(\hat{A})\psi = f(\beta)\psi$$

11–17. Show that if $f(\hat{A})$ and $g(\hat{A})$ are two polynomial functions of $\hat{A}$ (see the previous problem), then $f(\hat{A})$ and $g(\hat{A})$ commute.

11–18. We can define $\exp(\hat{A})$ by its Maclaurin series

$$e^{\hat{A}} = \hat{I} + \hat{A} + \frac{1}{2!}\hat{A}^2 + \frac{1}{3!}\hat{A}^3 + \cdots$$

Under what conditions does $e^{\hat{A}+\hat{B}} = e^{\hat{A}} e^{\hat{B}}$?

11–19. Show that $[\hat{A}, \hat{B}\hat{C}] = [\hat{A}, \hat{B}]\hat{C} + \hat{B}[\hat{A}, \hat{C}]$.

11–20. Show that

$$\psi_0(x) = (1/\pi)^{1/4} e^{-x^2/2}$$
$$\psi_1(x) = (4/\pi)^{1/4} x e^{-x^2/2}$$
$$\psi_2(x) = (1/4\pi)^{1/4}(2x^2 - 1) e^{-x^2/2}$$

are orthonormal over the interval $(-\infty, \infty)$.

11–21. Show that the set of functions $\{(2/a)^{1/2} \cos(n\pi x/a)\}$, $n = 0, 1, 2, \ldots$, is orthonormal over the interval $(0, a)$.

CHAPTER 12

Functions of Several Variables

In Chapters 1 and 2, we reviewed some of the essential features of the calculus of functions of a single variable. As the title to this chapter implies, we shall now discuss functions of more than one variable. Many physical quantities depend upon more than one variable. For example, the pressure of a fixed quantity of a gas depends upon the temperature and the volume. Even just a cursory look at any book on thermodynamics shows that the formulas of thermodynamics abound in partial derivatives. A significant difference between functions of a single variable and functions of more than one variable is that you can form partial derivatives of functions of several variables, which leads to mixed partial higher derivatives and a variety of chain rules.

After discussing partial derivatives and total derivatives, we then go on to discuss maxima and minima of functions of two variables. This topic is a little more involved than for functions of a single variable. The nature of a critical point (a point where the first partial derivatives are equal to zero) depends upon relative values of the second partial derivatives, $\partial^2 f/\partial x^2$, $\partial^2 f/\partial y^2$, and $\partial^2 f/\partial x \partial y$, and can give rise to a maximum, a minimum, or a saddle point. Finally, in the last section we discuss multiple integrals.

12.1 Partial Derivatives

The evaluation of a partial derivative of a function of several variables is similar to the evaluation of the derivative of a function of a single variable. Suppose we have a function $f(x, y)$. Then the partial derivative of $f(x, y)$ with respect to x is defined by

$$\frac{\partial f}{\partial x} = \lim_{\Delta x \to 0} \frac{f(x + \Delta x, y) - f(x, y)}{\Delta x} \tag{12.1}$$

with a similar equation for $\partial f/\partial y$.

Equation 12.1 says that you can determine partial derivatives by differentiating with respect to one variable while treating the other one as a constant. For example, if $f(x, y) = e^x \sin xy$, then

$$f_x = \frac{\partial f}{\partial x} = e^x \sin xy + ye^x \cos xy$$

and

$$f_y = \frac{\partial f}{\partial y} = xe^x \cos xy$$

where we have introduced the common notation f_x for $\partial f/\partial x$ and f_y for $\partial f/\partial y$. This notation is often convenient, but you have to be careful not to confuse f_x and f_y with the components of a vector. Sometimes we write $(\partial f/\partial x)_y$ to emphasize that y is held constant when we take the derivative of f with respect to x. Usually, however, it will be clear from the context which variables are held constant.

EXAMPLE 12–1

The van der Waals equation is an approximate equation for the pressure of a gas as a function of its temperature and volume. The van der Waals equation for one mole of a gas is

$$P = \frac{RT}{V - b} - \frac{a}{V^2} \qquad (12.2)$$

where R is the molar gas constant and a and b are constants that are characteristic of the particular gas. Determine $\partial P/\partial T$ and $\partial P/\partial V$.

SOLUTION:

$$\left(\frac{\partial P}{\partial T} \right)_V = \frac{R}{V - b}$$

and

$$\left(\frac{\partial P}{\partial V} \right)_T = -\frac{RT}{(V - b)^2} + \frac{2a}{V^3}$$

Physically, these equations govern how the pressure varies as we change the temperature at constant volume, and how the pressure changes as we change the volume at constant temperature. Notice that $(\partial P/\partial V)_T$ is a function of both T and V.

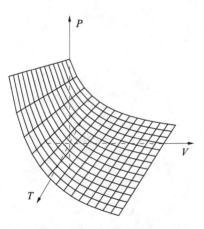

Figure 12.1. The pressure surface for the van der Waals equation. The pressure of the gas is equal to the height of the surface above the TV plane.

Partial derivatives have a nice geometric interpretation. Figure 12.1 shows the surface corresponding to Equation 12.2. Let's choose some fixed temperature T_0. The condition $T_0 = $ constant is described by a plane parallel to the PV plane and intersecting the T axis at T_0. Then the partial derivative $(\partial P/\partial V)_{T_0}$ is the slope of the line formed by the intersection of the pressure surface in Figure 12.1 with

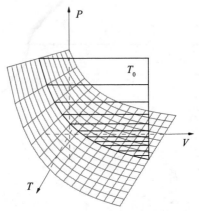

P

T_0

V

T

Figure 12.2. The intersection of the T_0 plane and the pressure surface shown in Figure 12.1. The slope of this line is $(\partial P/\partial V)_{T_0}$.

the T_0 = constant plane. Figure 12.2 shows the curve formed by the intersection of the T_0 plane and the pressure surface. The slope of the curve is $(\partial P/\partial V)_{T_0}$. Similarly, the partial derivative $(\partial P/\partial T)_V$ is the slope of the curve that is formed by the intersection of the pressure surface and a plane that is parallel to the PT plane; in other words, a plane with $V = V_0$ = constant.

You should realize that f_x and f_y themselves can be functions of x and y. For example, if $f(x, y) = y^2 e^x$, then $f_x(x, y) = y^2 e^x$ and $f_y(x, y) = 2ye^x$. Therefore, we can form partial derivatives of $f_x(x, y)$ and $f_y(x, y)$ just as we did for $f(x, y)$. The second partial derivatives of $f(x, y)$ are

$$\frac{\partial}{\partial x}\left(\frac{\partial f}{\partial x}\right) = \frac{\partial f_x}{\partial x} = \frac{\partial^2 f}{\partial x^2} = f_{xx}$$

$$\frac{\partial}{\partial y}\left(\frac{\partial f}{\partial y}\right) = \frac{\partial f_y}{\partial y} = \frac{\partial^2 f}{\partial y^2} = f_{yy}$$

(12.3)

EXAMPLE 12–2

Show that $V(x, y, z) = (x^2 + y^2 + z^2)^{-1/2}$ satisfies the equation

$$\frac{\partial^2 V}{\partial x^2} + \frac{\partial^2 V}{\partial y^2} + \frac{\partial^2 V}{\partial z^2} = 0$$

everywhere except for the origin $(0, 0, 0)$, where it diverges. This equation is called Laplace's equation. Among other things, Laplace's equation determines the electrostatic potential $V(x, y, z)$ in a charge-free region. Observe that $V(x, y, z) = (x^2 + y^2 + z^2)^{-1/2} = 1/r$, the Coulomb potential (within a multiplicative constant) due to a charge situated at the origin.

SOLUTION:

$$V_x = -\frac{x}{(x^2 + y^2 + z^2)^{3/2}}$$

$$V_{xx} = \frac{2x^2 - y^2 - z^2}{(x^2 + y^2 + z^2)^{5/2}}$$

Because $V(x, y, z)$ is symmetric in x, y, and z, we can obtain V_{yy} by interchanging x and y in V_{xx}, and we can obtain V_{zz} by interchanging x and z. Therefore,

$$V_{xx} + V_{yy} + V_{zz} = \frac{(2x^2 - y^2 - z^2) + (2y^2 - x^2 - z^2) + (2z^2 - x^2 - y^2)}{(x^2 + y^2 + z^2)^{5/2}}$$

$$= 0$$

The types of derivatives in Equation 12.3 are similar to the second derivative of a function of a single variable, but functions of more than one variable admit *mixed partial second derivatives* as well:

$$\frac{\partial}{\partial x}\left(\frac{\partial f}{\partial y}\right) = \frac{\partial^2 f}{\partial x \partial y} = f_{yx}$$

$$\frac{\partial}{\partial y}\left(\frac{\partial f}{\partial x}\right) = \frac{\partial^2 f}{\partial y \partial x} = f_{xy}$$

(12.4)

For example, if $f(x, y) = xy^2 + e^{x^2 y}$, then

$$f_x = y^2 + 2xye^{x^2 y} \qquad\qquad f_y = 2xy + x^2 e^{x^2 y}$$

$$f_{xy} = 2y + 2x(1 + x^2 y)e^{x^2 y} \qquad f_{yx} = 2y + 2x(1 + x^2 y)e^{x^2 y}$$

Note that $f_{xy} = f_{yx}$. This will always be true if f_{xy} and f_{yx} are continuous, which is the case for almost all the functions that we deal with in physical applications. Problem 12–3 has you show that $\partial^2 P / \partial V \partial T = \partial^2 P / \partial T \partial V$ for the van der Waals equation.

The equality of mixed partial second derivatives is used often in thermodynamics. For example, thermodynamics tells us that the entropy (S) and the pressure (P) of a substance can be expressed as partial derivatives of the Helmholtz energy, $A = A(V, T)$, which is a function of the volume and the temperature:

$$S = -\left(\frac{\partial A}{\partial T}\right)_V \qquad \text{and} \qquad P = -\left(\frac{\partial A}{\partial V}\right)_T$$

Using the relation

$$\frac{\partial^2 A}{\partial V \partial T} = \left[\frac{\partial}{\partial V}\left(\frac{\partial A}{\partial T}\right)_V\right]_T = \left[\frac{\partial}{\partial T}\left(\frac{\partial A}{\partial V}\right)_T\right]_V$$

$$= \frac{\partial^2 A}{\partial T \partial V}$$

we see that

$$\left(\frac{\partial S}{\partial V}\right)_T = \left(\frac{\partial P}{\partial T}\right)_V$$

(12.5)

Equation 12.5 is known as a Maxwell relation in thermodynamics. The derivation of Equation 12.5 is a typical thermodynamic manipulation. Equation 12.5 is an important and useful equation because it allows us to calculate the entropy of a substance (which is not a directly measurable quantity) in terms of the pressure-volume-temperature (P-V-T) dependence of the substance, which is readily measurable.

12.2 Total Differentials

The partial derivatives given in Example 12–1 indicate how P changes with one independent variable, keeping the other one fixed. We often want to know how a dependent variable changes with a change in the values of both (or more) of its independent variables. Using the example $P = P(T, V)$ (for one mole), we write

$$\Delta P = P(T + \Delta T, V + \Delta V) - P(T, V)$$

If we add and subtract $P(T, V + \Delta V)$ to this equation, we obtain

$$\Delta P = [P(T + \Delta T, V + \Delta V) - P(T, V + \Delta V)]$$
$$+ [P(T, V + \Delta V) - P(T, V)]$$

Multiply the first two terms in brackets by $\Delta T/\Delta T$ and the second two terms by $\Delta V/\Delta V$ to get

$$\Delta P = \left[\frac{P(T + \Delta T, V + \Delta V) - P(T, V + \Delta V)}{\Delta T} \right] \Delta T$$
$$+ \left[\frac{P(T, V + \Delta V) - P(T, V)}{\Delta V} \right] \Delta V$$

Now let $\Delta T \to 0$ and $\Delta V \to 0$, in which case we have

$$dP = \lim_{\Delta T \to 0} \left[\frac{P(T + \Delta T, V) - P(T, V)}{\Delta T} \right] \Delta T$$
$$+ \lim_{\Delta V \to 0} \left[\frac{P(T, V + \Delta V) - P(T, V)}{\Delta V} \right] \Delta V \qquad (12.6)$$

The first limit gives $(\partial P/\partial T)_V$ (by definition) and the second gives $(\partial P/\partial V)_T$, so that Equation 12.6 gives our desired result:

$$dP = \left(\frac{\partial P}{\partial T} \right)_V dT + \left(\frac{\partial P}{\partial V} \right)_T dV \qquad (12.7)$$

Equation 12.7 is called the *total differential* of P. It simply says that the change in P is given by how P changes with T (keeping V constant) times the infinitesimal change in T plus how P changes with V (at constant T) times the infinitesimal change in V.

EXAMPLE 12–3

We can use Equation 12.7 to estimate the change in pressure when both the temperature and the volume are changed slightly. To this end, for finite ΔT and ΔV, we write Equation 12.7 as

$$\Delta P \approx \left(\frac{\partial P}{\partial T} \right)_V \Delta T + \left(\frac{\partial P}{\partial V} \right)_T \Delta V$$

Use this equation to estimate the change in pressure of one mole of an ideal gas if the temperature is changed from 273.15 K to 274.00 K and the volume is changed from 10.00 L to 9.90 L.

SOLUTION: We first need

$$\left(\frac{\partial P}{\partial T} \right)_V = \left[\frac{\partial}{\partial T} \left(\frac{RT}{V} \right) \right]_V = \frac{R}{V}$$

and

$$\left(\frac{\partial P}{\partial V}\right)_T = \left[\frac{\partial}{\partial V}\left(\frac{RT}{V}\right)\right]_V = -\frac{RT}{V^2}$$

so that

$$\Delta P \approx \frac{R}{V}\Delta T - \frac{RT}{V^2}\Delta V$$

$$= \frac{(8.314 \text{ J·K}^{-1}\cdot\text{mol}^{-1})}{(10.00 \text{ L·mol}^{-1})}(0.85 \text{ K})$$

$$- \frac{(8.314 \text{ J·K}^{-1}\cdot\text{mol}^{-1})(273.15 \text{ K})}{(10.00 \text{ L·mol}^{-1})^2}(-0.10 \text{ L·mol}^{-1})$$

$$= 3.0 \text{ J·L}^{-1} = 3.0 \times 10^3 \text{ J·m}^{-3} = 3.0 \times 10^3 \text{ Pa} = 0.030 \text{ bar}$$

Incidentally, in this particularly simple case, we calculate the exact change in P from

$$\Delta P = \frac{RT_2}{V_2} - \frac{RT_1}{V_1}$$

$$= (8.314 \text{ J·K}^{-1}\cdot\text{mol}^{-1})\left(\frac{274.00 \text{ K}}{9.90 \text{ L·mol}^{-1}} - \frac{273.15 \text{ K}}{10.00 \text{ L·mol}^{-1}}\right)$$

$$= 3.0 \text{ J·L}^{-1} = 0.030 \text{ bar}$$

Example 12–1 gives P as a function of T and V, or $P = P(T, V)$ for one mole of a van der Waals gas. We can form the total differential of P by differentiating the right side of Equation 12.2 with respect to T and V to obtain

$$dP = \frac{R}{V - b}dT - \frac{RT}{(V - b)^2}dV + \frac{2a}{V^3}dV$$

$$= \frac{R}{V - b}dT + \left[-\frac{RT}{(V - b)^2} + \frac{2a}{V^3}\right]dV \tag{12.8}$$

We can see from Example 12–1 that Equation 12.8 is just Equation 12.7 written for the van der Waals equation. Suppose, however, that we are given an arbitrary expression for dP, say,

$$dP = \frac{RT}{V - b}dT + \left[\frac{RT}{(V - b)^2} - \frac{a}{TV^2}\right]dV \tag{12.9}$$

and are asked to determine the equation of state $P = P(T, V)$ that leads to Equation 12.9. In fact, a simpler question is to ask if there even is a function $P(T, V)$ whose total differential is given by Equation 12.9. How can we tell? If there is such a function $P(T, V)$, then its total differential is (Equation 12.7)

$$dP = \left(\frac{\partial P}{\partial T}\right)_V dT + \left(\frac{\partial P}{\partial V}\right)_T dV$$

If there is such an equation of state, then its cross derivatives,

$$\left(\frac{\partial^2 P}{\partial V \partial T}\right) = \left[\frac{\partial}{\partial V}\left(\frac{\partial P}{\partial T}\right)_V\right]_T \quad \text{and} \quad \left(\frac{\partial^2 P}{\partial T \partial V}\right) = \left[\frac{\partial}{\partial T}\left(\frac{\partial P}{\partial V}\right)_T\right]_V$$

must be equal. If we apply this requirement to Equation 12.9, we find that

$$\frac{\partial}{\partial T}\left[\frac{RT}{(V-b)^2} - \frac{a}{TV^2}\right] = \frac{R}{(V-b)^2} + \frac{a}{T^2V^2}$$

and

$$\frac{\partial}{\partial V}\left(\frac{RT}{V-b}\right) = -\frac{RT}{(V-b)^2}$$

Thus, we see that the cross derivatives are *not* equal, so the expression given by Equation 12.9 is not the total differential of any function $P(T, V)$.

The differential df of a function $f(x, y)$ is called an *exact differential*. Equation 12.8 is an example of an exact differential. It is obtained by explicitly forming the total differential of $P(V, T)$ for the van der Waals equation. Problem 12–3 shows that the cross derivatives are equal, as they must be for an exact differential. If an expression for df turns out not to be the differential of a function, then df is called an *inexact differential*. Equation 12.9 is an example of an inexact differential. The cross derivatives of an inexact differential are not equal.

Exact and inexact differentials play a significant role in physical chemistry. If dy is an exact differential, then

$$\int_1^2 dy = y_2 - y_1 \qquad \text{(exact differential)}$$

so the integral depends only upon the endpoints (1 and 2) and not upon the path from 1 to 2. This statement is not true for an inexact differential, however, so

$$\int_1^2 dy \neq y_2 - y_1 \qquad \text{(inexact differential)}$$

The integral in this case depends not only upon the endpoints but also upon the path from 1 to 2.

The variable y in the first case is called a *state function* because its value depends only upon the initial and final states of the system, designated by 1 and 2 respectively, and not upon how the system gets from one state to another. In the second case, y is not a state function because its value depends upon how the system gets from the initial state to the final state.

12.3 Chain Rules for Partial Differentiation

Suppose that $u = f(x, y)$, where x and y are functions of a single variable t. Then the composite function $f(x(t), y(t)) = u(t)$ is a function of a single variable t and

$$\frac{du}{dt} = \frac{\partial u}{\partial x}\frac{dx}{dt} + \frac{\partial u}{\partial y}\frac{dy}{dt} \tag{12.10}$$

provided $\partial u/\partial x$ and $\partial u/\partial y$ are continuous. Equation 12.10 is called the *chain rule of partial differentiation*. Equation 12.10 is readily extended to a function of more than two independent variables.

EXAMPLE 12–4

Use Equation 12.10 to evaluate du/dt if $u(x, y) = x^2y + xy^2$ and $x(t) = te^{-t}$ and $y(t) = e^{-t}$.

SOLUTION:

$$\frac{\partial u}{\partial x} = 2xy + y^2 \qquad \frac{\partial u}{\partial y} = x^2 + 2xy$$

$$\frac{dx}{dt} = (1 - t)e^{-t} \qquad \frac{dy}{dt} = -e^{-t}$$

and so

$$\frac{du}{dt} = (2xy + y^2)(1 - t)e^{-t} + (x^2 + 2xy)(-e^{-t})$$

$$= [(1 + 2t)(1 - t) - t(t + 2)]e^{-3t}$$

$$= (1 - t - 3t^2)e^{-3t}$$

Of course you get the same result by substituting $x(t)$ and $y(t)$ into $u(x, y)$ and then differentiating with respect to t (Problem 12–13).

Now suppose that $u = u(x, y)$ and that $x = x(s, t)$ and $y = y(s, t)$. In this case u is also a function of s and t. We can extend Equation 12.10 to write

$$\frac{\partial u}{\partial s} = \frac{\partial u}{\partial x}\frac{\partial x}{\partial s} + \frac{\partial u}{\partial y}\frac{\partial y}{\partial s} \qquad (12.11a)$$

and

$$\frac{\partial u}{\partial t} = \frac{\partial u}{\partial x}\frac{\partial x}{\partial t} + \frac{\partial u}{\partial y}\frac{\partial y}{\partial t} \qquad (12.11b)$$

When u is regarded as a function of s and t and is differentiated with respect to s or t, then the other variable is held constant during the partial differentiation. When u is regarded as a function of x and y and is differentiated with respect to x or y, then the other variable is held constant during the partial differentiation.

EXAMPLE 12–5

If $u(x, y) = ye^{-x} + xy$ and $x(s, t) = s^2t$ and $y(s, t) = e^{-s} + t$, evaluate $\partial u/\partial s$ and $\partial u/\partial t$.

SOLUTION:

$$\frac{\partial u}{\partial s} = \frac{\partial u}{\partial x}\frac{\partial x}{\partial s} + \frac{\partial u}{\partial y}\frac{\partial y}{\partial s}$$

$$= (-ye^{-x} + y)(2st) + (e^{-x} + x)(-e^{-s})$$

$$= 2st[-(e^{-s} + t)e^{-s^2t} + e^{-s} + t] - (e^{-s^2t} + s^2t)e^{-s}$$

$$= 2st(e^{-s} + t)(1 - e^{-s^2t}) - (e^{-s^2t} + s^2t)e^{-s}$$

$$\frac{\partial u}{\partial t} = \frac{\partial u}{\partial x}\frac{\partial x}{\partial t} + \frac{\partial u}{\partial y}\frac{\partial y}{\partial t}$$

$$= (-ye^{-x} + y)s^2 + (e^{-x} + x)$$

$$= (e^{-s^2t} + s^2t) + s^2(e^{-s} + t)(1 - e^{-s^2t})$$

Problem 12–14 has you show that you get the same result by substituting $x = s^2t$ and $y = e^{-s} + t$ into $u(x, y)$ and then forming $\partial u/\partial s$ and $\partial u/\partial t$ directly.

There is another form of the chain rule that is very useful in thermodynamics. If $u = u(x, y)$ and $y = y(x, z)$, then

$$\left(\frac{\partial u}{\partial x}\right)_z = \left(\frac{\partial u}{\partial x}\right)_y + \left(\frac{\partial u}{\partial y}\right)_x \left(\frac{\partial y}{\partial x}\right)_z \tag{12.12}$$

Although we do not often use this equation directly in physical chemistry, it serves as a justification for a formal procedure that is used often. (A mathematical procedure is called formal if symbols are manipulated without regard to rigor.) We can "derive" Equation 12.12 by starting with the total differential of u,

$$du = \left(\frac{\partial u}{\partial x}\right)_y dx + \left(\frac{\partial u}{\partial y}\right)_x dy$$

and then dividing by dx while keeping z constant to obtain

$$\left(\frac{\partial u}{\partial x}\right)_z = \left(\frac{\partial u}{\partial x}\right)_y + \left(\frac{\partial u}{\partial y}\right)_x \left(\frac{\partial y}{\partial x}\right)_z$$

This is the same result as Equation 12.12, which serves to justify our formal procedure.

Let's see how we use this procedure in thermodynamics. Consider the thermodynamic equation

$$dU = TdS - PdV \tag{12.13}$$

which is just a statement of the first law of thermodynamics. It is common practice to say "divide (Equation 12.13) through by dV at constant T" to write

$$\left(\frac{\partial U}{\partial V}\right)_T = T\left(\frac{\partial S}{\partial V}\right)_T - P \tag{12.14}$$

Using Equation 12.5, Equation 12.14 becomes

$$\left(\frac{\partial U}{\partial V}\right)_T = T\left(\frac{\partial P}{\partial T}\right)_V - P \tag{12.15}$$

which is a standard equation of thermodynamics. It's easy to see that $(\partial U/\partial V)_T = 0$ for an ideal gas, which says that the thermodynamic energy of an ideal gas does not change in an isothermal expansion. Problems 12–28 through 12–32 involve deriving some other equations that you meet in thermodynamics.

12.4 Euler's Theorem

There is a theorem called *Euler's theorem* that is useful in thermodynamics and a number of other fields. First we define a *homogeneous function of degree p* as one that has the property that

$$f(\lambda x_1, \lambda x_2, \ldots, \lambda x_n) = \lambda^p f(x_1, x_2, \ldots, x_n) \tag{12.16}$$

where λ is a parameter. For example, $f(x, y, z) = x^2 z + yz^2 + xyz$ is homogeneous of degree 3 since

$$f(\lambda x, \lambda y, \lambda z) = (\lambda x)^2(\lambda z) + (\lambda y)(\lambda z)^2 + (\lambda x)(\lambda y)(\lambda z)$$
$$= \lambda^3(x^2 z + yz^2 + xyz) = \lambda^3 f(x, y, z)$$

Not every independent variable has to appear in Equation 12.16. The function $f(x, y, z, w)$ given by

$$f(x, y, z, w) = xy \sin z + \frac{x^3}{y}e^{-w^2}$$

is homogeneous of degree 2 in the independent variables x and y because

$$f(\lambda x, \lambda y) = \lambda^2 f(x, y) \tag{12.17}$$

The independent variables z and w are simply suppressed in Equation 12.17.

Euler's theorem says that

If $f(\lambda x, \lambda y) = \lambda^p f(x, y)$, then

$$pf(x, y) = x\frac{\partial f}{\partial x} + y\frac{\partial f}{\partial y} \tag{12.18}$$

The proof of Euler's theorem goes as follows. Start with

$$f(\lambda x, \lambda y) = \lambda^p f(x, y)$$

and let $u = \lambda x$ and $v = \lambda y$. Now differentiate both sides of

$$f(\lambda x, \lambda y) = \lambda^p f(x, y)$$

with respect to λ to get

$$\frac{\partial f}{\partial u}\frac{\partial u}{\partial \lambda} + \frac{\partial f}{\partial v}\frac{\partial v}{\partial \lambda} = p\lambda^{p-1} f(x, y)$$

But $\partial u/\partial\lambda = x$ and $\partial v/\partial\lambda = y$, so

$$p\lambda^{p-1}f(x, y) = x\frac{\partial f}{\partial u} + y\frac{\partial f}{\partial v}$$

Because this equation is true for any value of λ, it is true for $\lambda = 1$ and so $u = x$ and $v = y$ and

$$pf(x, y) = x\frac{\partial f}{\partial x} + y\frac{\partial f}{\partial y}$$

The usefulness of Euler's theorem in thermodynamics stems from the fact that if the variables x and y are extensive thermodynamic variables, then $f(\lambda x, \lambda y) = \lambda f(x, y)$. The following Example gives a concrete illustration of this result.

EXAMPLE 12–6

The thermodynamic energy (U) of a system can be expressed as a function of the entropy (S), the volume (V), and the number of moles (n). Use Euler's theorem to derive

$$U = S\left(\frac{\partial U}{\partial S}\right)_{V,n} + V\left(\frac{\partial U}{\partial V}\right)_{S,n} + n\left(\frac{\partial U}{\partial n}\right)_{S,V}$$

Do you recognize the resulting equation?

SOLUTION: The entropy, volume, and number of moles are all extensive thermodynamic quantities, and so $U(\lambda S, \lambda V, \lambda n) = \lambda U(S, V, n)$. Equation 12.18 says that

$$U = S\left(\frac{\partial U}{\partial S}\right)_{V,n} + V\left(\frac{\partial U}{\partial V}\right)_{S,n} + n\left(\frac{\partial U}{\partial n}\right)_{S,V}$$

This result is equivalent to the equation $G = \mu n = U - TS + PV = H - TS$ when you recognize that $(\partial U/\partial S)_{V,n} = T$, $(\partial U/\partial V)_{S,n} = -P$, and $(\partial U/\partial n)_{S,V} = \mu$ (the chemical potential).

Problems 12–17 and 12–18 provide other applications of Euler's theorem to thermodynamics.

12.5 Maxima and Minima

A function $f(x, y)$ has a local maximum value at the point (a, b) if $f(a, b)$ is greater than $f(x, y)$ evaluated at its neighboring points, and $f(x, y)$ has a minimum value if $f(a, b)$ is less than $f(x, y)$ evaluated at its neighboring points. This observation has the following geometrical interpretation. Let the equation $z = f(x, y)$ describe a smooth surface in three-dimensional space. If $f(x, y)$ has a local maximum value at (a, b) and if we denote this maximum value by $c = f(a, b)$, then the plane $z = c = f(a, b)$ is the horizontal tangent plane to the surface at the point

$x = a, y = b, z = c$, and all the points on the surface $z = f(x, y)$ in the neighborhood of this point lie below the plane (Figure 12.3). Similarly, if $f(x, y)$ has a local minimum value at (a, b), then all the points on the surface $z = f(x, y)$ in the neighborhood of (a, b) lie above the horizontal tangent plane $z = c = f(a, b)$ (Figure 12.4).

We can express the fact that $f(x, y)$ has a local maximum value at (a, b) mathematically by noting that $f(a \pm h, b \pm k) < f(a, b)$ for small values of h and k. Now, if we let $k = 0$, then this definition says that $f(x, y)$ is a local maximum at (a, b) if $f(a + h, b) < f(a, b)$. But this statement simply says that $f(x, b)$ is a local maximum at $x = a$ where $f(x, y)$ is considered to be a function of x with y held constant at $y = b$. Thus, the condition that a function of a single variable be a maximum at $x = a$, namely, $f'(x) = 0$ at $x = a$, becomes

$$\frac{\partial f}{\partial x} = 0 \quad \text{at } x = a, \ y = b \quad \text{or} \quad f_x(a, b) = 0 \quad (12.19)$$

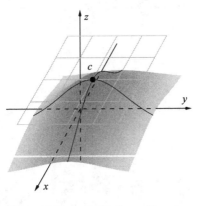

Figure 12.3. A graph of $z = f(x, y)$ in the neighborhood of its maximum value c at (a, b) and the plane $z = c = f(a, b)$.

The same argument with $h = 0$ instead of $k = 0$ gives

$$\frac{\partial f}{\partial y} = 0 \quad \text{at } x = a, \ y = b \quad \text{or} \quad f_y(a, b) = 0 \quad (12.20)$$

Furthermore, the same argument applies to the case in which $f(x, y)$ is a local minimum at (a, b), so Equations 12.19 and 12.20 must be satisfied for $f(x, y)$ to have a local extremum at (a, b). The point (a, b) is said to be a *critical point* of $f(x, y)$.

Just as in the case of a function of a single variable, however, Equations 12.19 and 12.20 are *necessary conditions*, but not sufficient conditions, that $f(x, y)$ be a local extremum at (a, b). A good example of the fact that Equations 12.19 and 12.20 are necessary but not sufficient conditions that $f(x, y)$ be an extremum at (a, b) is provided by $f(x, y) = x^2 - y^2$. We see that $f_x = 2x$ and $f_y = -2y$ are both equal to zero at the point $(0, 0)$, so Equations 12.19 and 12.20 are satisfied. Yet, considered as a function of x with y held constant at $y = 0$, $f(x, 0)$ has a minimum at $(0, 0)$ because $f_{xx} = 2 > 0$ at $(0, 0)$, while considered as a function of y with x held constant at $x = 0$, $f(0, y)$ has a maximum at $(0, 0)$ because $f_{yy} = -2 < 0$ at $(0, 0)$. Thus, the surface $z = f(x, y)$ has a maximum in the yz plane and a minimum in the xz plane, as shown in Figure 12.5. The critical point in this case is called a *saddle point* for obvious reasons.

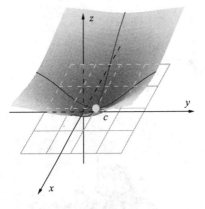

Figure 12.4. A graph of $z = f(x, y)$ in the neighborhood of its minimum value c at (a, b) and the plane $z = c = f(a, b)$.

EXAMPLE 12–7
Find the critical points for

$$f(x, y) = x^3 + y^3 - x - 6y + 10$$

SOLUTION: The equations for the critical points are

$$\frac{\partial f}{\partial x} = 3x^2 - 1 = 0 \quad \text{and} \quad \frac{\partial f}{\partial y} = 3y^2 - 6 = 0$$

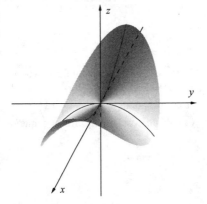

Figure 12.5. The saddle point of $z = x^2 - y^2$ at the point $(0, 0)$.

which yields the critical points $x = \pm 1/\sqrt{3}$ and $y = \pm\sqrt{2}$. The critical points are at

$$\left(\frac{1}{\sqrt{3}}, \sqrt{2}\right), \ \left(-\frac{1}{\sqrt{3}}, \sqrt{2}\right), \ \left(\frac{1}{\sqrt{3}}, -\sqrt{2}\right), \ \text{and} \ \left(-\frac{1}{\sqrt{3}}, -\sqrt{2}\right)$$

If $f'(a) = 0$ and $f''(a) \neq 0$ for a function of a single variable, it is the sign of $f''(a)$ that determines if $f(a)$ is a maximum or a minimum. Similarly, the signs of the second partial derivatives of $f(x, y)$ determine whether $f(a, b)$ is a maximum, a minimum, or a saddle point when $f_x(a, b) = f_y(a, b) = 0$. We first must define the quantity

$$D(a, b) = f_{xx}(a, b)f_{yy}(a, b) - f_{xy}^2(a, b)$$

Now, if

$$f_x(a, b) = 0 \qquad \text{and} \qquad f_y(a, b) = 0$$

then $f(a, b)$ will be

1. a local maximum if $f_{xx}(a, b) < 0$ and $D(a, b) > 0$
2. a local minimum if $f_{xx}(a, b) > 0$ and $D(a, b) > 0$
3. a saddle point if $D(a, b) < 0$

If $D(a, b) = 0$, higher partial derivatives must be considered.

Let's look at the four critical points in Example 12–7. First we note that

$$f_{xx} = 6x, \qquad f_{yy} = 6y, \qquad f_{xy} = 0, \qquad \text{and} \qquad D = 36xy$$

Thus, at

$$\left(\frac{1}{\sqrt{3}}, \sqrt{2}\right): \ D > 0, \ f_{xx} > 0; \ \text{a local minimum}$$

$$\left(-\frac{1}{\sqrt{3}}, \sqrt{2}\right): \ D < 0; \ \text{a saddle point}$$

$$\left(\frac{1}{\sqrt{3}}, -\sqrt{2}\right): \ D < 0; \ \text{a saddle point}$$

$$\left(-\frac{1}{\sqrt{3}}, -\sqrt{2}\right): \ D > 0, \ f_{xx} < 0; \ \text{a local maximum}$$

You can see these critical points in Figure 12.6.

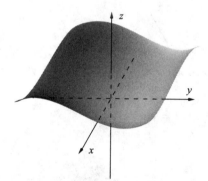

Figure 12.6. The graph of the function $z(x, y) = x^3 + y^3 - x - 6y + 10$ around its critical points at $\left(\pm\frac{1}{\sqrt{3}}, \pm\sqrt{2}\right)$. Can you identify each of the critical points?

EXAMPLE 12–8
Investigate the critical points of

$$f(x, y) = \ln(x^2 + y^2 + 2)$$

SOLUTION: The equations for the critical points are

$$f_x = \frac{2x}{x^2 + y^2 + 2} = 0 \qquad f_y = \frac{2y}{x^2 + y^2 + 2} = 0$$

from which we see that there is a critical point at $(0, 0)$. The second partial derivatives evaluated at the critical point are $f_{xx}(0, 0) = 1$; $f_{yy}(0, 0) = 1$; $f_{xy}(0, 0) = 0$. Therefore, $D(0, 0) = 1 > 0$ and $f_{xx}(0, 0) = 1 > 0$, and so the critical point is a local minimum (Figure 12.7).

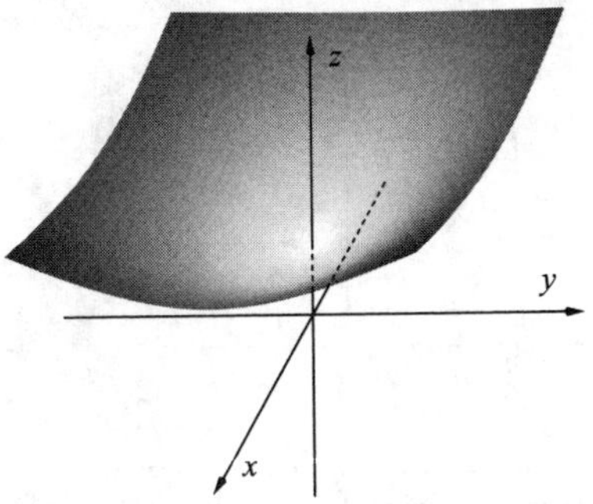

Figure 12.7. The behavior of the function $z(x, y) = \ln(x^2 + y^2 + 2)$, showing that there is a minimum at its critical point at $(0, 0)$.

EXAMPLE 12–9

Investigate the critical points of

$$f(x, y) = \frac{1}{2}x^2 - xy$$

SOLUTION: The critical points are given by

$$f_x = x - y = 0 \qquad f_y = -x = 0$$

and so we see that there is a critical point at $(0, 0)$. The second partial derivatives evaluated at the critical point are $f_{xx}(0, 0) = 1 > 0$; $f_{yy}(0, 0) = 0$; and $f_{xy}(0, 0) = -1$. Therefore, $D(0, 0) = -1 < 0$, and so the critical point is a saddle point (Figure 12.8).

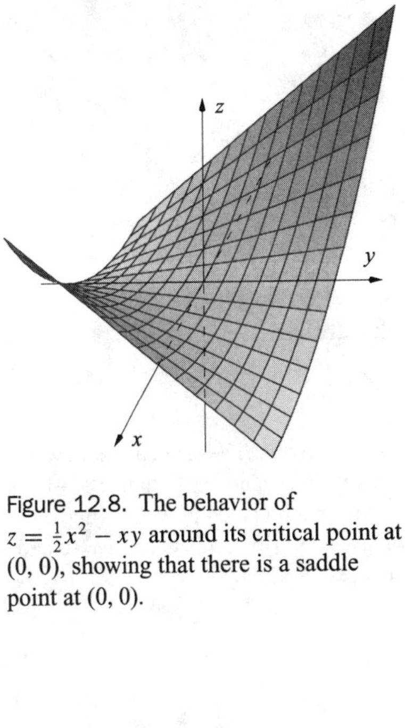

12.6 Multiple Integrals

For our first multiple integral, let's consider

$$I = \int_0^\infty \int_0^\infty \int_0^\infty e^{-\alpha x^2} e^{-\beta y^2} e^{-\gamma z^2} \, dx\,dy\,dz \qquad (12.21)$$

which occurs in the kinetic theory of gases and in statistical thermodynamics, among other places. The key thing to notice here is that even though Equation 12.21 is written as a triple integral, it is actually just the product of three single integrals

$$I = \int_0^\infty e^{-\alpha x^2} \, dx \int_0^\infty e^{-\beta y^2} \, dy \int_0^\infty e^{-\gamma z^2} \, dz$$

$$= \left(\frac{\pi}{4\alpha}\right)^{1/2} \left(\frac{\pi}{4\beta}\right)^{1/2} \left(\frac{\pi}{4\gamma}\right)^{1/2}$$

because the integrand itself is a product of a function of x, a function of y, and a function of z and the limits do not involve x, y, or z.

Let's go on now and discuss multiple integrals that do not separate into products of single integrals. The double integral of $f(x, y)$ over some region, R, in the xy plane is denoted by

$$I = \iint_R f(x, y) \, dA \qquad (12.22)$$

Figure 12.8. The behavior of $z = \frac{1}{2}x^2 - xy$ around its critical point at $(0, 0)$, showing that there is a saddle point at $(0, 0)$.

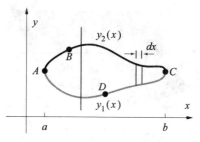

Figure 12.9. An illustration of a two-dimensional region R where any line parallel to the y axis crosses the boundary of R at two points at the most.

where dA is an element of area in the xy plane. Just as the single integral $\int_a^b f(x)\,dx$ has a geometric interpretation of being equal to the net area between $f(x)$ and the x axis over the interval (a, b), a double integral has a geometric interpretation of being the net volume between the surface $f(x, y)$ and the xy plane over the region R.

Consider the region shown in Figure 12.9, where any line parallel to the y axis crosses the boundary of R at two points at the most. Suppose that the top boundary of R (ABC in the figure) is described by $y_2(x)$ and the lower boundary (ADC) by $y_1(x)$, where $y_1(x)$ and $y_2(x)$ are continuous in the interval $a \le x \le b$. In this case, we can evaluate the integral in Equation 12.22 by letting $dA = dx\,dy$ and integrating over x and y in turn. Consider a vertical strip of width dx in Figure 12.9. The contribution to I of this strip between the curves ABC and ADB is given by

$$d\sigma(x) = dx \int_{y_1(x)}^{y_2(x)} dy\, f(x, y)$$

Note that $d\sigma(x)$ is a function of x. We can find I over the region in Figure 12.9 by adding the contributions from all the vertical strips between $x = a$ and $x = b$, which amounts to integrating $d\sigma(x)$ over x from a to b, or in an equation

$$I = \int_a^b \left\{ \int_{y_1(x)}^{y_2(x)} f(x, y)\,dy \right\} dx \tag{12.23}$$

Equation 12.23 is called an *iterated integral*: the y integration in brackets produces a function of x, which is then integrated between the limits $a \le x \le b$.

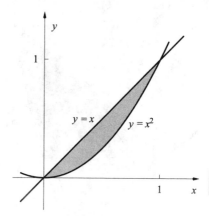

Figure 12.10. The shaded region whose area is determined in Example 12–10. It is the area bounded by $y = x$ (upper) and $y = x^2$ (lower) between $x = 0$ and $x = 1$.

EXAMPLE 12–10
Use Equation 12.23 to evaluate the area of the shaded region shown in Figure 12.10.

SOLUTION: In this case $f(x, y) = 1$ in Equation 12.22. The shaded region is bounded by $y = x$ (upper) and $y = x^2$ (lower). Equation 12.23 gives

$$I = \int_0^1 \left\{ \int_{x^2}^x dy \right\} dx$$
$$= \int_0^1 (x - x^2)\,dx = \frac{1}{6}$$

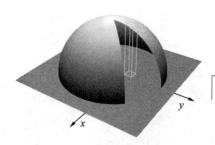

Figure 12.11. A pictorial illustration of the integration scheme used in Example 12–11 to determine the volume of a hemisphere.

EXAMPLE 12–11
Use Equation 12.23 to determine the volume of the hemisphere described by $x^2 + y^2 + z^2 = a^2$ for $z > 0$.

SOLUTION: In this case, the function $f(x, y)$ represents the height $z = f(x, y) = (a^2 - x^2 - y^2)^{1/2}$ of the hemispherical dome above the xy plane (Figure 12.11) and the integration variables x and y vary over the unit circle $x^2 + y^2 = a^2$ in the $z = 0$ plane. We'll calculate 1/4 of the volume of the hemisphere by restricting x and y to take on only positive values. The boundary of the hemisphere in the $z = 0$ plane is the circle given by $x^2 + y^2 = a^2$, and so y in Equation 12.23 varies from 0 to $(a^2 - x^2)^{1/2}$ and x varies from 0 to a (Figure 12.11). Therefore, we write

$$\frac{V}{4} = \int_0^a \left\{ \int_0^{(a^2-x^2)^{1/2}} (a^2 - x^2 - y^2)^{1/2} dy \right\} dx$$

$$= \int_0^a \left\{ \frac{1}{2} \left[y(a^2 - x^2 - y^2)^{1/2} + (a^2 - x^2)\sin^{-1} \frac{y}{(a^2 - x^2)^{1/2}} \right]_0^{(a^2-x^2)^{1/2}} \right\} dx$$

$$= \int_0^a \left\{ \frac{(a^2 - x^2)}{2} \frac{\pi}{2} \right\} dx = \frac{\pi}{4} \frac{2a^3}{3} = \frac{\pi a^3}{6}$$

Four times this result, $2\pi a^3/3$, is the volume of a hemisphere of radius a.

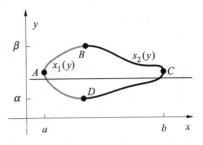

Figure 12.12. The region shown in Figure 12.9 where any line parallel to the x axis crosses the boundary of R at two points at the most.

In Equation 12.23, we integrate over y first and then over x. We can reverse the order of integration and integrate over x first and then over y, and write

$$I = \int_\alpha^\beta \left\{ \int_{x_1(y)}^{x_2(y)} f(x, y)\, dx \right\} dy \qquad (12.24)$$

where $x_1(y)$ and $x_2(y)$ bound the region in Figure 12.9 in a horizontal sense (Figure 12.12). Notice that the x integration yields a function of y that is then integrated over y between α and β in Figure 12.12. Let's use Equation 12.24 to evaluate the integral

$$I = \iint_R x\, dx dy$$

where R is the region in the first quadrant bounded by the curves $y = x^2$ and $y = x$ (Figure 12.13a). In this case, the limits of the x integration are $x = y$ to $x = y^{1/2}$ and the limits of the y integration are 0 to 1. Therefore,

$$I = \int_0^1 \left\{ \int_y^{y^{1/2}} x\, dx \right\} dy = \int_0^1 \left[\frac{y}{2} - \frac{y^2}{2} \right] dy = \frac{1}{12}$$

We could also have used Equation 12.23 to evaluate I. In this case, the limits of the y integration are x^2 to x and the limits of the x integration are 0 to 1 (Figure 12.13b), and so

$$I = \int_0^1 \left\{ x \int_{x^2}^x dy \right\} dx = \int_0^1 x(x - x^2)\, dx = \frac{1}{12}$$

In the first case, we find the contributions from horizontal strips and then add them up in the vertical direction (Figure 12.13a), and in the second case, we find the contributions from vertical strips and then add them up in the horizontal direction (Figure 12.13b).

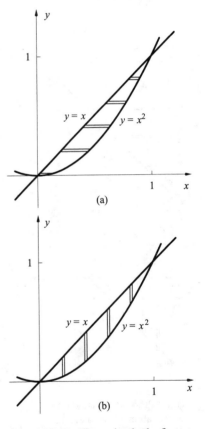

Figure 12.13. The region in the first quadrant bounded by the curves $y = x^2$ and $y = x$. (a) The integration is over x first (y to $y^{1/2}$) and then over y (0 to 1). (b) The integration is over y first (x^2 to x) and then over x (0 to 1).

160

In Equations 12.23 and 12.24, we used curly brackets to emphasize which variable is integrated first. This is not standard notation. Equation 12.23 is often written as

$$I = \int_a^b \int_{y_1(x)}^{y_2(x)} f(x, y)\, dy\, dx \tag{12.25}$$

with the understanding that the inner integration is performed first. A better notation, however, is to write Equation 12.23 as

$$I = \int_a^b dx \int_{y_1(x)}^{y_2(x)} dy\; f(x, y) \tag{12.26}$$

where the y integration is thought of as an operator that acts on $f(x, y)$ that produces a function of x followed by the x integration as an operator. The order of the two operators is from right to left as usual. For example, if $f(x, y) = xy$ and $y_2(x) = 2x$ and $y_1(x) = x$ in Equation 12.26, then

$$I = \int_a^b dx\, x \int_x^{2x} dy\, y$$

$$= \int_a^b dx\, x \left(\frac{4x^2 - x^2}{2}\right) = \frac{3}{2}\int_a^b dx\, x^3 = \frac{3}{8}(b^4 - a^4)$$

The key point here is to realize that you perform the integrations sequentially from right to left; you wait for the y integration to produce its result *before* you integrate over x. This notation is very convenient and well worth using.

It is often beneficial to reverse the order of integration in a double integral. A double integral such as

$$I = \int_0^x du \int_0^u dt\; v(t) \tag{12.27}$$

occurs in the statistical mechanics of fluids. Let's reverse the order of integration and integrate over u first. It is always helpful to draw a picture illustrating the integration region. Figure 12.14a shows this region for Equation 12.27. We integrate over t from 0 to the line $t = u$ for some arbitrary value of u (the horizontal strips in Figure 12.14a) and then over u from 0 to x (we add up the horizontal strips). Figure 12.14b illustrates the integration over the same region in the reverse order. We integrate over u from t to x (the vertical strips in Figure 12.14b) and then over t from 0 to x (add up the vertical strips). In either case, we sweep out the same region. Thus,

$$I = \int_0^x du \int_0^u dt\; v(t) = \int_0^x dt\; v(t) \int_t^x du = \int_0^x dt (x - t) v(t) \tag{12.28}$$

We were able to reduce Equation 12.27 to a single integral by reversing the order of integration.

Equations 12.23 and 12.24 are readily extended to three dimensions. For example,

$$I = \int_a^b \left\{ \int_{y_1(x)}^{y_2(x)} \left[\int_{g_1(x,y)}^{g_2(x,y)} f(x, y, z)\, dz \right] dy \right\} dx \tag{12.29}$$

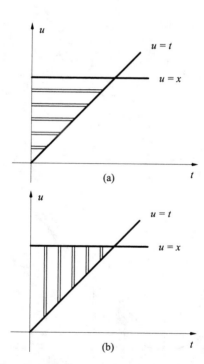

Figure 12.14. Pictorial aids to the evaluation of the integral in Equation 12.27 by reversing the orders of integration. In (a), we integrate over t first, and in (b), we integrate over u first.

or in operator notation

$$I = \int_a^b dx \int_{y_1(x)}^{y_2(x)} dy \int_{g_1(x,y)}^{g_2(x,y)} dz \; f(x, y, z) \qquad (12.30)$$

Notice that in either Equation 12.29 or 12.30, the z integration yields a function of x and y, then the y integration yields a function of x, and finally the x integration between a and b yields I.

Most of the computer algebra systems that are available can evaluate multiple integrals. For example, the command

```
Integrate [ Sqrt [ a^2 - x^2 - y^2 ], {x, 0, a}, {y, 0, Sqrt [a^2 - x^2] } ]
```

in *Mathematica* gives

$$I = \int_0^a dx \int_0^{(a^2-x^2)^{1/2}} dy \, (a^2 - x^2 - y^2)^{1/2} = \frac{\pi a^3}{6}$$

as we found in Example 12–11.

Problems

12–1. Determine all the partial derivatives up to second order of $f(x, y) =$
    (a) $xe^y + y$ (b) $y \sin x + x^2$ (c) $e^{-(x^2+y^2)}$

12–2. Show that $f_{xy} = f_{yx}$ for (a) $f(x, y) = x^2 e^{-y^2}$, (b) $f(x, y) = e^{-y} \cos xy$, and
    (c) $f(x, y) = \sin xy$.

12–3. Show that $\partial^2 P / \partial V \partial T = \partial^2 P / \partial T \partial V$ for the van der Waals equation for one mole of a gas, Equation 12.2.

12–4. Show that $c(x, t) = (4\pi Dt)^{-1/2} e^{-x^2/4Dt}$ satisfies the equation $\dfrac{\partial c}{\partial t} = D \dfrac{\partial^2 c}{\partial x^2}$.

12–5. Show that $f(x, y) = \sinh ax \cos ay$ satisfies Laplace's equation
    $\dfrac{\partial^2 f}{\partial x^2} + \dfrac{\partial^2 f}{\partial y^2} = 0$.

12–6. Show that

$$\left(\frac{\partial V}{\partial T} \right)_{n,P} = \frac{1}{\left(\dfrac{\partial T}{\partial V} \right)_{n,P}}$$

for an ideal gas and for a gas whose equation of state is $P = nRT/(V - nb)$, where b is a constant. This relation is generally true and is called the *reciprocal identity*. Notice that the same variables must be held fixed on both sides of the equation.

12–7. The thermodynamic equation $\left(\dfrac{\partial U}{\partial V} \right)_T = T \left(\dfrac{\partial P}{\partial T} \right)_V - P$ shows how the energy U of a system varies with the volume in terms of pressure, volume, and (kelvin) temperature of the system. Evaluate $(\partial U / \partial V)_T$ for one mole of an ideal gas $(PV = RT)$ and for one mole of a van der Waals gas $\left[\left(P + \dfrac{a}{V^2} \right) (V - b) = RT \right]$, where a and b are constants.

12–8. Given that the heat capacity at constant volume is defined by $C_V = (\partial U/\partial T)_V$ and given the expression in Problem 12–7, derive the equation $(\partial C_V/\partial V)_T = T(\partial^2 P/\partial T^2)_V$.

12–9. Thermodynamics tells us that the difference between the heat capacity at constant pressure and the heat capacity at constant volume is given by $C_P - C_V = T (\partial P/\partial T)_V (\partial V/\partial T)_P$. Show that $C_P - C_V = R$ for one mole of an ideal gas.

12–10. Use the expression in Problem 12–8 to determine $(\partial C_V/\partial V)_T$ for one mole of an ideal gas and for one mole of a van der Waals gas.

12–11. Is $dV = \pi r^2 dh + 2\pi rh dr$ an exact differential?

12–12. Is $dx = C_V(T) \, dT + \dfrac{RT}{V} dV$ an exact differential? What about dx/T?

12–13. Verify that you get the same result for Example 12–4 if you substitute $x(t)$ and $y(t)$ into $u(x, y)$ and then differentiate with respect to t.

12–14. Verify that you get the same result for Example 12–5 if you substitute $x(s, t)$ and $y(s, t)$ directly into $u(x, y)$ and then take partial derivatives.

12–15. Use the chain rule to evaluate du/dt if $u(x, y, z) = x^2 + ze^y$, where $x(t) = t$, $y(t) = t^2$, and $z(t) = t^3$. Verify your result by substituting $x = t$, $y = t^2$, and $z = t^3$ directly into u.

12–16. Evaluate $\partial u/\partial s$ and $\partial u/\partial t$ if $u(x, y) = e^{x+y}$, $x(t, s) = te^s$, and $y(s) = \sin s$.

12–17. Let Y be any extensive property. Use Euler's theorem to show that $Y(n_1, n_2, \ldots, T, P) = \sum n_j \overline{Y}_j$, where $\overline{Y}_j = (\partial Y/\partial n_j)_{T,P,n_{k \neq j}}$. What are the $\overline{Y}_j$ called? Can you interpret this equation physically?

12–18. The Helmholtz energy (A) of a system can be expressed as a function of the temperature (T), the volume (V), and the number of moles (n). Apply Euler's theorem to $A = A(T, V, n)$. Do you recognize the resulting equation?

12–19. Determine whether $f(x, y)$ has a maximum, a minimum, or a saddle point at the given critical points.
   (a) $f(x, y) = 2x^2 + 8xy + y^4$ at $(4, -2)$ and $(-4, 2)$
   (b) $f(x, y) = 2x - x^2 + 2y^2 - y^4$ at $(1, 1)$ and $(1, -1)$
   (c) $f(x, y) = 4 + x + y - x^2 - xy - y^2/2$ at $(0, 1)$
   (d) $f(x, y) = e^{2x-4y-x^2-y^2}$ at $(1, -2)$

12–20. Find the critical points and determine whether $f(x, y)$ has a maximum, a minimum, or a saddle point at them.
   (a) $f(x, y) = x^2 + y^2 + 2x - 4y + 8$
   (b) $f(x, y) = x^2 - y^2 + 2x - 4y + 8$
   (c) $f(x, y) = x^2 - 2x + y^2 - 2y + 3$

12–21. Find the critical points and determine whether $f(x, y)$ has a maximum, a minimum, or a saddle point at them.
   (a) $f(x, y) = xy + 6$
   (b) $f(x, y) = x^2 + y^2 - 6x + 2y + 5$
   (c) $f(x, y) = 3x^2 + 12x + 4y^3 - 6y^2 + 5$

12–22. Determine the area between the two parabolas $y^2 = a - x$ and $y^2 = a - ax$. Take $a > 1$.

12–23. Determine the volume under the surface described by $z = 1 - x^2 - y^2$ and over the square with vertices $(\pm 1, 0)$ and $(0, \pm 1)$ in the $z = 0$ plane. (See Figure 12.15.)

12–24. Evaluate the following integrals by reversing the order of integration:

(a) $\displaystyle\int_0^1 dy \int_y^1 dx \frac{ye^x}{x}$ (b) $\displaystyle\int_0^{\pi/2} dx \int_x^{\pi/2} du \frac{\sin u}{u}$

12–25. Show that $\displaystyle\int_0^1 dy \int_y^1 dx\, ye^{x^3} = \frac{1}{6}(e - 1)$. *Hint*: Reverse the order of integration.

12–26. Use any CAS to evaluate $\int_0^1 dx \int_0^{x^3} dy\, e^{y/x}$.

12–27. Use any CAS to evaluate $\int_0^2 dy \int_0^y dx\, \sqrt{x^2 + y^2}$.

The next five problems involve the derivations of various thermodynamic relations.

12–28. Starting with $dU = TdS - PdV$, add $d(PV)$ and subtract $d(TS)$ to both sides to obtain

$$d(U + PV - TS) = dG = -SdT + VdP$$

where G is the Gibbs energy. Use this expression to derive the Maxwell relation

$$\left(\frac{\partial S}{\partial P}\right)_T = -\left(\frac{\partial V}{\partial T}\right)_P$$

12–29. Subtract $d(TS)$ from both sides of $dU = TdS - PdV$ to obtain $d(U - TS) = dA = -SdT - PdV$. Now derive Equation 12.5.

12–30. We derived Equation 12.15 formally in Section 12.3, but we shall develop a more rigorous derivation here. Because we want to derive an expression for $(\partial U/\partial V)_T$, we start with $U = U(V, T)$. Now write the total differential of U,

$$dU = \left(\frac{\partial U}{\partial V}\right)_T dV + \left(\frac{\partial U}{\partial T}\right)_V dT \qquad (1)$$

Now eliminate dS from $dU = TdS - PdV$ by using the total differential of $S = S(V, T)$ to obtain

$$dU = \left[T\left(\frac{\partial S}{\partial V}\right)_T - P\right]dV + T\left(\frac{\partial S}{\partial T}\right)_V dT \qquad (2)$$

Compare equations 1 and 2 to obtain

$$\left(\frac{\partial U}{\partial V}\right)_T = T\left(\frac{\partial S}{\partial V}\right)_T - P \qquad (3)$$

and

$$\left(\frac{\partial S}{\partial T}\right)_V = \frac{1}{T}\left(\frac{\partial U}{\partial T}\right)_V = \frac{C_V}{T} \qquad (4)$$

Now substitute Equation 12.5 into equation 3 to obtain Equation 12.15. Comment on the utility of equations 3 and 4.

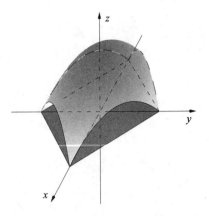

Figure 12.15. An illustration of the volume to be determined in Problem 12–23.

12–31. Use the procedure of the previous problem to derive

$$\left(\frac{\partial H}{\partial P}\right)_T = V - T\left(\frac{\partial V}{\partial T}\right)_P \tag{1}$$

and

$$\left(\frac{\partial S}{\partial T}\right)_P = \frac{C_P}{T} \tag{2}$$

Evaluate $(\partial H/\partial P)_T$ for an ideal gas.

12–32. We derive here what many students think is a bizarre thermodynamic relation. Use Equation 12.5 and equation 4 of Problem 12–30 to show that the total differential of $S = S(T, V)$ can be written as

$$dS = \frac{C_V}{T}dT + \left(\frac{\partial P}{\partial T}\right)_V dV$$

Now eliminate dV from this result by inserting the total differential of $V = V(T, P)$ to obtain

$$dS = \left[\frac{C_V}{T} + \left(\frac{\partial P}{\partial T}\right)_V \left(\frac{\partial V}{\partial T}\right)_P\right] dT + \left(\frac{\partial P}{\partial T}\right)_V \left(\frac{\partial V}{\partial P}\right)_T dP$$

Finally, compare this result to the total differential of $S = S(T, P)$ and use the results of Problems 12–28 and 12–31 to get

$$C_P - C_V = T\left(\frac{\partial P}{\partial T}\right)_V \left(\frac{\partial V}{\partial T}\right)_P \tag{1}$$

and

$$\left(\frac{\partial P}{\partial V}\right)_V \left(\frac{\partial V}{\partial P}\right)_T = -\left(\frac{\partial V}{\partial T}\right)_P \tag{2}$$

First, show that $C_P - C_V = nR$ for an ideal gas. Now use the result of Problem 12–6 to obtain

$$\left(\frac{\partial T}{\partial V}\right)_P \left(\frac{\partial V}{\partial P}\right)_T \left(\frac{\partial P}{\partial T}\right)_V = -1 \tag{3}$$

Verify this result for an ideal gas.

12–33. In this problem, we'll show how to determine a function given its total differential. Suppose $dz = (2x + y)\,dx + (x + y)\,dy$. Show that this expression is an exact differential. To determine the functional form of $z(x, y)$, we integrate $\partial z/\partial x$ *partially* with respect to x to obtain

$$z(x, y) = \int \frac{\partial z}{\partial x}\,dx = \int (2x + y)\,dx = x^2 + xy + f(y)$$

where $f(y)$ is a function of y to be determined. We obtain a function of y in this case because we are integrating with respect to x partially. We can determine $f(y)$ by differentiating $z(x, y)$ with respect to y and equating the result to $\partial z/\partial y$ in the total differential. This step gives

$$\frac{\partial z}{\partial y} = \frac{\partial}{\partial y}[x^2 + xy + f(y)] = x + \frac{df}{dy} = x + y$$

or $df/dy = y$, or $f(y) = y^2/2 +$ constant. Thus, $z(x, y) = x^2 + xy + y^2/z +$ constant. Use this procedure to determine $z(x, y)$ given that $dz = (x^2 + \sin y)\,dx + (x \cos y - 2y)\,dy$.

12–34. Use the procedure described in the previous problem to determine $z(x, y)$, given that

$$dz = [2x \sin y + (1 + y)e^x]\,dx + (x^2 \cos y + 2y + e^x)\,dy$$

12–35. We'll learn in Chapter 15 that the linear partial differential equation

$$\frac{\partial^2 u}{\partial x^2} = \frac{1}{v^2}\frac{\partial^2 u}{\partial t^2} \tag{1}$$

where $u = u(x, t)$ and v is a constant, is called the *one-dimensional wave equation*. We're going to find a new pair of independent variables that transform this equation into a simpler form. Let

$$\xi = x + at \quad \text{and} \quad \eta = x + bt$$

where a and b are constants to be determined. Using Equation 12.11, show that

$$\frac{\partial^2 u}{\partial x^2} = \frac{\partial^2 u}{\partial \xi^2} + 2\frac{\partial^2 u}{\partial \eta \partial \xi} + \frac{\partial^2 u}{\partial \eta^2} \tag{2}$$

and

$$\frac{\partial^2 u}{\partial t^2} = a^2\frac{\partial^2 u}{\partial \xi^2} + 2ab\frac{\partial^2 u}{\partial \eta \partial \xi} + b^2\frac{\partial^2 u}{\partial \eta^2} \tag{3}$$

Substitute equations 2 and 3 into equation 1 to get

$$\left(1 - \frac{a^2}{v^2}\right)\frac{\partial^2 u}{\partial \xi^2} + 2\left(1 - \frac{ab}{v^2}\right)\frac{\partial^2 u}{\partial \eta \partial \xi} + \left(1 - \frac{b^2}{v^2}\right)\frac{\partial^2 u}{\partial \eta^2} = 0 \tag{4}$$

The parameters a and b are at our disposal, so let's choose $a = \pm v$ and $b = \pm v$. Now show that with an appropriate choice of signs, equation 4 becomes

$$\frac{\partial^2 u}{\partial \eta \partial \xi} = 0 \tag{5}$$

where

$$\xi = x + vt \quad \text{and} \quad \eta = x - vt \tag{6}$$

Equations 5 and 6 are equivalent to equation 1.

12–36. Show that the solution to equation 5 in the previous problem can be written as

$$u(x, t) = f(\xi) + g(\eta) = f(x + vt) + g(x - vt) \tag{1}$$

where $f(\xi)$ and $g(\eta)$ are (almost) arbitrary functions. It turns out that f and g must be twice-differentiable, in which case, equation 1 is the general solution of the one-dimensional wave equation of the previous problem.

CHAPTER 13

Vectors

A vector is a quantity that has both magnitude and direction. Examples of vectors are position, force, velocity, and momentum. We specify the position of something, for example, by giving not only its distance from a certain point but also its direction from that point. We shall review how to add and subtract vectors to get new vectors and how to express them in terms of the unit vectors of some coordinate system. We'll then discuss the two ways that vectors can be multiplied together to give scalar products and vector products and illustrate a few of their applications. In the final section, we shall introduce vector operators, the gradient operator and the divergence operator, and then show how they arise in physical problems.

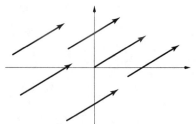

Figure 13.1. All the vectors in this figure are equal because they have the same length and the same direction.

13.1 Representation of Vectors

A vector can be represented geometrically by an arrow from the origin of a coordinate system. The length of the arrow represents the magnitude of the vector, and the direction of the arrow represents the direction of the vector. Vectors that have the same length and the same direction are equal. Thus, all the vectors shown in Figure 13.1 are equal. It makes no difference where the tail is located, although we often locate it at the origin of a coordinate system for convenience.

Two vectors can be added together to get a new vector. Consider the two vectors **u** and **v** in Figure 13.2. (We denote vectors by boldface symbols.) To find **w** = **u** + **v**, we place the tail of **u** at the tip of **v** and then draw **w** from the tail of **v** to the tip of **u**, as shown in the figure. We could also place the tail of **v** at the tip of **u** and draw **w** from the tail of **u** to the tip of **v**. As Figure 13.2 indicates, we get

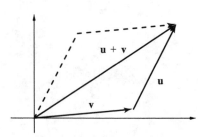

Figure 13.2. An illustration of the (commutative) addition of two vectors, **u** + **v** = **v** + **u** = **w**.

the same result either way, so we see that

$$\mathbf{w} = \mathbf{u} + \mathbf{v} = \mathbf{v} + \mathbf{u} \tag{13.1}$$

Vector addition is commutative.

To subtract two vectors, we draw one of them in the opposite direction and then add it to the other. Writing a vector in its opposite direction is equivalent to forming the vector $-\mathbf{v}$ (Figure 13.3). Thus, mathematically we have

$$\mathbf{t} = \mathbf{u} - \mathbf{v} = \mathbf{u} + (-\mathbf{v}) \tag{13.2}$$

Generally, a number a times a vector is a new vector that is parallel to $\mathbf{u}$ but whose length is a times the length of $\mathbf{u}$. If a is positive, then $a\mathbf{u}$ lies in the same direction as $\mathbf{u}$, but if a is negative, then $a\mathbf{u}$ lies in the opposite direction.

A useful set of vectors are the vectors that are of unit length and point along the positive x, y, and z axes of a cartesian coordinate system. These *unit vectors* (unit length), which we designate by $\mathbf{i}, \mathbf{j}$, and $\mathbf{k}$, respectively, are shown in Figure 13.4. We shall always draw a cartesian coordinate system so that it is right-handed. A *right-handed coordinate system* is such that when you curl the four fingers of your right hand from $\mathbf{i}$ to $\mathbf{j}$, your thumb points along $\mathbf{k}$ (Figure 13.5). Any three-dimensional vector $\mathbf{u}$ can be described in terms of these unit vectors,

$$\mathbf{u} = u_x\,\mathbf{i} + u_y\,\mathbf{j} + u_z\,\mathbf{k} \tag{13.3}$$

where, for example, $u_x\,\mathbf{i}$ is u_x units long and lies in the direction of $\mathbf{i}$. The quantities u_x, u_y, and u_z in Equation 13.3 are the *components* of $\mathbf{u}$. They are the projections of $\mathbf{u}$ along the respective cartesian axes (Figure 13.6). In terms of components, the sum or difference of two vectors is given by

$$\mathbf{u} \pm \mathbf{v} = (u_x \pm v_x)\,\mathbf{i} + (u_y \pm v_y)\,\mathbf{j} + (u_z \pm v_z)\,\mathbf{k} \tag{13.4}$$

Figure 13.6 shows that the length of $\mathbf{u}$ is given by

$$u = |\mathbf{u}| = (u_x^2 + u_y^2 + u_z^2)^{1/2} \tag{13.5}$$

We shall often denote the magnitude of $\mathbf{u}$ by u.

EXAMPLE 13–1
If $\mathbf{u} = 2\mathbf{i} - \mathbf{j} + 3\mathbf{k}$ and $\mathbf{v} = -\mathbf{i} + 2\mathbf{j} - \mathbf{k}$, then what is the length of $\mathbf{u} + \mathbf{v}$?

SOLUTION: Using Equation 13.4, we have

$$\mathbf{u} + \mathbf{v} = (2 - 1)\mathbf{i} + (-1 + 2)\mathbf{j} + (3 - 1)\mathbf{k} = \mathbf{i} + \mathbf{j} + 2\mathbf{k}$$

and using Equation 13.5 gives

$$|\mathbf{u} + \mathbf{v}| = (1^2 + 1^2 + 2^2)^{1/2} = \sqrt{6}$$

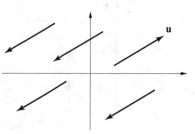

Figure 13.3. The vector $-\mathbf{u}$ points in the opposite direction as $\mathbf{u}$. All the vectors pointing downward in the figure are equal to $-\mathbf{u}$.

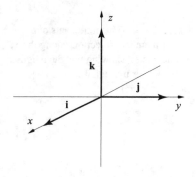

Figure 13.4. The fundamental unit vectors $\mathbf{i}, \mathbf{j}$, and $\mathbf{k}$ of a cartesian coordinate system.

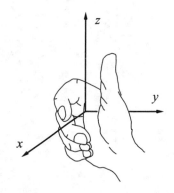

Figure 13.5. An illustration of a right-handed cartesian coordinate system.

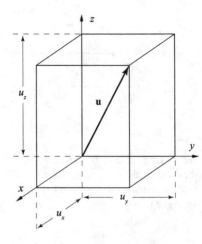

Figure 13.6. The components of a vector **u** are its projections along the x, y, and z axes, showing that the length of the vector **u** is equal to $u = |\mathbf{u}| = (u_x^2 + u_y^2 + u_z^2)^{1/2}$.

13.2 Products of Vectors

There are two ways to form the product of two vectors, and both have many applications in physical chemistry. One way yields a scalar quantity (in other words, just a number), and the other yields a vector. Not surprisingly, we call the result of the first method a *scalar product* and the result of the second method a *vector product*.

Scalar Products

The scalar product of two vectors **u** and **v** is defined as

$$\mathbf{u} \cdot \mathbf{v} = |\mathbf{u}||\mathbf{v}| \cos\theta \tag{13.6}$$

where θ is the angle between **u** and **v**. Note from the definition that

$$\mathbf{u} \cdot \mathbf{v} = \mathbf{v} \cdot \mathbf{u} \tag{13.7}$$

Taking a scalar product is a *commutative operation*. The dot between **u** and **v** is such a standard notation that $\mathbf{u} \cdot \mathbf{v}$ is often called the *dot product* of **u** and **v**. The dot products of the unit vectors **i**, **j**, and **k** are

$$\mathbf{i} \cdot \mathbf{i} = \mathbf{j} \cdot \mathbf{j} = \mathbf{k} \cdot \mathbf{k} = |1||1| \cos 0° = 1$$
$$\tag{13.8}$$
$$\mathbf{i} \cdot \mathbf{j} = \mathbf{j} \cdot \mathbf{i} = \mathbf{i} \cdot \mathbf{k} = \mathbf{k} \cdot \mathbf{i} = \mathbf{j} \cdot \mathbf{k} = \mathbf{k} \cdot \mathbf{j} = |1||1| \cos 90° = 0$$

We can use Equations 13.8 to evaluate the dot product of two vectors:

$$\begin{aligned} \mathbf{u} \cdot \mathbf{v} &= (u_x \mathbf{i} + u_y \mathbf{j} + u_z \mathbf{k}) \cdot (v_x \mathbf{i} + v_y \mathbf{j} + v_z \mathbf{k}) \\ &= u_x v_x \, \mathbf{i} \cdot \mathbf{i} + u_x v_y \, \mathbf{i} \cdot \mathbf{j} + u_x v_z \, \mathbf{i} \cdot \mathbf{k} \\ &\quad + u_y v_x \, \mathbf{j} \cdot \mathbf{i} + u_y v_y \, \mathbf{j} \cdot \mathbf{j} + u_y v_z \, \mathbf{j} \cdot \mathbf{k} \\ &\quad + u_z v_x \, \mathbf{k} \cdot \mathbf{i} + u_z v_y \, \mathbf{k} \cdot \mathbf{j} + u_z v_z \, \mathbf{k} \cdot \mathbf{k} \end{aligned}$$

and so

$$\mathbf{u} \cdot \mathbf{v} = u_x v_x + u_y v_y + u_z v_z \tag{13.9}$$

Note that if $\mathbf{v} = \mathbf{u}$, then $\mathbf{u} \cdot \mathbf{u} = u_x^2 + u_y^2 + u_z^2 = u^2$.

EXAMPLE 13–2
Find the scalar product of $\mathbf{u} = 2\mathbf{i} - \mathbf{j} + 3\mathbf{k}$ and $\mathbf{v} = -\mathbf{i} + 2\mathbf{j} - \mathbf{k}$.

SOLUTION: Equation 13.9 gives

$$\mathbf{u} \cdot \mathbf{v} = -2 - 2 - 3 = -7$$

EXAMPLE 13–3
Find the angle between the two vectors $\mathbf{u} = \mathbf{i} + 3\mathbf{j} - \mathbf{k}$ and $\mathbf{v} = \mathbf{j} - \mathbf{k}$.

SOLUTION: We use Equation 13.6, but first we must find

$$|\mathbf{u}| = (\mathbf{u} \cdot \mathbf{u})^{1/2} = (1 + 9 + 1)^{1/2} = \sqrt{11}$$
$$|\mathbf{v}| = (\mathbf{v} \cdot \mathbf{v})^{1/2} = (0 + 1 + 1)^{1/2} = \sqrt{2}$$

and

$$\mathbf{u} \cdot \mathbf{v} = 0 + 3 + 1 = 4$$

Therefore,

$$\cos\theta = \frac{\mathbf{u} \cdot \mathbf{v}}{|\mathbf{u}||\mathbf{v}|} = \frac{4}{\sqrt{22}} = 0.8528$$

or $\theta = 31.48°$.

Because $\cos 90° = 0$, the dot product of any two vectors that are perpendicular to each other is equal to zero. For example, the dot products between the $\mathbf{i}$, $\mathbf{j}$, and $\mathbf{k}$ cartesian unit vectors are equal to zero, as Equations 13.8 say.

EXAMPLE 13–4
Show that the vectors $\mathbf{v}_1 = \frac{1}{\sqrt{3}}\mathbf{i} + \frac{1}{\sqrt{3}}\mathbf{j} + \frac{1}{\sqrt{3}}\mathbf{k}$, $\mathbf{v}_2 = \frac{1}{\sqrt{6}}\mathbf{i} - \frac{2}{\sqrt{6}}\mathbf{j} + \frac{1}{\sqrt{6}}\mathbf{k}$, and $\mathbf{v}_3 = -\frac{1}{\sqrt{2}}\mathbf{i} + \frac{1}{\sqrt{2}}\mathbf{k}$ are of unit length and are mutually perpendicular.

SOLUTION: The lengths are given by

$$v_1 = (\mathbf{v}_1 \cdot \mathbf{v}_1)^{1/2} = \left(\frac{1}{3} + \frac{1}{3} + \frac{1}{3}\right)^{1/2} = 1$$

$$v_2 = (\mathbf{v}_2 \cdot \mathbf{v}_2)^{1/2} = \left(\frac{1}{6} + \frac{4}{6} + \frac{1}{6}\right)^{1/2} = 1$$

$$v_3 = (\mathbf{v}_3 \cdot \mathbf{v}_3)^{1/2} = \left(\frac{1}{2} + 0 + \frac{1}{2}\right)^{1/2} = 1$$

The dot products between the different vectors are

$$\mathbf{v}_1 \cdot \mathbf{v}_2 = \frac{1}{\sqrt{18}} - \frac{2}{\sqrt{18}} + \frac{1}{\sqrt{18}} = 0$$

$$\mathbf{v}_1 \cdot \mathbf{v}_3 = -\frac{1}{\sqrt{6}} + 0 + \frac{1}{\sqrt{6}} = 0$$

$$\mathbf{v}_2 \cdot \mathbf{v}_3 = -\frac{1}{\sqrt{12}} + 0 + \frac{1}{\sqrt{12}} = 0$$

None of the vector operations that we have used so far are limited to two or three dimensions. We can easily generalize Equation 13.9 to N dimensions by writing

$$\mathbf{u} \cdot \mathbf{v} = \sum_{j=1}^{N} u_j v_j \tag{13.10}$$

The length of an N-dimensional vector is given by

$$u = (\mathbf{u} \cdot \mathbf{u})^{1/2} = \left(\sum_{j=1}^{N} u_j^2 \right)^{1/2} \tag{13.11}$$

If the dot product of two N-dimensional vectors is equal to zero, then we say that the two vectors are *orthogonal*. Thus, the term *orthogonal* is just a generalization of *perpendicular*. Furthermore, if the length of a vector is equal to 1, then the vector is said to be *normalized*. A set of mutually orthogonal vectors that are also normalized is said to be *orthonormal*. The set of vectors in Example 13–4 are orthonormal. It is common notation to represent N-dimensional vectors by just listing their components within parentheses. Problem 13–5 has you show that the set of vectors $(1/\sqrt{3}, 1/\sqrt{3}, 0, 1/\sqrt{3})$, $(1/\sqrt{3}, -1/\sqrt{3}, 1/\sqrt{3}, 0)$, $(0, 1/\sqrt{3}, 1/\sqrt{3}, -1/\sqrt{3})$, and $(1/\sqrt{3}, 0, -1/\sqrt{3}, -1/\sqrt{3})$ is orthonormal.

One application of a dot product involves the definition of work. Recall that work is defined as force times distance, where *force* means the component of force that lies in the same direction as the displacement. If we let $\mathbf{F}$ be the force and $\mathbf{d}$ be the displacement, then *work* is defined as

$$\text{work} = \mathbf{F} \cdot \mathbf{d} \tag{13.12}$$

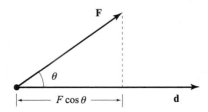

Figure 13.7. Work is defined as $w = \mathbf{F} \cdot \mathbf{d}$, or $(F \cos \theta)d$, where $F \cos \theta$ is the component of $\mathbf{F}$ along $\mathbf{d}$.

We can write Equation 13.12 as $(F \cos \theta)(d)$ to emphasize that $F \cos \theta$ is the component of $\mathbf{F}$ in the direction of $\mathbf{d}$ (Figure 13.7).

Another important application of a dot product involves the interaction of a dipole moment with an electric field. The separation of opposite charges in a molecule gives rise to a dipole moment, which we indicate by an arrow pointing from the negative charge to the positive charge. For example, because a chlorine atom is more electronegative than a hydrogen atom, HCl has a dipole moment, which we indicate by writing $\overleftarrow{\text{HCl}}$. (Some authors define the direction of a dipole moment to be opposite the one that we are using, but most physical chemistry and physics books use the convention shown here.) A dipole moment is a vector quantity whose magnitude is equal to the product of the positive charge and the distance between the positive and negative charges and whose direction is from the negative charge to the positive charge. Thus, for the two separated charges illustrated in Figure 13.8, the dipole moment $\boldsymbol{\mu}$ is equal to

$$\boldsymbol{\mu} = q \, \mathbf{r}$$

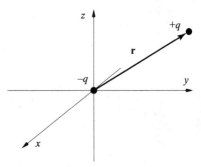

Figure 13.8. A dipole moment is a vector that points from a negative charge, $-q$, to a positive charge, $+q$, and whose magnitude is qr.

If we apply an electric field $\mathbf{E}$ to a dipole moment, then the potential energy of interaction will be

$$V = -\boldsymbol{\mu} \cdot \mathbf{E} \tag{13.13}$$

Equation 13.13 is used frequently in physical chemistry.

Vector Products

The vector product of two vectors is a vector defined by

$$\mathbf{u} \times \mathbf{v} = |\mathbf{u}||\mathbf{v}|\,\mathbf{c}\sin\theta \tag{13.14}$$

where θ is the angle between $\mathbf{u}$ and $\mathbf{v}$, and $\mathbf{c}$ is a unit vector perpendicular to the plane formed by $\mathbf{u}$ and $\mathbf{v}$ (Figure 13.9). The direction of $\mathbf{c}$ is given by the right-hand rule: If the four fingers of your right hand curl from $\mathbf{u}$ to $\mathbf{v}$, then $\mathbf{c}$ lies along the direction of your thumb. (See Figure 13.5 for a similar construction.) The notation given in Equation 13.14 is so commonly used that the vector product is often called the *cross product*. Because the direction of $\mathbf{c}$ is given by the right-hand rule, the cross product operation is not commutative, and, in particular,

$$\mathbf{u} \times \mathbf{v} = -\mathbf{v} \times \mathbf{u} \tag{13.15}$$

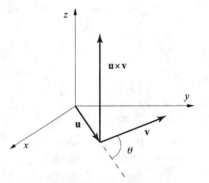

Figure 13.9. An illustration of the vector product $\mathbf{u} \times \mathbf{v}$. The direction of $\mathbf{u} \times \mathbf{v}$ is given by the right-hand rule.

The cross products of the cartesian unit vectors are

$$\mathbf{i} \times \mathbf{i} = \mathbf{j} \times \mathbf{j} = \mathbf{k} \times \mathbf{k} = |1||1|\,\mathbf{c}\sin 0° = 0$$

$$\mathbf{i} \times \mathbf{j} = -\mathbf{j} \times \mathbf{i} = |1||1|\,\mathbf{k}\sin 90° = \mathbf{k}$$

$$\mathbf{j} \times \mathbf{k} = -\mathbf{k} \times \mathbf{j} = \mathbf{i} \tag{13.16}$$

$$\mathbf{k} \times \mathbf{i} = -\mathbf{i} \times \mathbf{k} = \mathbf{j}$$

In terms of components of $\mathbf{u}$ and $\mathbf{v}$, we have (Problem 13–8)

$$\mathbf{u} \times \mathbf{v} = (u_y v_z - u_z v_y)\,\mathbf{i} + (u_z v_x - u_x v_z)\,\mathbf{j} + (u_x v_y - u_y v_x)\,\mathbf{k} \tag{13.17}$$

Equation 13.17 can be conveniently expressed as a determinant:

$$\mathbf{u} \times \mathbf{v} = \begin{vmatrix} \mathbf{i} & \mathbf{j} & \mathbf{k} \\ u_x & u_y & u_z \\ v_x & v_y & v_z \end{vmatrix} \tag{13.18}$$

(See Chapter 17 for a discussion of determinants.) Equations 13.17 and 13.18 are equivalent.

EXAMPLE 13–5

Given $\mathbf{u} = -2\mathbf{i} + \mathbf{j} + \mathbf{k}$ and $\mathbf{v} = 3\mathbf{i} - \mathbf{j} + \mathbf{k}$, determine $\mathbf{w} = \mathbf{u} \times \mathbf{v}$.

SOLUTION: Using Equation 13.17, we have

$$\mathbf{w} = [(1)(1) - (1)(-1)]\,\mathbf{i} + [(1)(3) - (-2)(1)]\,\mathbf{j} + [(-2)(-1) - (1)(3)]\,\mathbf{k}$$

$$= 2\mathbf{i} + 5\mathbf{j} - \mathbf{k}$$

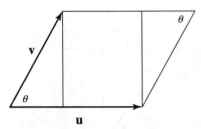

Figure 13.10. An illustration of the geometric interpretation of $\mathbf{u} \times \mathbf{v}$.

The magnitude of the cross product of two vectors has a nice geometric interpretation. Figure 13.10 shows a parallelogram whose sides are the vectors $\mathbf{u}$ and $\mathbf{v}$. The area of this paralellelogram is equal to the areas of the two triangles plus the area of the rectangle, or

$$A = 2\left(\frac{1}{2}\right)(v\cos\theta)(v\sin\theta) + (v\sin\theta)(u - v\cos\theta) = uv\sin\theta \qquad (13.19)$$

Notice that Equation 13.19 says that the area is equal to the length of one side times the height. But $uv\sin\theta$ is simply the magnitude of $\mathbf{u} \times \mathbf{v}$, so we can write

$$A = |\mathbf{u} \times \mathbf{v}| = uv\sin\theta \qquad (13.20)$$

The cross product results in a vector, so now let's consider the *triple scalar product* $\mathbf{u} \cdot (\mathbf{v} \times \mathbf{w})$. Using Equation 13.17, we have

$$\mathbf{u} \cdot (\mathbf{v} \times \mathbf{w}) = u_x(v_y w_z - v_z w_y) + u_y(v_z w_x - v_x w_z) + u_z(v_x w_y - v_y w_x)$$

This expression looks just like Equation 13.17 with $\mathbf{i}$, $\mathbf{j}$, and $\mathbf{k}$ replaced by u_x, u_y, and u_z, so $\mathbf{u} \cdot (\mathbf{v} \times \mathbf{w})$ can be expressed in the form

$$\mathbf{u} \cdot (\mathbf{v} \times \mathbf{w}) = \begin{vmatrix} u_x & u_y & u_z \\ v_x & v_y & v_z \\ w_x & w_y & w_z \end{vmatrix} \qquad (13.21)$$

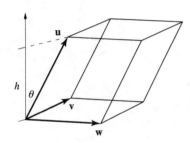

Figure 13.11. A parallelepiped with sides $\mathbf{u}$, $\mathbf{v}$, and $\mathbf{w}$.

which is just Equation 13.18 with $\mathbf{i}$, $\mathbf{j}$, and $\mathbf{k}$ replaced by u_x, u_y, and u_z.

Just as $|\mathbf{u} \times \mathbf{v}|$ is the area of the parallelogram with sides $\mathbf{u}$ and $\mathbf{v}$, $|\mathbf{u} \cdot (\mathbf{v} \times \mathbf{w})|$ is the volume of the parallelepiped with sides $\mathbf{u}$, $\mathbf{v}$, and $\mathbf{w}$ (Figure 13.11). To see that this is so, note that $|\mathbf{u} \cdot (\mathbf{v} \times \mathbf{w})| = u|\mathbf{v} \times \mathbf{w}|\cos\theta$ and that $|\mathbf{v} \times \mathbf{w}|$ is the area of the base of the parallelepiped shown in Figure 13.11 and $u\cos\theta$ is its height.

One physically important application of a cross product involves the definition of angular momentum. If a particle has a momentum $\mathbf{p} = m\mathbf{v}$ at a position $\mathbf{r}$ from a fixed point (as in Figure 13.12), then its *angular momentum* is defined by

$$\mathbf{l} = \mathbf{r} \times \mathbf{p} \qquad (13.22)$$

Note that the angular momentum is a vector perpendicular to the plane formed by $\mathbf{r}$ and $\mathbf{p}$ (Figure 13.13). In terms of components, $\mathbf{l}$ is equal to (see Equation 13.17)

$$\mathbf{l} = (yp_z - zp_y)\mathbf{i} + (zp_x - xp_z)\mathbf{j} + (xp_y - yp_x)\mathbf{k} \qquad (13.23)$$

Angular momentum plays an important role in quantum mechanics.

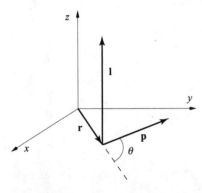

Figure 13.12. The angular momentum of a particle of momentum $\mathbf{p}$ and position $\mathbf{r}$ from a fixed center is a vector perpendicular to the plane formed by $\mathbf{r}$ and $\mathbf{p}$ and in the direction of $\mathbf{r} \times \mathbf{p}$.

Another example that involves a cross product is the equation that gives the force $\mathbf{F}$ on a particle of charge q moving with velocity $\mathbf{v}$ through a magnetic field $\mathbf{B}$:

$$\mathbf{F} = q(\mathbf{v} \times \mathbf{B})$$

Note that the force is perpendicular to $\mathbf{v}$, and so the effect of $\mathbf{B}$ is to cause the motion of the particle to curve, not to speed up or slow down.

EXAMPLE 13–6

The force acting on a particle of charge q moving with velocity $\mathbf{v}$ in a magnetic field $\mathbf{B}$ is $\mathbf{F} = q\mathbf{v} \times \mathbf{B}$. Determine the motion if $\mathbf{B} = B\mathbf{k}$, where B is a constant.

SOLUTION: Newton's equations of motion are

$$m\frac{d\mathbf{v}}{dt} = q\mathbf{v} \times \mathbf{B} = q \begin{vmatrix} \mathbf{i} & \mathbf{j} & \mathbf{k} \\ v_x & v_y & v_z \\ 0 & 0 & B \end{vmatrix}$$

$$= \mathbf{i}\, qv_y B - \mathbf{j}\, qv_x B$$

or

$$m\frac{dv_x}{dt} = qv_y B \qquad m\frac{dv_y}{dt} = -qv_x B \qquad m\frac{dv_z}{dt} = 0$$

Problem 13–13 takes you through the solution to these equations, and the final result is

$$x(t) = x_0 + \frac{v_{0y}}{\omega} + \frac{v_{0x}}{\omega}\sin\omega t - \frac{v_{0y}}{\omega}\cos\omega t$$

$$y(t) = y_0 - \frac{v_{0x}}{\omega} + \frac{v_{0y}}{\omega}\sin\omega t + \frac{v_{0x}}{\omega}\cos\omega t \qquad (1)$$

$$z(t) = z_0 + v_{0z}t$$

where $\omega = qB/m$. These equations for $x(t)$ and $y(t)$ give

$$\left(x(t) - x_0 - \frac{v_{0y}}{\omega}\right)^2 + \left(y(t) - y_0 + \frac{v_{0x}}{\omega}\right)^2 = \frac{v_{0x}^2 + v_{0y}^2}{\omega^2}$$

which is the equation of a circle of radius $(v_{0x}^2 + v_{0y}^2)^{1/2}/\omega$ centered at $x = x_0 + v_{0y}/\omega$ and $y = y_0 - v_{0x}/\omega$. Thus, the motion in the xy direction is a circle of constant radius. According to equations 1, the motion in the z direction is uniform, and so we see that equations 1 describe a trajectory that is a helix with a uniform speed along the z axis (Figure 13.14). The radius of the helix is

$$R = \frac{(v_{0x}^2 + v_{0y}^2)^{1/2}}{\omega} = \frac{mv_0^2}{qB}$$

and the frequency with which the particle revolves in the xy direction, $\omega = qB/m$, is called the *Larmour frequency*, or the *cyclotron frequency*.

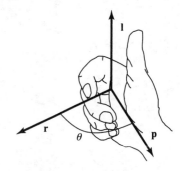

Figure 13.13. Angular momentum is a vector quantity that lies perpendicular to the plane formed by $\mathbf{r}$ and $\mathbf{p}$ and is directed such that the vectors $\mathbf{r}$, $\mathbf{p}$, and $\mathbf{l}$ form a right-handed coordinate system.

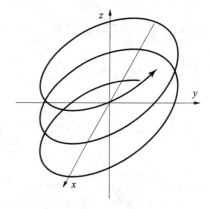

Figure 13.14. The helix described in Example 13–6. The motion in the xy direction is circular, and the motion in the z direction is uniform (constant velocity).

13.3 Vector Calculus

We saw in Example 13–6 that we can take derivatives of vectors. Suppose that the components of momentum, $\mathbf{p}$, depend upon time. Then

$$\frac{d\mathbf{p}(t)}{dt} = \frac{dp_x(t)}{dt}\mathbf{i} + \frac{dp_y(t)}{dt}\mathbf{j} + \frac{dp_z(t)}{dt}\mathbf{k} \qquad (13.24)$$

(There are no derivatives of **i**, **j**, and **k** because they are fixed in space.) Newton's equation of motion is

$$\frac{d\mathbf{p}}{dt} = \mathbf{F} \tag{13.25}$$

Equation 13.25 is actually three separate equations, one for each component. (See Example 13–6.) Because $\mathbf{p} = m\mathbf{v}$, we can write Newton's equations as

$$m\frac{d\mathbf{v}}{dt} = \mathbf{F}$$

if m is a constant. Furthermore, because $\mathbf{v} = d\mathbf{r}/dt$, we can also express Newton's equations as

$$m\frac{d^2\mathbf{r}}{dt^2} = \mathbf{F} \tag{13.26}$$

Once again, Equation 13.26 represents a set of three equations, one for each component.

There are a couple of differential vector operators that occur frequently in chemical and physical problems. One of these is the *gradient operator*, or simply the *gradient*, which is defined by

$$\boldsymbol{\nabla} f(x, y, z) = \text{grad} f(x, y, z) = \mathbf{i}\,\frac{\partial f}{\partial x} + \mathbf{j}\,\frac{\partial f}{\partial y} + \mathbf{k}\,\frac{\partial f}{\partial z} \tag{13.27}$$

Note that the gradient operator, $\boldsymbol{\nabla}$, operates on a scalar function. The vector $\boldsymbol{\nabla} f$ is called the *gradient vector* of $f(x, y, z)$. Consider a set of contour lines on a topographical map or a set of isotherms or isobars on a weather map or a set of equipotentials in a potential energy diagram. Those lines are collectively called *level curves*. If a surface is described by $z = f(x, y)$, then the level curves are given by $z = $ constant (Figure 13.15). The path traced out by $\boldsymbol{\nabla} f$ in Figure 13.15 is normal (perpendicular) to each level curve that it crosses and follows the direction of steepest descent. For a set of equipotentials, for example, $\boldsymbol{\nabla} f$ represents the corresponding electric field and traces out the direction of accalaration that a positively charged particle will follow (Figure 13.16).

Many physical laws are expressed in terms of a gradient vector. For example, Fick's law of diffusion says that the flux of a solute is proportional to the gradient of its concentration, or if $c(x, y, z, t)$ is the concentration of solute at the point (x, y, z) at time t, then

$$\text{flux of solute} = -D\boldsymbol{\nabla}c(x, y, z, t)$$

where D is called the diffusion constant. Similarly, Fourier's law of heat flow says that the flux of heat is described by

$$\text{flux of heat} = -\lambda\boldsymbol{\nabla}T(x, y, z, t)$$

where T is the temperature and λ is the thermal conductivity. If $V(x, y, z)$ is a mechanical potential energy experienced by a body, then the force on the body is

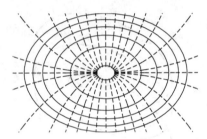

Figure 13.15. A set of level curves (solid) for the surface $z = f(x, y)$ and the path $\boldsymbol{\nabla} f(x, y)$ (dotted), which follows the direction of steepest descent.

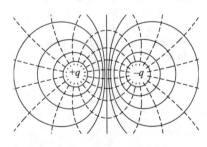

Figure 13.16. The equipotentials (solid) and the electric field (dotted) of an electric dipole formed by equal and opposite charges.

given by

$$\mathbf{F}(x, y, z) = -\nabla V(x, y, z) \qquad (13.28)$$

In addition, if $\phi(x, y, z)$ is an electrostatic potential, then the electric field associated with that potential is given by

$$\mathbf{E}(x, y, z) = -\nabla \phi(x, y, z) \qquad (13.29)$$

EXAMPLE 13–7

Suppose that a particle experiences a potential energy

$$V(x, y, z) = \frac{k_x x^2}{2} + \frac{k_y y^2}{2} + \frac{k_z z^2}{2}$$

where the k's are constants. Derive an expression for the force acting on the particle.

SOLUTION: We use Equation 13.28 to write

$$\begin{aligned}
\mathbf{F}(x, y, z) &= -\mathbf{i}\frac{\partial V}{\partial x} - \mathbf{j}\frac{\partial V}{\partial y} - \mathbf{k}\frac{\partial V}{\partial z} \\
&= -\mathbf{i}\, k_x x - \mathbf{j}\, k_y y - \mathbf{k}\, k_z z
\end{aligned}$$

Another differential vector operator that occurs frequently in physical problems is the *divergence operator*, or simply the *divergence*, which is defined by

$$\text{div}\mathbf{A}(x, y, z) = \nabla \cdot \mathbf{A}(x, y, z) = \frac{\partial A_x}{\partial x} + \frac{\partial A_y}{\partial y} + \frac{\partial A_z}{\partial z} \qquad (13.30)$$

Note that the divergence operator acts upon a vector $\mathbf{A}(x, y, z)$. The divergence is a measure of the flow rate per unit volume out of a small region surrounding the point (x, y, z).

EXAMPLE 13–8

Show that div grad $\phi = \nabla^2 \phi$.

SOLUTION: From Equation 13.27,

$$\text{grad}\,\phi = \mathbf{i}\frac{\partial \phi}{\partial x} + \mathbf{j}\frac{\partial \phi}{\partial y} + \mathbf{k}\frac{\partial \phi}{\partial z}$$

and from Equation 13.30,

$$\text{div grad}\,\phi = \frac{\partial^2 \phi}{\partial x^2} + \frac{\partial^2 \phi}{\partial y^2} + \frac{\partial^2 \phi}{\partial z^2} \qquad (13.31)$$

As we have pointed out before, the operator

$$\nabla^2 = \frac{\partial^2}{\partial x^2} + \frac{\partial^2}{\partial y^2} + \frac{\partial^2}{\partial z^2} \tag{13.32}$$

is called the Laplacian operator.

Problems

13–1. Find the lengths of the vectors $\mathbf{v} = 2\,\mathbf{i} - \mathbf{j} + 3\,\mathbf{k}$ and $\mathbf{v} = x\,\mathbf{i} + y\,\mathbf{j} + z\,\mathbf{k}$.

13–2. Show that the vectors $\mathbf{u} = 2\,\mathbf{i} - 4\,\mathbf{j} - 2\,\mathbf{k}$ and $\mathbf{v} = 3\,\mathbf{i} + 4\,\mathbf{j} - 5\,\mathbf{k}$ are orthogonal.

13–3. Show that the vector $\mathbf{v} = 2\,\mathbf{i} - 3\,\mathbf{k}$ lies entirely in a plane perpendicular to the y axis.

13–4. Find the angle between the two vectors $\mathbf{u} = -\mathbf{i} + 2\,\mathbf{j} + \mathbf{k}$ and $\mathbf{v} = 3\,\mathbf{i} - \mathbf{j} + 2\,\mathbf{k}$.

13–5. Show that the set of vectors $(1/\sqrt{3}, 1/\sqrt{3}, 0, 1/\sqrt{3})$, $(1/\sqrt{3}, -1/\sqrt{3}, 1/\sqrt{3}, 0)$, $(0, 1/\sqrt{3}, 1/\sqrt{3}, -1/\sqrt{3})$, and $(1/\sqrt{3}, 0, -1/\sqrt{3}, -1/\sqrt{3})$ is orthonormal.

13–6. Determine $\mathbf{w} = \mathbf{u} \times \mathbf{v}$ given that $\mathbf{u} = -\mathbf{i} + 2\,\mathbf{j} + \mathbf{k}$ and $\mathbf{v} = 3\,\mathbf{i} - \mathbf{j} + 2\,\mathbf{k}$. What is $\mathbf{v} \times \mathbf{u}$ equal to?

13–7. Show that $\mathbf{u} \times \mathbf{u} = \mathbf{0}$.

13–8. Using Equations 13.16, show that $\mathbf{u} \times \mathbf{v}$ is given by Equation 13.17.

13–9. Show that the angular momentum $l = |\mathbf{l}| = mvr$ for motion in a circle centered at the origin.

13–10. Show that

$$\frac{d}{dt}(\mathbf{u} \cdot \mathbf{v}) = \frac{d\mathbf{u}}{dt} \cdot \mathbf{v} + \mathbf{u} \cdot \frac{d\mathbf{v}}{dt}$$

and

$$\frac{d}{dt}(\mathbf{u} \times \mathbf{v}) = \frac{d\mathbf{u}}{dt} \times \mathbf{v} + \mathbf{u} \times \frac{d\mathbf{v}}{dt}$$

13–11. Using the results of the previous problem, show that

$$\mathbf{u} \times \frac{d^2\mathbf{u}}{dt^2} = \frac{d}{dt}\left(\mathbf{u} \times \frac{d\mathbf{u}}{dt}\right)$$

13–12. Starting with Newton's equations

$$m\frac{d^2\mathbf{r}}{dt^2} = \mathbf{F}(x, y, z)$$

operate from the left by $\mathbf{r} \times$ and use the result of Problem 13–11 to show that

$$m\frac{d}{dt}\left(\mathbf{r} \times \frac{d\mathbf{r}}{dt}\right) = \mathbf{r} \times \mathbf{F}$$

Because momentum is defined as $\mathbf{p} = m\,\mathbf{v} = m\dfrac{d\mathbf{r}}{dt}$, the above expression reads

$$\frac{d}{dt}(\mathbf{r} \times \mathbf{p}) = \mathbf{r} \times \mathbf{F}$$

But $\mathbf{r} \times \mathbf{p} = \mathbf{l}$, the angular momentum, and so we have for constant m

$$\frac{d\mathbf{l}}{dt} = \mathbf{r} \times \mathbf{F}$$

This is the form of Newton's equation for a rotating system. Notice that $d\mathbf{l}/dt = 0$, or that angular momentum is conserved if $\mathbf{r} \times \mathbf{F} = 0$. Can you identify $\mathbf{r} \times \mathbf{F}$?

13–13.   We'll solve the differential equations that occur in Example 13–6 in this problem. First write them as

$$\frac{dv_x}{dt} = \omega v_y \qquad \frac{dv_y}{dt} = -\omega v_x \qquad \frac{dv_z}{dt} = 0 \qquad (1)$$

where $\omega = qB/m$. The equation for dv_z/dt is easy and gives $z(t) = z_0 + v_{0z}t$. Differentiate the equations for dv_x/dt and dv_y/dt and use equations 1 to get

$$\frac{d^2 v_x}{dt^2} + \omega^2 v_x = 0 \qquad \text{and} \qquad \frac{d^2 v_y}{dt^2} + \omega^2 v_y = 0$$

Show that

$$v_x = A \cos \omega t + B \sin \omega t \qquad \text{and} \qquad v_y = C \cos \omega t + D \sin \omega t \qquad (2)$$

are their solutions. Substitute this result into equations 1 to show that $C = B = v_{0y}$ and $D = -A = -v_{0x}$. Finally, integrate equations 2 to obtain

$$x(t) = x_0 + \frac{v_{0y}}{\omega} + \frac{v_{0x}}{\omega} \sin \omega t - \frac{v_{0y}}{\omega} \cos \omega t$$

and

$$y(t) = y_0 - \frac{v_{0x}}{\omega} + \frac{v_{0y}}{\omega} \sin \omega t + \frac{v_{0x}}{\omega} \cos \omega t$$

Problem 6–22 shows how to solve equations 1 using complex numbers.

13–14.   Find the gradient of $f(x, y, z) = x^2 - yz + xz^2$ at the point $(1, 1, 1)$.

13–15.   Show that $\nabla(\mathbf{c} \cdot \mathbf{r}) = \mathbf{c}$ if $\mathbf{c}$ is a constant vector.

13–16.   Find div $\mathbf{A}$ for (a) $\mathbf{A} = xy^2\,\mathbf{i} + 2xyz\,\mathbf{j} - x^2z\,\mathbf{k}$ and
(b) $\mathbf{A} = (x - \cos yz)\,\mathbf{i} + (y - \cos xz)\,\mathbf{j} + (z - \cos xy)\,\mathbf{k}$.

For the next five problems, take $\mathbf{r} = x\,\mathbf{i} + y\,\mathbf{j} + z\,\mathbf{k}$.

13–17.   Show that $\operatorname{grad}(1/r) = -\mathbf{r}/r^3$ and that $\operatorname{grad}(1/r^3) = -3\,\mathbf{r}/r^5$ (provided $\mathbf{r} \neq \mathbf{0}$).

13–18.   Show that $\nabla \cdot \mathbf{r} = 3$ (provided $\mathbf{r} \neq \mathbf{0}$).

13–19.   Show that div $(\mathbf{r}/r^3) = 0$ (provided $\mathbf{r} \neq \mathbf{0}$).

13–20.   Show that $\nabla^2 \left(\dfrac{1}{r} \right) = 0$ (provided $r \neq 0$).

13–21.   The electrostatic potential produced by a dipole moment $\boldsymbol{\mu}$ located at the origin and directed along the x axis is given by

$$\phi(x, y, z) = \frac{\mu x}{(x^2 + y^2 + z^2)^{3/2}} \qquad (x, y, z \neq 0)$$

Derive an expression for the electric field associated with this potential. If the dipole moment is directed in an arbitrary direction, then

$$\phi = \frac{\boldsymbol{\mu} \cdot \mathbf{r}}{r^3}$$

First show that this expression reduces to the one above if $\boldsymbol{\mu}$ is directed along the x axis. Now show that the electric field in this case is given by

$$\mathbf{E} = -\boldsymbol{\nabla}\phi = \frac{3\,(\boldsymbol{\mu}\cdot\mathbf{r})\,\mathbf{r}}{r^5} - \frac{\boldsymbol{\mu}}{r^3}$$

Finally, show that this expression reduces to the one that you obtained above for the case where $\boldsymbol{\mu}$ points along the x axis. *Hint*: Use the results of Problems 13–15 and 13–17.

13–22. (a) Prove that div $\phi\mathbf{v} = \phi\boldsymbol{\nabla}\cdot\mathbf{v} + \mathbf{v}\cdot\boldsymbol{\nabla}\phi$. (b) Use this result to evaluate div $\phi\mathbf{v}$ if $\phi = xy$ and $\mathbf{v} = y^2\,\mathbf{i} + xz\,\mathbf{k}$. (c) Evaluate it by applying div directly to $\phi\mathbf{v}$ and compare your result.

13–23. You proved the *Schwarz inequality* for complex numbers in Problem 5–22. For vectors, the Schwarz inequality takes the form

$$|\mathbf{u}\cdot\mathbf{v}| \le |\mathbf{u}|\,|\mathbf{v}|$$

To prove the Schwarz inequality for vectors, start with

$$(\mathbf{u} + \lambda\,\mathbf{v})\cdot(\mathbf{u} + \lambda\,\mathbf{v}) \ge 0$$

where λ is an arbitrary number. Expand this expression as a quadratic form in λ, and then let $\lambda = -\mathbf{u}\cdot\mathbf{v}/v^2$ to obtain $(\mathbf{u}\cdot\mathbf{v})^2 \le u^2 v^2$, or $|\mathbf{u}\cdot\mathbf{v}| \le |\mathbf{u}|\,|\mathbf{v}|$. Give a geometric interpretation to this inequality. Do you see a parallel between this result for two-dimensional vectors and the complex number version?

13–24. The inequality

$$|\mathbf{u} + \mathbf{v}| \le |\mathbf{u}| + |\mathbf{v}|$$

is called the *triangle inequality*. Prove this inequality by starting with

$$|\mathbf{u} + \mathbf{v}|^2 = |\mathbf{u}|^2 + |\mathbf{v}|^2 + 2\,\mathbf{u}\cdot\mathbf{v}$$

and then using the Schwarz inequality (previous problem). Why do you think this is called the triangle inequality?

13–25. We shall discuss the rotation of vectors in Chapter 18, but it should be clear at this point that the length of a vector does not change when it is rotated. In an equation, if we let $\mathbf{u}_0$ be the original vector and $\mathbf{u}_R$ be the rotated vector, then $u_R = u_0$. We say that the length of a vector is *invariant under rotation*. Now show that the scalar product of two vectors is invariant under rotation. Interpret this result physically. *Hint*: Start with $\mathbf{w}_0 = \mathbf{u}_0 + \mathbf{v}_0$ and $\mathbf{w}_R = \mathbf{u}_R + \mathbf{v}_R$ and use the invariance property of $\mathbf{w}$, $\mathbf{u}$, and $\mathbf{v}$.

CHAPTER 14

Plane Polar Coordinates and Spherical Coordinates

Although cartesian, or rectangular, coordinates are the first ones we learn to use and use most often, they are not always the most convenient. For example, if the system has a natural center of symmetry, as in the case of an atom with its massive nucleus at its center, it is much more convenient to use spherical coordinates, which are constructed with exactly such systems in mind. There are many examples of problems that become much easier by using the appropriate choice of coordinate system. Usually, the symmetry of the system will suggest which of a number of available coordinate systems to use. In this chapter, we'll study polar coordinates and spherical coordinates and learn how to express vector quantities, such as the gradient and the divergence, in these coordinate systems.

14.1 Plane Polar Coordinates

Instead of locating a point in a plane by the two coordinates (x, y), we can locate it equally well by specifying its distance r from the origin, and the angle θ that the line from the origin to the point makes with the positive x axis (Figure 14.1). The coordinates r and θ are called *polar coordinates*. We shall restrict r to $r \geq 0$ and allow θ to vary from 0 to 2π. You can see from Figure 14.1 that the relation between rectangular coordinates and polar coordinates is given by

$$x = r \cos \theta \quad \text{and} \quad y = r \sin \theta \qquad (14.1)$$

and

$$r^2 = x^2 + y^2 \quad \text{and} \quad \theta = \tan^{-1} \frac{y}{x} \qquad (14.2)$$

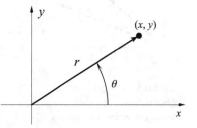

Figure 14.1. The specification of the location of a point in a plane by polar coordinates (r, θ).

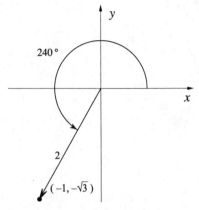

240°

2

$(-1, -\sqrt{3})$

Figure 14.2. The point $x = -1$, $y = -\sqrt{3}$.

Thus, the point $(1, 1)$ in rectangular coordinates becomes the point $(\sqrt{2}, \pi/4)$ in polar coordinates. When calculating θ from the arctangent formula in Equation 14.2, you must bear in mind in which quadrant the point lies. Using Equation 14.2 blindly for the point $(x = -1, y = -1)$ gives $\theta = \pi/4$, but realize also that $\tan 5\pi/4 = \tan(225°) = 1$, which is the correct result.

EXAMPLE 14–1

Equations 14.1 give

$$\theta = \cos^{-1}\frac{x}{r} \qquad \text{and} \qquad \theta = \sin^{-1}\frac{y}{r}$$

and Equation 14.2 says that

$$\theta = \tan^{-1}\frac{y}{x}$$

Use each of these relations to calculate θ for the point $(x = -1, y = -\sqrt{3})$ (Figure 14.2).

SOLUTION: In this case, $r = (x^2 + y^2)^{1/2} = 2$. Using a hand calculator, you'll find that

$$\theta = \cos^{-1}\left(\frac{-1}{2}\right) = 120°$$

$$\theta = \sin^{-1}\left(\frac{-\sqrt{3}}{2}\right) = -60°$$

and

$$\theta = \tan^{-1}\left(\frac{-\sqrt{3}}{-1}\right) = 60°$$

none of which is correct! The point lies in the third quadrant, and the correct answer is $180° + \cos^{-1}(1/2) = 240°$. The problem here is that the arccosine, arcsine, and arctangent are multivalued functions.

Consider an integral of the form

$$I = \iint\limits_{R} dxdy\, f(x, y) \tag{14.3}$$

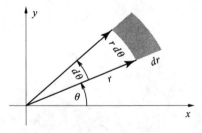

Figure 14.3. A geometrical construction of the differential area element in polar coordinates. The differential area element is given by the product of the differential arclength, $r\, d\theta$, and the differential thickness dr of the shaded area, or $r\, dr d\theta$.

over some region R in the xy plane. What is the form of the corresponding integral in polar coordinates? Let's realize at the outset that $dxdy$ does *not* become simply $drd\theta$. To see just what it does become, refer to Figure 14.3, where you can see the differential area element that is mapped out when r changes by dr and θ changes by $d\theta$. Because dr and $d\theta$ are infinitesimally small, the area is essentially rectangular, with sides dr and $r\, d\theta$ (the arclength) as indicated in Figure 14.3. Thus, the correct transformation of a differential area element in cartesian coordinates to plane polar coordinates is $dxdy \rightarrow rdrd\theta$, and so we have

$$dA = rdrd\theta \qquad (14.4)$$

in polar coordinates. Therefore, Equation 14.3 becomes

$$I = \iint_R rdrd\theta \, f(r, \theta) \qquad (14.5)$$

in polar coordinates.

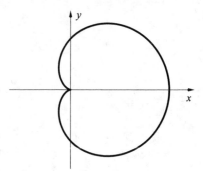

Figure 14.4. A cardioid, $r = a(1 + \cos\theta)$.

Let's use Equation 14.5 to determine the area bounded by a closed curve (in which case $f(r, \theta) = 1$). Figure 14.4 shows a curve called a *cardioid*. (Many curves in polar coordinates have colorful names. The School of Mathematics at the University of St. Andrews, Scotland, has an outstanding tutorial website— *www-history.mcs.st-and.ac.uk/history/*— with a "Famous Curves Index" that gives interactive access to almost 100 "famous curves," with names such as Freeth's nephroid, Fermat's spiral, and the conchoid of de Sluze.) Anyway, the equation for the cardioid shown in Figure 14.4 is $r = a(1 + \cos\theta)$. Because we are given $r(\theta)$, Equation 14.5 becomes

$$I = \int_0^{2\pi} d\theta \int_0^{r(\theta)} dr\, r = \frac{1}{2}\int_0^{2\pi} d\theta\, r^2(\theta) = \frac{a^2}{2}\int_0^{2\pi} d\theta\,(1 + \cos\theta)^2$$

$$= \frac{a^2}{2}\left(2\pi + \frac{2\pi}{2}\right) = \frac{3\pi a^2}{2}$$

The integral

$$I = \int_0^\infty dx\, e^{-ax^2} \qquad (14.6)$$

is not easy to evaluate by elementary methods, but Equation 14.5 provides a standard trick to evaluate it. First write I^2 as

$$I^2 = \int_0^\infty dx\, e^{-ax^2} \int_0^\infty dy\, e^{-ay^2} = \int_0^\infty \int_0^\infty dxdy\, e^{-a(x^2+y^2)} \qquad (14.7)$$

Notice that we did *not* write I^2 as $\int_0^\infty dx \int_0^\infty dx\, e^{-2ax^2}$, whatever that notation would mean. We must use two different dummy integration variables when we express I^2 as a double integral, as in Equation 14.7. Now we convert Equation 14.7 to plane polar coordinates and obtain

$$I^2 = \iint_R rdrd\theta \, e^{-ar^2} \qquad (14.8)$$

Let's look at the integration limits of r and θ. The region of integration in Equation 14.7 is the entire first quadrant ($0 \le x < \infty$ and $0 \le y < \infty$). Therefore, the limits of r and θ are $0 \le r < \infty$ and $0 \le \theta \le \pi/2$ because these limits map out the first quadrant. Using these limits, Equation 14.8 becomes

$$I^2 = \int_0^\infty dr\, r \int_0^{\pi/2} d\theta\, e^{-ar^2} = \frac{\pi}{2}\int_0^\infty dr\, re^{-ar^2}$$

But this remaining integrand is just of the form $e^{-u}\, du$ (with $u = ar^2$), and so we get $I^2 = \int_0^\infty e^{-u}du/2a = \pi/4a$, or $I = (\pi/4a)^{1/2}$.

There is one final topic involving polar coordinates that we should discuss here. The two-dimensional Laplacian operator is

$$\nabla^2 = \frac{\partial^2}{\partial x^2} + \frac{\partial^2}{\partial y^2}$$

When dealing with two-dimensional problems involving a center of symmetry so that we use plane polar coordinates, we express ∇^2 in terms of polar coordinates rather than cartesian coordinates. The conversion of ∇^2 from cartesian coordinates to polar coordinates can be carried out starting with Equations 14.1 and is a good exercise involving partial derivatives. The final result is (see Problems 14–15 and 14–16)

$$\nabla^2 = \frac{\partial^2}{\partial r^2} + \frac{1}{r}\frac{\partial}{\partial r} + \frac{1}{r^2}\frac{\partial^2}{\partial \theta^2} \tag{14.9}$$

14.2 Spherical Coordinates

Instead of locating a point in space by specifying the cartesian coordinates x, y, and z, we can equally well locate the same point by specifying the spherical coordinates r, θ, and ϕ. From Figure 14.5, we can see that the relations between the two sets of coordinates are given by

$$\begin{aligned} x &= r\sin\theta\cos\phi \\ y &= r\sin\theta\sin\phi \\ z &= r\cos\theta \end{aligned} \tag{14.10}$$

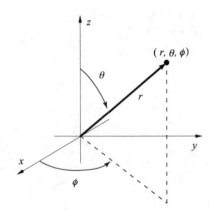

Figure 14.5. A representation of a point in a spherical coordinate system. A point is specified by the spherical coordinates r, θ, and ϕ.

This coordinate system is called a *spherical coordinate system* because the graph of the equation $r = c = $ constant is a sphere of radius c centered at the origin. Occasionally, we need to know r, θ, and ϕ in terms of x, y, and z. These relations are given by

$$\begin{aligned} r &= \left(x^2 + y^2 + z^2\right)^{1/2} \\ \cos\theta &= \frac{z}{(x^2 + y^2 + z^2)^{1/2}} \\ \tan\phi &= \frac{y}{x} \end{aligned} \tag{14.11}$$

Any point on the surface of a sphere of unit radius can be specified by the values of θ and ϕ. The angle θ represents the declination from the North Pole, and hence $0 \leq \theta \leq \pi$. The angle ϕ represents the angle about the equator, and so $0 \leq \phi \leq 2\pi$. Although there is a natural zero value for θ (along the North Pole), there is none for ϕ. Conventionally, the angle ϕ is measured from the x axis, as illustrated in Figure 14.5. Note that r, being the distance from the origin, is intrinsically a positive quantity. In mathematical terms, $0 \leq r < \infty$.

We frequently encounter integrals involving spherical coordinates. The differential volume element in cartesian coordinates is $dxdydz$, but it is not quite so simple in spherical coordinates. Figure 14.6 shows a differential volume element

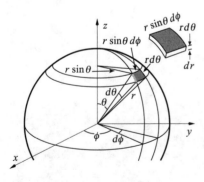

Figure 14.6. A geometrical construction of the differential volume element in spherical coordinates. The differential volume is the product of the three differential lengths, dr, $r\,d\theta$, and $r\sin\theta\,d\phi$.

in spherical coordinates, which can be seen to be

$$dV = (r\sin\theta d\phi)(rd\theta)dr = r^2\sin\theta drd\theta d\phi \qquad (14.12)$$

Let's use Equation 14.12 to evaluate the volume of a sphere of radius a. In this case, $0 \le r \le a$, $0 \le \theta \le \pi$, and $0 \le \phi \le 2\pi$. Therefore,

$$V = \int_0^a r^2 dr \int_0^\pi \sin\theta d\theta \int_0^{2\pi} d\phi = \left(\frac{a^3}{3}\right)(2)(2\pi) = \frac{4\pi a^3}{3}$$

Similarly, if we integrate only over θ and ϕ, then we obtain

$$dV = r^2 dr \int_0^\pi \sin\theta d\theta \int_0^{2\pi} d\phi = 4\pi r^2 dr \qquad (14.13)$$

This quantity is the volume of a spherical shell of radius r and thickness dr (Figure 14.7). The factor $4\pi r^2$ represents the surface area of the spherical shell, and dr is its thickness.

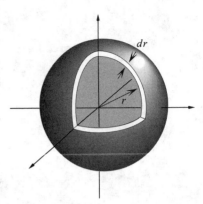

Figure 14.7. A spherical shell of radius r and thickness dr. The volume of such a shell is $4\pi r^2 dr$, which is its area $(4\pi r^2)$ times its thickness (dr).

The quantity

$$dA = r^2\sin\theta \, d\theta d\phi \qquad (14.14)$$

is the differential area on the surface of a sphere of radius r. (See Figure 14.8.) If we integrate Equation 14.14 over all values of θ and ϕ, then we obtain $A = 4\pi r^2$, the area of the surface of a sphere of radius r.

Often, the integral we need to evaluate will be of the form

$$I = \int_0^\infty dr\, r^2 \int_0^\pi d\theta\, \sin\theta \int_0^{2\pi} d\phi\, F(r,\theta,\phi) \qquad (14.15)$$

Recall from Chapter 12 that each integral "acts on" everything that lies to its right; in other words, we first integrate $F(r,\theta,\phi)$ over ϕ from 0 to 2π, then multiply the result by $\sin\theta$ and integrate over θ from 0 to π, and finally multiply that result by r^2 and integrate over r from 0 to ∞. The advantage of the notation in Equation 14.15 is that the integration variable and its associated limits are always unambiguous. As an example of the application of this notation, let's evaluate Equation 14.15 with

$$F(r,\theta,\phi) = \frac{1}{32\pi}r^2 e^{-r}\sin^2\theta\cos^2\phi$$

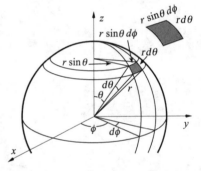

Figure 14.8. A pictorial illustration that the differential area element in spherical coordinates is $dA = r^2\sin\theta\, d\theta d\phi$.

(This function is the square of a $2p_x$ hydrogen atomic orbital.) If we substitute $F(r,\theta,\phi)$ into Equation 14.15, we obtain

$$I = \frac{1}{32\pi}\int_0^\infty dr\, r^2 \int_0^\pi d\theta\, \sin\theta \int_0^{2\pi} d\phi\, r^2 e^{-r}\sin^2\theta\cos^2\phi$$

The integral over ϕ gives

$$\int_0^{2\pi} d\phi\, \cos^2\phi = \pi$$

so that

$$I = \frac{1}{32}\int_0^\infty dr\, r^2 \int_0^\pi d\theta\, \sin\theta\, r^2 e^{-r}\sin^2\theta \qquad (14.16)$$

The integral over θ, I_θ, is

$$I_\theta = \int_0^\pi d\theta \, \sin^3 \theta$$

It is often convenient to change variables and let $x = \cos\theta$ in integrals involving θ. Then $\sin\theta \, d\theta$ becomes $-dx$ and the limits become $+1$ to -1, so in this case we have

$$I_\theta = \int_0^\pi d\theta \, \sin^3 \theta = -\int_1^{-1} dx \, (1 - x^2) = \int_{-1}^1 dx \, (1 - x^2) = 2 - \frac{2}{3} = \frac{4}{3}$$

Using this result in Equation 14.17 gives

$$I = \frac{1}{24} \int_0^\infty dr \, r^4 e^{-r} = \frac{1}{24}(4!) = 1$$

This final result for I simply shows that our above expression for a $2p_x$ hydrogen atomic orbital is normalized.

Frequently, the integrand in Equation 14.15 will be a function only of r, in which case we say that the integrand is spherically symmetric. Let's look at Equation 14.15 when $F(r, \theta, \phi) = f(r)$:

$$I = \int_0^\infty dr \, r^2 \int_0^\pi d\theta \, \sin\theta \int_0^{2\pi} d\phi f(r) \qquad (14.17)$$

Because $f(r)$ is independent of θ and ϕ, we can integrate over ϕ to get 2π and then integrate over θ to get 2:

$$\int_0^\pi \sin\theta \, d\theta = \int_{-1}^1 dx = 2$$

Therefore, Equation 14.17 becomes

$$I = \int_0^\infty f(r) 4\pi r^2 \, dr \qquad (14.18)$$

The point here is that if $F(r, \theta, \phi) = f(r)$, then Equation 14.15 becomes effectively a one-dimensional integral with a factor of $4\pi r^2 dr$ multiplying the integrand. The quantity $4\pi r^2 dr$ is the volume of a spherical shell of radius r and thickness dr (Figure 14.7).

EXAMPLE 14–2
A $1s$ hydrogen atomic orbital is given by

$$f(r) = \frac{1}{(\pi a_0^3)^{1/2}} e^{-r/a_0}$$

Show that the square of this function is normalized.

SOLUTION: Realize that $f(r)$ is a spherically symmetric function of x, y, and z, where $r = (x^2 + y^2 + z^2)^{1/2}$. Therefore, we use Equation 14.18

and write

$$I = \int_0^\infty f^2(r)\, 4\pi r^2\, dr = \frac{4\pi}{\pi a_0^3} \int_0^\infty r^2 e^{-2r/a_0} dr$$

$$= \frac{4}{a_0^3} \cdot \frac{2}{(2/a_0)^3} = 1$$

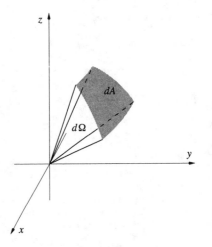

If we restrict ourselves to the surface of a sphere of unit radius, then the angular part of Equation 14.14 gives us the differential surface area

$$dA = \sin\theta\, d\theta d\phi \qquad (14.19)$$

If we integrate over the entire spherical surface ($0 \leq \theta \leq \pi, 0 \leq \phi \leq 2\pi$), then

$$A = \int_0^\pi \sin\theta\, d\theta \int_0^{2\pi} d\phi = 4\pi \qquad (14.20)$$

which is the area of a sphere of unit radius.

We call the solid enclosed by the surface that connects the origin and the area dA a *solid angle*, as shown in Figure 14.9. Because of Equation 14.20, we say that a complete solid angle is 4π, just as we say that a complete angle of a circle is 2π. We often denote a solid angle by $d\Omega$, so that we sometimes write

Figure 14.9. The solid angle, $d\Omega$, subtended by the differential area element $dA = \sin\theta d\theta d\phi$.

$$d\Omega = \sin\theta\, d\theta d\phi \qquad (14.21)$$

and Equation 14.20 becomes

$$\int_{\text{sphere}} d\Omega = 4\pi \qquad (14.22)$$

In discussing the quantum theory of a hydrogen atom, we frequently encounter angular integrals of the form

$$I = \int_0^\pi d\theta\, \sin\theta \int_0^{2\pi} d\phi\, F(\theta, \phi) \qquad (14.23)$$

Note that we are integrating $F(\theta, \phi)$ over the surface of a sphere. For example, we encounter the integral

$$I = \frac{15}{8\pi} \int_0^{2\pi} d\phi \int_0^\pi d\theta\, \sin^2\theta \cos^2\theta \sin\theta$$

The value of this integral is

$$I = \frac{15}{8\pi} \int_0^\pi d\theta\, \sin\theta \cos^2\theta \sin^2\theta \int_0^{2\pi} d\phi$$

$$= \frac{15}{4} \int_{-1}^1 (1 - x^2) x^2\, dx = \frac{15}{4} \left[\frac{2}{3} - \frac{2}{5} \right] = 1$$

EXAMPLE 14–3
Show that

$$I = \int_0^\pi d\theta \, \sin\theta \int_0^{2\pi} d\phi \, Y_1^1(\theta, \phi)^* Y_1^{-1}(\theta, \phi) = 0$$

where

$$Y_1^1(\theta, \phi) = -\left(\frac{3}{8\pi}\right)^{1/2} e^{i\phi} \sin\theta$$

and

$$Y_1^{-1}(\theta, \phi) = \left(\frac{3}{8\pi}\right)^{1/2} e^{-i\phi} \sin\theta$$

SOLUTION:

$$I = -\frac{3}{8\pi} \int_0^\pi d\theta \, \sin^3\theta \int_0^{2\pi} d\phi \, e^{-2i\phi}$$

The integral over ϕ is an integral over a complete cycle of $\sin 2\phi$ and $\cos 2\phi$ and therefore $I = 0$. We say that $Y_1^1(\theta, \phi)$ and $Y_1^{-1}(\theta, \phi)$ are orthogonal over the surface of a unit sphere.

When dealing with problems involving a center of symmetry, we need to express the Laplacian operator, ∇^2, in terms of spherical coordinates rather than cartesian coordinates. The conversion of ∇^2 from cartesian coordinates to spherical coordinates can be carried out starting with Equations 14.10, but it is a long, tedious exercise involving partial derivatives that perhaps you should do once, but probably never again. The final result is (Problem 14–22)

$$\nabla^2 = \frac{1}{r^2}\frac{\partial}{\partial r}\left(r^2\frac{\partial}{\partial r}\right) + \frac{1}{r^2 \sin\theta}\frac{\partial}{\partial\theta}\left(\sin\theta\frac{\partial}{\partial\theta}\right) + \frac{1}{r^2\sin^2\theta}\frac{\partial^2}{\partial\phi^2} \qquad (14.24)$$

EXAMPLE 14–4
Show that $u(r, \theta, \phi) = 1/r$ is a solution to $\nabla^2 u = 0$. (This equation is called *Laplace's equation*.)

SOLUTION: The fact that u depends only upon r means that $\nabla^2 u$ reduces to

$$\nabla^2 u = \frac{1}{r^2}\frac{\partial}{\partial r}\left(r^2\frac{\partial u}{\partial r}\right)$$

If we substitute $u = 1/r$ into this expression, we find that $r^2 \partial u/\partial r = -1$ and that $\nabla^2 u = 0$.

There is one final topic involving spherical coordinates that we shall discuss here. We often deal with Fourier transforms in three dimensions. The extension of

Equation 10.7 to three dimensions is

$$\hat{F}(k_x, k_y, k_z) = \frac{1}{(2\pi)^{3/2}} \int\!\!\!\int\!\!\!\int_{-\infty}^{\infty} dx\,dy\,dz\, f(x, y, z)e^{-i\mathbf{k}\cdot\mathbf{r}} \qquad (14.25)$$

where $\mathbf{k}\cdot\mathbf{r} = k_x x + k_y y + k_z z$. The inverse transform, in a more common notation, is

$$f(\mathbf{r}) = \frac{1}{(2\pi)^{3/2}} \int\!\!\!\int\!\!\!\int d^3 k\, \hat{F}(\mathbf{k})e^{i\mathbf{k}\cdot\mathbf{r}} \qquad (14.26)$$

where $f(\mathbf{r})$ denotes $f(x, y, z)$, $\hat{F}(\mathbf{k})$ denotes $\hat{F}(k_x, k_y, k_z)$, and $d^3k = dk_x\,dk_y\,dk_z$. Suppose now that $f(\mathbf{r})$ depends only upon $|\mathbf{r}| = (x^2 + y^2 + z^2)^{1/2}$, as is often the case. To evaluate $\hat{F}(\mathbf{k})$ when $f(\mathbf{r}) = f(|\mathbf{r}|)$, we introduce spherical coordinates into Equation 14.25 and write

$$\hat{F}(\mathbf{k}) = \frac{1}{(2\pi)^{3/2}} \int_0^\infty dr\, r^2 \int_0^\pi d\theta\, \sin\theta \int_0^{2\pi} d\phi\, f(r)e^{-i\mathbf{k}\cdot\mathbf{r}} \qquad (14.27)$$

Choose the z axis (the polar axis) of our spherical coordinate system to point along $\mathbf{k}$, so that $\mathbf{k}\cdot\mathbf{r} = kr\cos\theta$. We can then integrate over θ and ϕ in Equation 14.27 to get (Problem 14–19)

$$\hat{F}(k) = \left(\frac{2}{\pi}\right)^{1/2} \int_0^\infty dr\, f(r)\frac{r\sin kr}{k} \qquad (14.28)$$

The inverse formula is

$$f(r) = \left(\frac{2}{\pi}\right)^{1/2} \int_0^\infty dk\, \hat{F}(k)\frac{k\sin kr}{r} \qquad (14.29)$$

Notice that $\hat{F}(\mathbf{k}) = \hat{F}(|\mathbf{k}|) = \hat{F}(k)$ when $f(\mathbf{r}) = f(|\mathbf{r}|) = f(r)$; in other words, the Fourier transform of $f(r)$ is a function of only the magnitude of $\mathbf{k}$ when $f(r)$ is spherically symmetric.

EXAMPLE 14–5

Determine the Fourier transform of

$$f(r) = \frac{Z^3}{\pi}e^{-2Zr}$$

given that r is the radial coordinate of a spherical coordinate system.

SOLUTION: We use Equation 14.28:

$$\hat{F}(k) = \left(\frac{2}{\pi}\right)^{1/2} \frac{Z^3}{\pi k} \int_0^\infty re^{-2Zr}\sin kr\, dr$$

$$= \left(\frac{2}{\pi}\right)^{1/2} \frac{4Z^4/\pi}{(k^2 + 4Z^2)^2}$$

Problems

14–1. Calculate the value of θ in plane polar coordinates for the points
(a) $(x = -1, y = \sqrt{3})$, (b) $(x = -1, y = -1)$, (c) $(x = 1, y = 1)$,
(d) $(x = \sqrt{3}, y = -1)$.

14–2. Calculate the area enclosed by the curve described by $r = a \sin \theta$. What does this curve look like when plotted? Remember that $r \geq 0$.

14–3. Calculate the area enclosed by the curve described by $r = 2 \cos 2\theta$ (lemniscate of Bernoulli). Be sure to sketch the curve first in order to determine the limits on θ. Remember that $r \geq 0$.

14–4. Express the following points given in cartesian coordinates in terms of spherical coordinates: (x, y, z) : $(1, 0, 0)$; $(0, 1, 0)$; $(0, 0, 1)$; $(0, 0, -1)$.

14–5. Describe the graphs of the following equations in spherical coordinates:
(a) $r = 5$ (b) $\theta = \pi/4$ (c) $\phi = \pi/2$

14–6. Use Equation 14.12 to determine the volume of a hemisphere of radius a.

14–7. Use Equation 14.14 to determine the surface area of a hemisphere of radius a.

14–8. Evaluate the integral

$$I = \int_0^\pi \cos^2 \theta \sin^3 \theta \, d\theta$$

by letting $x = \cos \theta$.

14–9. A $2p_y$ hydrogen atom orbital is given by

$$\psi_{2p_y} = \frac{1}{4\sqrt{2\pi}} r e^{-r/2} \sin \theta \sin \phi$$

Show that ψ_{2p_y} is normalized. (Don't forget to square ψ_{2p_y} first.)

14–10. A $2s$ hydrogen atomic orbital is given by

$$\psi_{2s} = \frac{1}{4\sqrt{2\pi}} (2 - r) e^{-r/2}$$

Show that ψ_{2s} is normalized. (Don't forget to square ψ_{2s} first.)

14–11. Show that

$$Y_1^0(\theta, \phi) = \left(\frac{3}{4\pi} \right)^{1/2} \cos \theta$$

$$Y_1^1(\theta, \phi) = -\left(\frac{3}{8\pi} \right)^{1/2} e^{i\phi} \sin \theta$$

and

$$Y_1^{-1}(\theta, \phi) = \left(\frac{3}{8\pi} \right)^{1/2} e^{-i\phi} \sin \theta$$

are orthonormal over the surface of a unit sphere.

14–12. Evaluate the integral of $\cos \theta$ and $\cos^2 \theta$ over the surface of a unit sphere.

14–13. We frequently use the notation $d\mathbf{r}$ to represent the volume element in spherical coordinates. Evaluate the integral

$$I = \int d\mathbf{r} \, e^{-r} \cos^2 \theta$$

where the integral is over all space (in other words, over all possible values of r, θ, and ϕ).

14-14. Show that the two functions $f_1(r, \theta) = e^{-r} \cos \theta$ and $f_2(r, \theta) = (2 - r)e^{-r/2} \cos \theta$ are orthogonal over all space (in other words, over all possible values of r, θ, and ϕ).

14-15. We shall transform ∇^2 from two-dimensional cartesian coordinates to polar coordinates in this problem and the next problem. If a function $f(r, \theta)$ depends upon the polar coordinates r and θ, then the chain rule of partial differentiation says that

$$\left(\frac{\partial f}{\partial x}\right)_y = \left(\frac{\partial f}{\partial r}\right)_\theta \left(\frac{\partial r}{\partial x}\right)_y + \left(\frac{\partial f}{\partial \theta}\right)_r \left(\frac{\partial \theta}{\partial x}\right)_y \tag{1}$$

and that

$$\left(\frac{\partial f}{\partial y}\right)_x = \left(\frac{\partial f}{\partial r}\right)_\theta \left(\frac{\partial r}{\partial y}\right)_x + \left(\frac{\partial f}{\partial \theta}\right)_r \left(\frac{\partial \theta}{\partial y}\right)_x \tag{2}$$

For simplicity, we shall assume that r is equal to a constant, l, so that we can ignore terms involving derivatives with respect to r. Using equations 1 and 2, show that

$$\left(\frac{\partial f}{\partial x}\right)_y = -\frac{\sin \theta}{l} \left(\frac{\partial f}{\partial \theta}\right)_r \quad \text{and} \quad \left(\frac{\partial f}{\partial y}\right)_x = \frac{\cos \theta}{l} \left(\frac{\partial f}{\partial \theta}\right)_r \tag{3}$$

Now apply equation 1 again to show that

$$\left(\frac{\partial^2 f}{\partial x^2}\right)_y = \left[\frac{\partial}{\partial x}\left(\frac{\partial f}{\partial x}\right)_y\right] = \left[\frac{\partial}{\partial \theta}\left(\frac{\partial f}{\partial x}\right)_y\right]_r \left(\frac{\partial \theta}{\partial x}\right)_y$$

$$= \left\{\frac{\partial}{\partial \theta}\left[-\frac{\sin \theta}{l}\left(\frac{\partial f}{\partial \theta}\right)_r\right]\right\}_r \left(-\frac{\sin \theta}{l}\right)$$

$$= \frac{\sin \theta \cos \theta}{l^2}\left(\frac{\partial f}{\partial \theta}\right) + \frac{\sin^2 \theta}{l^2}\left(\frac{\partial^2 f}{\partial \theta^2}\right)$$

Similarly, show that

$$\left(\frac{\partial^2 f}{\partial y^2}\right)_x = -\frac{\sin \theta \cos \theta}{l^2}\left(\frac{\partial f}{\partial \theta}\right) + \frac{\cos^2 \theta}{l^2}\left(\frac{\partial^2 f}{\partial \theta^2}\right)$$

and that

$$\nabla^2 f = \frac{\partial^2 f}{\partial x^2} + \frac{\partial^2 f}{\partial y^2} \longrightarrow \frac{1}{l^2}\left(\frac{\partial^2 f}{\partial \theta^2}\right)$$

14-16. Generalize Problem 14-15 to the case of a particle moving in a plane under the influence of a central force; in other words, convert

$$\nabla^2 = \frac{\partial^2}{\partial x^2} + \frac{\partial^2}{\partial y^2}$$

to plane polar coordinates, this time without assuming that r is a constant (see Equation 14.9).

14–17. Show that $u(r, \theta, \phi) = r \sin \theta \cos \phi$ satisfies Laplace's equation in spherical coordinates, $\nabla^2 u = 0$.

14–18. Show that $u(r, \theta, \phi) = r^2 \sin^2 \theta \cos 2\phi$ satisfies Laplace's equation in spherical coordinates, $\nabla^2 u = 0$.

14–19. Derive Equations 14.28 and 14.29.

14–20. Determine the Fourier transform of $f(r) = \dfrac{1}{(4\pi\alpha)^{3/2}} e^{-r^2/4\alpha}$, where α is a constant.

14–21. We determined the conversion of the differential area element from cartesian coordinates to plane polar coordinates $(dxdy \to r\,drd\theta)$ geometrically using Figure 14.3. There is an analytic method that we can use to make the transformation, which is given by

$$dxdy = \begin{vmatrix} \partial x/\partial r & \partial x/\partial \theta \\ \partial y/\partial r & \partial y/\partial \theta \end{vmatrix} drd\theta \tag{1}$$

where the determinant on the right side here is called a *Jacobian determinant*. (See Chapter 17 for a discussion of determinants.) If we substitute Equations 14.1 into equation 1, we find

$$\begin{vmatrix} \partial x/\partial r & \partial x/\partial \theta \\ \partial y/\partial r & \partial y/\partial \theta \end{vmatrix} = \begin{vmatrix} \cos\theta & -r\sin\theta \\ \sin\theta & r\cos\theta \end{vmatrix} = r(\cos^2\theta + \sin^2\theta) = r$$

in agreement with Equation 14.4. This procedure is readily extended to three dimensions. Use the three-dimensional extension of equation 1 to derive Equation 14.12.

14–22. Derive Equation 14.24.

CHAPTER 15

The Classical Wave Equation

In 1925, Erwin Schrödinger and Werner Heisenberg independently formulated a general quantum theory. At first sight, the two methods appeared to be different because Heisenberg's method is formulated in terms of matrices, whereas Schrödinger's method is formulated in terms of partial differential equations. Just a year later, however, the two formulations were shown to be mathematically equivalent. Because most students of physical chemistry are not familiar with matrix algebra, quantum theory is customarily presented according to Schrödinger's formulation, the central feature of which is a partial differential equation now known as the *Schrödinger equation*. In this chapter, we're going to discuss a simpler partial differential equation called the *classical wave equation*. The classical wave equation describes various wave phenomena such as a vibrating string, a vibrating drum head, ocean waves, and acoustic waves. Not only does the classical wave equation provide a physical background to the Schrödinger equation, but, in addition, the methods that we use to solve it are similar to those that we use to solve the Schrödinger equation.

15.1 A Vibrating String

Consider a uniform string stretched between two fixed points, as shown in Figure 15.1. The maximum displacement of the string from its equilibrium horizontal position is called its *amplitude*. If we let $u(x, t)$ be the displacement of the string, then $u(x, t)$ satisfies the equation

$$\frac{\partial^2 u}{\partial x^2} = \frac{1}{v^2}\frac{\partial^2 u}{\partial t^2} \tag{15.1}$$

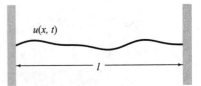

Figure 15.1. A vibrating string whose ends are fixed at 0 and l. The displacement of the vibration at position x and time t is $u(x, t)$.

where v is the speed with which a disturbance moves along the string. (You derive Equation 15.1 in Problem 15–12.) Equation 15.1 is the *classical wave equation*. Equation 15.1 is a *partial differential equation* because the unknown, $u(x, t)$ in this case, occurs in partial derivatives. The variables x and t are said to be the *independent variables* and $u(x, t)$, which depends upon x and t, is said to be the *dependent variable*. Equation 15.1 is a *linear partial differential equation* because $u(x, t)$ and its derivatives appear only to the first power and there are no cross terms.

In addition to having to satisfy Equation 15.1, the displacement $u(x, t)$ must satisfy certain physical conditions as well. Because the ends of the string are held fixed, the displacement at these two points is always zero, and so we have the requirement that

$$u(0, t) = 0 \quad \text{and} \quad u(l, t) = 0 \quad \text{(for all } t) \tag{15.2}$$

These two conditions are called *boundary conditions* because they specify the behavior of $u(x, t)$ at the boundaries. Generally, a partial differential equation must be solved subject to certain boundary conditions, the nature of which will be apparent on physical grounds.

15.2 The Method of Separation of Variables

The classical wave equation, as well as the Schrödinger equation and many other partial differential equations that arise in physical chemistry, can be solved by a method called *separation of variables*. We shall use the problem of a vibrating string to illustrate this method.

The key step in the method of separation of variables is to assume that $u(x, t)$ factors into a function of x, $X(x)$, times a function of t, $T(t)$, or that

$$u(x, t) = X(x)T(t) \tag{15.3}$$

If we substitute Equation 15.3 into Equation 15.1, we obtain

$$T(t)\frac{d^2X(x)}{dx^2} = \frac{1}{v^2}X(x)\frac{d^2T(t)}{dt^2} \tag{15.4}$$

Now we divide both sides of Equation 15.4 by $u(x, t) = X(x)T(t)$ and obtain

$$\frac{1}{X(x)}\frac{d^2X(x)}{dx^2} = \frac{1}{v^2T(t)}\frac{d^2T(t)}{dt^2} \tag{15.5}$$

The left side of Equation 15.5 is a function of x only, and the right side is a function of t only. Because x and t are independent variables, each side of Equation 15.5 can be varied independently. The only way for the equality of the two sides to be maintained under any variation of x and t is for each side to be equal to a constant. If we let this constant be K, we can write

$$\frac{1}{X(x)}\frac{d^2X(x)}{dx^2} = K \tag{15.6}$$

and

$$\frac{1}{v^2 T(t)} \frac{d^2 T(t)}{dt^2} = K \tag{15.7}$$

where K is called the *separation constant* and will be determined later. Equations 15.6 and 15.7 can be written as

$$\frac{d^2 X(x)}{dx^2} - K X(x) = 0 \tag{15.8}$$

and

$$\frac{d^2 T(t)}{dt^2} - K v^2 T(t) = 0 \tag{15.9}$$

Thus, the method of separation of variables has reduced the partial differential equation in two independent variables to two ordinary differential equations. In this particular case, the resulting differential equations are linear with constant coefficients and so are easy to solve (see Section 6.2).

The value of K in Equations 15.8 and 15.9 is yet to be determined. We do not know right now whether K is positive, negative, or even zero. Let's first assume that $K = 0$. In this case, Equations 15.8 and 15.9 can be integrated immediately to give

$$X(x) = a_1 x + b_1 \tag{15.10}$$

and

$$T(t) = a_2 t + b_2 \tag{15.11}$$

where the a's and b's are just integration constants, which can be determined by using the boundary conditions given in Equations 15.2. In terms of $X(x)$ and $T(t)$, the boundary conditions are

$$u(0, t) = X(0)T(t) = 0 \quad \text{and} \quad u(l, t) = X(l)T(t) = 0$$

Because $T(t)$ certainly does not vanish for all t, we must have that

$$X(0) = 0 \quad \text{and} \quad X(l) = 0 \tag{15.12}$$

which is how the boundary conditions affect $X(x)$. Going back to Equation 15.10, we conclude that the only way to satisfy Equations 15.12 is for $a_1 = b_1 = 0$, which means that $X(x) = 0$ and that $u(x, t) = 0$ for all x. This is called a *trivial solution* to Equation 15.1 and is of no physical interest. (Throwing away solutions to mathematical equations should not disturb you. What we know is that every physically acceptable solution $u(x, t)$ must satisfy Equation 15.1, not that every solution to the equation is physically acceptable.)

Now let's assume that $K > 0$ in Equation 15.8. To this end, write K as k^2, where k is real. This assures that K is positive because it is the square of a real number. In this case, Equation 15.8 becomes

$$\frac{d^2 X(x)}{dx^2} - k^2 X(x) = 0 \tag{15.13}$$

whose general solution is

$$X(x) = c_1 e^{kx} + c_2 e^{-kx} \tag{15.14}$$

Applying the boundary conditions given by Equations 15.12 to Equation 15.14 gives

$$c_1 + c_2 = 0 \quad \text{and} \quad c_1 e^{kl} + c_2 e^{-kl} = 0$$

The only way to satisfy these conditions for $k > 0$ is with $c_1 = c_2 = 0$, and so once again, we find only a trivial solution (Problem 15–1). So far, we have found only a trivial solution to Equation 15.1 if $K = 0$ or $K > 0$.

Let's hope that assuming K to be negative gives us something interesting. If we set $K = -\beta^2$, then K is negative if β is real. In this case, Equation 15.8 is

$$\frac{d^2 X(x)}{dx^2} + \beta^2 X(x) = 0 \tag{15.15}$$

whose general solution is

$$X(x) = A \cos \beta x + B \sin \beta x \tag{15.16}$$

The boundary condition that $X(0) = 0$ implies that $A = 0$. The condition at the boundary $x = l$ says that

$$X(l) = B \sin \beta l = 0 \tag{15.17}$$

Equation 15.17 can be satisfied in two ways. One is that $B = 0$, but this along with the fact that $A = 0$ yields a trivial solution. The other way is to require that $\sin \beta l = 0$. Because $\sin \theta = 0$ when $\theta = 0, \pi, 2\pi, 3\pi, \ldots$, Equation 15.17 implies that

$$\beta_n l = n\pi \qquad n = 1, 2, 3, \ldots \tag{15.18}$$

where we have omitted the $n = 0$ case because it leads to $\beta = 0$ and a trivial solution. We have also subscripted β with an n to emphasize that its value depends upon n. Equation 15.18 determines the parameter β_n and hence the separation constant $K = -\beta_n^2$. So far, then, we have that

$$X(x) = B \sin \frac{n\pi x}{l} \tag{15.19}$$

Remember that we have Equation 15.9 to solve also. Because $K = -\beta_n^2$, Equation 15.9 can be written as

$$\frac{d^2 T(t)}{dt^2} + \beta_n^2 v^2 T(t) = 0 \tag{15.20}$$

where Equation 15.18 says that $\beta_n = n\pi/l$. The general solution to Equation 15.20 is

$$T(t) = D \cos \omega_n t + E \sin \omega_n t \tag{15.21}$$

where $\omega_n = \beta_n v = n\pi v/l$. We have no conditions to specify D and E yet, so the displacement $u(x, t)$ is given by (see Equation 15.3)

$$u(x, t) = X(x)T(t)$$
$$= \left(B \sin \frac{n\pi x}{l}\right)(D \cos \omega_n t + E \sin \omega_n t)$$
$$= (a \cos \omega_n t + b \sin \omega_n t) \sin \frac{n\pi x}{l} \qquad n = 1, 2, \ldots$$

where we have let $a = DB$ and $b = EB$. Because there is a $u(x, t)$ for each integer n and because the values of a and b may depend on n, we should write $u(x, t)$ as

$$u_n(x, t) = (a_n \cos \omega_n t + b_n \sin \omega_n t) \sin \frac{n\pi x}{l} \qquad n = 1, 2, \ldots \qquad (15.22)$$

EXAMPLE 15–1
Show that Equation 15.22 is a solution to Equation 15.1.

SOLUTION: The second partial derivatives of $u_n(x, t)$ are

$$\frac{\partial^2 u_n(x, t)}{\partial x^2} = -\frac{n^2 \pi^2}{l^2}(a_n \cos \omega_n t + b_n \sin \omega_n t) \sin \frac{n\pi x}{l}$$
$$= -\frac{n^2 \pi^2}{l^2} u_n(x, t)$$

$$\frac{\partial^2 u_n(x, t)}{\partial t^2} = -\omega_n^2 u_n(x, t)$$

Using the fact that $\omega_n = n\pi v/l$, we see that $\omega_n^2 = v^2(n^2\pi^2/l^2)$, and that Equation 15.1 is satisfied.

Because each $u_n(x, t)$ in Equation 15.22 is a solution to the linear differential equation, Equation 15.1, their sum is also a solution of Equation 15.1 and is, in fact, the general solution. Therefore, we have

$$u(x, t) = \sum_{n=1}^{\infty} (a_n \cos \omega_n t + b_n \sin \omega_n t) \sin \frac{n\pi x}{l} \qquad n = 1, 2, \ldots \qquad (15.23)$$

No matter how the string is plucked initially, its shape will evolve according to Equation 15.23. We can easily verify that Equation 15.23 is a solution to Equation 15.1 by direct substitution (Problem 15–2).

Students are sometimes wary of the method of separation of variables when they first meet it, particularly of the assumption in Equation 15.3. You might wonder if there are other solutions satisfying the same boundary conditions but describing some other type of vibration that cannot be written in the form $u(x, t) = X(x)T(t)$. It turns out that the theory of partial differential equations tells us that if $u(x, t)$ satisfies Equations 15.1 and 15.2, then it is the only function that does so. Therefore, once we find a solution, by whatever means, we can be assured that it is the only solution.

15.3 Superposition of Normal Modes

We can write the general solution given by Equation 15.23 in another form. We showed in Section 6.2 that $a \cos \omega t + b \sin \omega t$ can be written as $A \cos (\omega t + \phi)$, where A and ϕ are constants expressible in terms of a and b (see Equation 6.23). Using this relation, we can write Equation 15.23 in the form

$$u(x, t) = \sum_{n=1}^{\infty} A_n \cos (\omega_n t + \phi_n) \sin \frac{n\pi x}{l} = \sum_{n=1}^{\infty} u_n(x, t) \qquad (15.24)$$

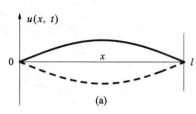

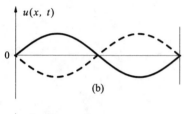

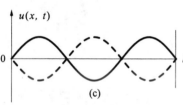

Equation 15.24 has a nice physical interpretation. Each $u_n(x, t)$ is called a *normal mode*, and the time dependence of each normal mode represents harmonic motion of frequency

$$\nu_n = \frac{\omega_n}{2\pi} = \frac{\nu n}{2l} \qquad (15.25)$$

where we have used the fact that $\omega_n = \beta_n v = n\pi v/l$ (see Equation 15.18). The corresponding wavelength is given by v/ν_n, or by $\lambda_n = 2l/n$.

The spatial dependence of the first few terms in Equation 15.24 is shown in Figure 15.2. The first term, $u_1(x, t)$, called the *fundamental mode* or *first harmonic*, represents a cosinusoidal (harmonic) time dependence of frequency $v/2l$ of the motion depicted in Figure 15.2a. The wavelength in this case is $\lambda_1 = 2l$. The *second harmonic* or *first overtone*, $u_2(x, t)$, vibrates harmonically with frequency v/l and looks like the motion depicted in Figure 15.2b. The wavelength in this case is $\lambda_2 = l$. Note that the midpoint of this harmonic is fixed at zero for all t. Such a point is called a *node*, a concept that arises in quantum mechanics as well. Notice that $u(0)$ and $u(l)$ are also equal to zero. These terms are not nodes because their values are fixed by the boundary conditions. Note that the second harmonic oscillates with twice the frequency of the first harmonic. Figure 15.2c shows that the *third harmonic* or *second overtone* has two nodes. The waves shown in Figure 15.2 are called *standing waves* because the positions of the nodes are fixed in time. Between the nodes, the string oscillates up and down.

Figure 15.2. The first three normal modes of a vibrating string. Note that each normal mode is a standing wave and that the nth harmonic has $n - 1$ nodes. The wavelengths of the standing waves satisfy the condition $\lambda_n = 2l/n$.

Consider a simple case in which $u(x, t)$ consists of only the first two harmonics and is of the form (see Equation 15.24)

$$u(x, t) = \cos \omega_1 t \sin \frac{\pi x}{l} + \frac{1}{2} \cos \left(\omega_2 t + \frac{\pi}{2} \right) \sin \frac{2\pi x}{l} \qquad (15.26)$$

Equation 15.26 is illustrated in Figure 15.3. The left side of Figure 15.3 shows the time dependence of each mode separately. Notice that $u_2(x, t)$ has gone through one complete oscillation in the time depicted while $u_1(x, t)$ has gone through only one-half of a cycle, nicely illustrating that $\omega_2 = 2\omega_1$. The right side of Figure 15.3 shows the sum of the two harmonics, or the actual motion of the string, as a function of time. You can see how a superposition of the standing waves in the left side of the figure yields the traveling wave in the right side. The decomposition of any complicated, general wave motion into a sum or superposition of normal modes is

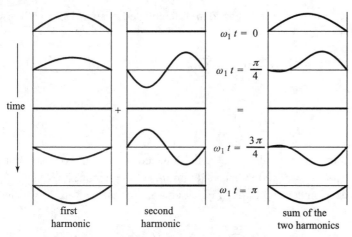

Figure 15.3. An illustration of how two standing waves can combine to give a traveling wave. In both parts, time increases downward. The left portion shows the independent motion of the first two harmonics. Both harmonics are standing waves; the first harmonic goes through half a cycle, and the second harmonic goes through one complete cycle in the time shown. The right side shows the sum of the two harmonics. The sum is not a standing wave. As shown, the sum is a traveling wave that travels back and forth between the fixed ends. The traveling wave has gone through one-half of a cycle in the time shown.

a fundamental property of oscillatory behavior and follows from the fact that the wave equation is a linear equation.

15.4 Fourier Series Solutions

We're not finished solving Equation 15.1. Equation 15.23 (or equivalently, Equation 15.24) still has two infinite sets of constants to be determined. Note that Equation 15.23 is of the form of a Fourier series, where the Fourier coefficients are $a_n \cos \omega_n t + b_n \sin \omega_n t$. We can evaluate all the a_n and b_n from two initial conditions. Suppose that we are given some initial displacement and some initial velocity of the string

$$u(x, 0) = f(x) \qquad \text{and} \qquad u_t(x, 0) = g(x)$$

Setting $t = 0$ in Equation 15.23 gives

$$u(x, 0) = f(x) = \sum_{n=1}^{\infty} a_n \sin \frac{n\pi x}{l} \qquad 0 \leq x \leq l \qquad (15.27)$$

Equation 15.27 is a Fourier series (Chapter 9). We can evaluate the a_n by using the orthogonality of $\{\sin n\pi x / l\}$ over the interval $(0, l)$, in which case we have

$$a_n = \frac{2}{l} \int_0^l f(x) \sin \frac{n\pi x}{l} dx \qquad n = 1, 2, \ldots \qquad (15.28)$$

Similarly, assuming that it is valid to differentiate Equation 15.23 with respect to t, we can write

$$u_t(x, 0) = g(x) = \sum_{n=1}^{\infty} \omega_n b_n \sin \frac{n\pi x}{l} \qquad 0 \le x \le l \qquad (15.29)$$

and

$$b_n = \frac{2}{\omega_n l} \int_0^l g(x) \sin \frac{n\pi x}{l} dx \qquad n = 1, 2, \ldots \qquad (15.30)$$

Let's use these results to solve the wave equation with boundary conditions $u(0, t) = u(l, t) = 0$ and initial conditions $u(x, 0) = u_0 \sin 2\pi x/l$ and $u_t(x, 0) = 0$. The development up through Equation 15.23 is unchanged, so we shall start with

$$u(x, t) = \sum_{n=1}^{\infty} (a_n \cos \omega_n t + b_n \sin \omega_n t) \sin \frac{n\pi x}{l}$$

where $\omega_n = n\pi v/l$. The initial condition $u_t(x, 0) = 0$ implies that the $b_n = 0$ (Equation 15.30). The a_n are given by the initial condition $u(x, 0) = u_0 \sin 2\pi x/l$,

$$a_n = \frac{2u_0}{l} \int_0^l \sin \frac{2\pi x}{l} \sin \frac{n\pi x}{l} dx = u_0 \delta_{n2}$$

where δ_{n2} is a Kronecker delta. The complete solution is

$$u(x, t) = u_0 \sin \frac{2\pi x}{l} \cos \frac{2\pi vt}{l} \qquad (15.31)$$

It's easy to show that this solution satisfies the wave equation along with the given boundary conditions and initial conditions (Problem 15–3).

EXAMPLE 15–2

Suppose a string is initially displaced a small distance h at the middle of the string and then released (in other words, plucked in the middle). Determine the subsequent motion of the string.

SOLUTION: The initial conditions translate into

$$u(x, 0) = \begin{cases} \dfrac{2hx}{l} & 0 < x < \dfrac{l}{2} \\ \dfrac{2h}{l}(l - x) & \dfrac{l}{2} < x < l \end{cases}$$

and $u_t(x, 0) = 0$. Equation 15.30 gives $b_n = 0$ and Equation 15.28 gives

$$a_n = \frac{2}{l} \left[\int_0^{l/2} \frac{2hx}{l} \sin \frac{n\pi x}{l} dx + \int_{l/2}^l \frac{2h}{l}(x - l) \sin \frac{n\pi x}{l} dx \right]$$

$$= \frac{8h}{n^2\pi^2} \sin \frac{n\pi}{2} = \begin{cases} \dfrac{(-1)^n 8h}{n^2\pi^2} & n \text{ odd} \\ 0 & n \text{ even} \end{cases}$$

Thus,

$$u(x, t) = \frac{8h}{\pi^2} \sum_{n=0}^{\infty} \frac{(-1)^n}{(2n+1)^2} \cos \frac{(2n+1)v\pi t}{l} \sin \frac{(2n+1)\pi x}{l}$$

Notice that only the odd harmonics are excited; none of the harmonics having a node at the center ($x = l/2$) are excited by this initial displacement. Figure 15.4 shows the displacement of the string as a function of time.

15.5 A Vibrating Square Membrane

The generalization of Equation 15.1 to two dimensions is

$$\frac{\partial^2 u}{\partial x^2} + \frac{\partial^2 u}{\partial y^2} = \frac{1}{v^2} \frac{\partial^2 u}{\partial t^2} \tag{15.32}$$

where $u = u(x, y, t)$ and x, y, and t are the independent variables. We will apply this equation to a rectangular membrane whose entire perimeter is clamped. By referring to the geometry in Figure 15.5, we see that the boundary conditions that $u(x, y, t)$ must satisfy because its four edges are clamped are

$$u(0, y) = u(a, y) = 0$$
$$u(x, 0) = u(x, b) = 0 \qquad \text{(for all } t\text{)} \tag{15.33}$$

By applying the method of separation of variables to Equation 15.32, we assume that $u(x, y, t)$ can be written as the product of a spatial part and a temporal part or that

$$u(x, y, t) = F(x, y)T(t) \tag{15.34}$$

We substitute Equation 15.34 into Equation 15.32 and divide both sides by $F(x, y)T(t)$ to find

$$\frac{1}{v^2 T(t)} \frac{d^2 T}{dt^2} = \frac{1}{F(x, y)} \left(\frac{\partial^2 F}{\partial x^2} + \frac{\partial^2 F}{\partial y^2} \right)$$

The right side of this equation is a function of x and y only, and the left side is a function of t only. The equality can be true for all t, x, and y only if both sides are equal to a constant. Anticipating that the separation constant will be negative, as it was in the previous section, we write it as $-\beta^2$ and obtain the two separate equations

$$\frac{d^2 T}{dt^2} + v^2 \beta^2 T(t) = 0 \tag{15.35}$$

and

$$\frac{\partial^2 F}{\partial x^2} + \frac{\partial^2 F}{\partial y^2} + \beta^2 F(x, y) = 0 \tag{15.36}$$

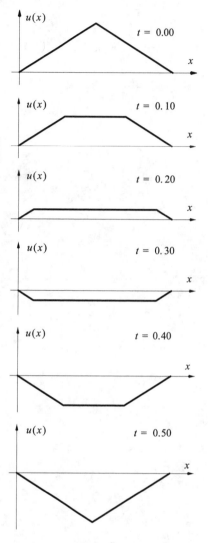

Figure 15.4. The time dependence of the displacement of the string described in Example 15–2.

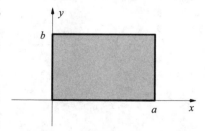

Figure 15.5. A rectangular membrane clamped along its perimeter.

Equation 15.36 is still a partial differential equation. To solve it, we once again use separation of variables. Substitute $F(x, y) = X(x)Y(y)$ into Equation 15.35 and divide both sides by $X(x)Y(y)$ to obtain

$$\frac{1}{X(x)}\frac{d^2 X}{dx^2} + \frac{1}{Y(y)}\frac{d^2 Y}{dy^2} + \beta^2 = 0 \tag{15.37}$$

Again we argue that because x and y are independent variables, the only way this equation can be valid is if

$$\frac{1}{X(x)}\frac{d^2 X}{dx^2} = -p^2 \tag{15.38}$$

and

$$\frac{1}{Y(y)}\frac{d^2 Y}{dy^2} = -q^2 \tag{15.39}$$

where p^2 and q^2 are separation constants, which according to Equation 15.37 must satisfy

$$p^2 + q^2 = \beta^2 \tag{15.40}$$

Equations 15.38 and 15.39 can be rewritten as

$$\frac{d^2 X}{dx^2} + p^2 X(x) = 0 \tag{15.41}$$

and

$$\frac{d^2 Y}{dy^2} + q^2 Y(y) = 0 \tag{15.42}$$

Note that Equation 15.32, a partial differential equation in three variables, has been reduced to three ordinary differential equations (Equations 15.35, 15.41, and 15.42). The boundary conditions, Equations 15.33, in terms of the functions $X(x)$ and $Y(y)$ are

$$X(0)Y(y) = X(a)Y(y) = 0$$

and

$$X(x)Y(0) = X(x)Y(b) = 0$$

which imply that

$$\begin{aligned} X(0) = X(a) = 0 \\ Y(0) = Y(b) = 0 \end{aligned} \tag{15.43}$$

The solutions to Equations 15.41 and 15.42 with these boundary conditions are

$$X_n(x) = B_n \sin\frac{n\pi x}{a} \qquad n = 1, 2, \ldots \tag{15.44}$$

and

$$Y_m(y) = D_m \sin\frac{m\pi y}{b} \qquad m = 1, 2, \ldots \tag{15.45}$$

Recalling that $p^2 + q^2 = \beta^2$, we see that

$$\beta_{nm} = \pi \left(\frac{n^2}{a^2} + \frac{m^2}{b^2} \right)^{1/2} \qquad \begin{array}{l} n = 1,\ 2, \ldots \\ m = 1,\ 2, \ldots \end{array} \qquad (15.46)$$

where we have subscripted β to emphasize that it depends on the two integers n and m.

Finally, now we solve Equation 15.35 for the time dependence:

$$\begin{aligned} T_{nm}(t) &= E_{nm} \cos \omega_{nm} t + F_{nm} \sin \omega_{nm} t \\ &= G_{nm} \cos(\omega_{nm} t + \phi_{nm}) \end{aligned} \qquad (15.47)$$

where

$$\omega_{nm} = v\beta_{nm} = v\pi \left(\frac{n^2}{a^2} + \frac{m^2}{b^2} \right)^{1/2} \qquad (15.48)$$

One solution to Equation 15.32 is given by the product $u_{nm}(x, y, t) = X_n(x)Y_m(y)T_{nm}(t)$, and the general solution is given by

$$\begin{aligned} u(x,\ y,\ t) &= \sum_{n=1}^{\infty} \sum_{m=1}^{\infty} u_{nm}(x,\ y,\ t) \\ &= \sum_{n=1}^{\infty} \sum_{m=1}^{\infty} A_{nm} \cos(\omega_{nm}t + \phi_{nm}) \sin \frac{n\pi x}{a} \sin \frac{m\pi y}{b} \end{aligned} \qquad (15.49)$$

where $A_{nm} = B_n D_m G_{nm}$.

As in the one-dimensional case of a vibrating string, we see that the general vibrational motion of a rectangular drum can be expressed as a superposition of normal modes, $u_{nm}(x, y, t)$. Some of these modes are shown in Figure 15.6. Note that in this two-dimensional problem we obtain *nodal lines*. In two-dimensional problems, the nodes are lines, as compared with points in one-dimensional problems.

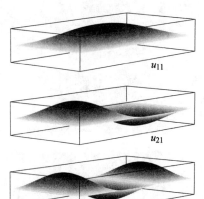

Figure 15.6. The first few normal modes of a rectangular membrane.

Problems

15–1. Show that the only solutions to the equations $c_1 + c_2 = 0$ and $c_1 e^{kl} + c_2 e^{-kl} = 0$ for $k > 0$ is the trivial solution $c_1 = c_2 = 0$. A perfectly satisfactory way to show this is by brute force. We'll learn in Chapter 17, however, that a requirement that there be a nontrivial solution to these equations is that the determinant of the coefficients must equal zero. If you are familiar with determinants, then show that the determinant associated with $c_1 + c_2 = 0$ and $c_1 e^{kl} + c_2 e^{-kl} = 0$ is equal to $e^{-kl} - e^{kl}$, which does not equal zero for $k > 0$.

15–2. Show that Equation 15.23 is a solution to Equation 15.1.

15–3. Show that Equation 15.31 satisfies the wave equation and the boundary conditions $u(0, t) = u(l, t) = 0$ and the initial conditions $u(x, 0) = u_0 \sin 2\pi x/l$ and $u_t(x, 0) = 0$.

15–4. Solve the one-dimensional wave equation subject to the conditions $u(0, t) = u(l, t) = 0$, $u(x, 0) = u_0 \sin(3\pi x/l)$, and $u_t(x, 0) = 0$.

15–5. If $u(x, 0) = u_0 \sin^3 \pi x/l$ in the previous problem, predict which normal modes will be excited.

15–6. Consider a particle constrained to move freely over the surface of a rectangle of sides a and b. The Schrödinger equation for this problem is

$$\frac{\partial^2 \psi}{\partial x^2} + \frac{\partial^2 \psi}{\partial y^2} + \left(\frac{8\pi^2 m E}{h^2}\right)\psi(x, y) = 0$$

with the boundary conditions

$$\psi(0, y) = \psi(a, y) = 0 \qquad \text{for all } y, \qquad 0 \le y \le b$$

$$\psi(x, 0) = \psi(x, b) = 0 \qquad \text{for all } x, \qquad 0 \le x \le a$$

Solve this equation for $\psi(x, y)$, apply the boundary conditions, and show that the energy is quantized according to

$$E_{n_x, n_y} = \frac{n_x^2 h^2}{8ma^2} + \frac{n_y^2 h^2}{8mb^2} \qquad \begin{aligned} n_x &= 1, \ 2, \ 3, \ldots \\ n_y &= 1, \ 2, \ 3, \ldots \end{aligned}$$

15–7. Extend the previous problem to three dimensions, where a particle is constrained to move freely throughout a rectangular box of sides a, b, and c. The Schrödinger equation for this system is

$$\frac{\partial^2 \psi}{\partial x^2} + \frac{\partial^2 \psi}{\partial y^2} + \frac{\partial^2 \psi}{\partial z^2} + \left(\frac{8\pi^2 m E}{h^2}\right)\psi(x, y, z) = 0$$

and the boundary conditions are that $\psi(x, y, z)$ vanishes over all the surfaces of the box. (We shall solve this problem in the next chapter.)

15–8. Show that $u(x, t) = A \sin\left[\dfrac{2\pi}{\lambda}(x - vt)\right]$ is a wave of wavelength λ and frequency $\nu = v/\lambda$ traveling to the right with a velocity v.

15–9. The previous problem introduces you to the idea of a traveling wave. Use the trigonometric identity $\sin \alpha \cos \beta = \dfrac{1}{2}\sin(\alpha + \beta) - \dfrac{1}{2}\sin(\alpha - \beta)$ to show that a standing wave such as $\cos \omega_n t \, \sin n\pi x/l$ can be expressed as the sum of two traveling waves, traveling in opposite directions with the same speed. Remember that $\omega_n = n\pi v/l$ (Equation 15.25).

15–10. The traveling waves in the previous problem are of the form $\sin\left[n\pi(x \pm vt)/l\right]$. Show that this form can be expressed as $\sin\left[\dfrac{2\pi}{\lambda_n}(x \pm vt)\right]$.

15–11. Show that the traveling waves in the previous two problems are solutions to the one-dimensional wave equation.

15–12. We shall derive the one-dimensional wave equation in this problem. Consider a perfectly flexible homogeneous string stretched to a uniform tension τ between two points. Let $u(x, t)$ be the displacement of the string from its horizontal position (see Figure 15.7). The quantities τ_1 and τ_2 in Figure 15.7 are the tensions at the points P and Q on the string. Both τ_1 and τ_2 are tangential to the curve of the string. Assuming that there is only vertical motion of the string, the

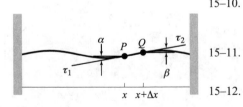

Figure 15.7. A vibrating string at an instant of time. The quantities shown in the figure are used in the derivation of the classical one-dimensional wave equation in Problem 15–12.

horizontal components of the tensions at all points along the string must be equal. Using the notation in Figure 15.7, we have that

$$\tau_1 \cos \alpha = \tau_2 \cos \beta = \tau = \text{constant} \tag{1}$$

There is a net force in the vertical direction that causes the vertical motion of the string. Show that

$$\text{net vertical force} = \tau_2 \sin \beta - \tau_1 \sin \alpha$$

By Newton's second law, this net force is equal to the mass $\rho \Delta x$ along the segment PQ times the acceleration of the string, $\partial^2 u / \partial t^2$. Thus, we write

$$\tau_2 \sin \beta - \tau_1 \sin \alpha = \rho \Delta x \frac{\partial^2 u}{\partial t^2} \tag{2}$$

Divide equation 1 by equation 2 to get

$$\tan \beta - \tan \alpha = \frac{\rho \Delta x}{\tau} \frac{\partial^2 u}{\partial t^2} \tag{3}$$

Now use the fact that $\tan \beta$ and $\tan \alpha$ are the slopes of the curve of the string at $x + \Delta x$ and at x, respectively, to get

$$u_x(x + \Delta x) - u_x(x) = \frac{\rho \Delta x}{\tau} \frac{\partial^2 u}{\partial t^2} \tag{4}$$

Now show that equation 4 becomes

$$\frac{\partial^2 u}{\partial x^2} = \frac{1}{v^2} \frac{\partial^2 u}{\partial t^2}$$

in the limit $\Delta x \to 0$, where $v = (\tau / \rho)^{1/2}$ has units of speed.

15–13.   In this problem, we'll derive an expression for the energy of a vibrating string. The kinetic energy part is easy: because the velocity at any point of the string is $\partial u / \partial t$, the kinetic energy, T, of the entire string is

$$T = \int_0^l \frac{1}{2} \rho \left(\frac{\partial u}{\partial t} \right)^2 dx$$

where ρ is the linear mass density of the string. The potential energy is found by considering the increase in length of the small arc PQ of length ds in Figure 15.7. The segment of the string along that arc has increased its length from dx to ds. Therefore, the potential energy associated with this increase is

$$V = \int_0^l \tau (ds - dx)$$

where τ is the (constant) tension in the string. Using the fact that $(ds)^2 = (dx)^2 + (du)^2$, show that

$$V = \int_0^l \tau \left\{ \left[1 + \left(\frac{\partial u}{\partial x} \right)^2 \right]^{1/2} - 1 \right\} dx$$

Using the fact that $(1 + x)^{1/2} \approx 1 + (x/2)$ for small x, show that

$$V = \frac{1}{2}\tau \int_0^l \left(\frac{\partial u}{\partial x}\right)^2 dx$$

for small displacements. The total energy of the vibrating string is the sum of T and V and so

$$E = \frac{\rho}{2} \int_0^l \left(\frac{\partial u}{\partial t}\right)^2 dx + \frac{\tau}{2} \int_0^l \left(\frac{\partial u}{\partial x}\right)^2 dx$$

for a constant linear mass density.

15–14. This problem shows that the intensity of a wave is proportional to the square of its amplitude. Equation 15.24 shows that the nth normal mode is

$$u_n(x, l) = A_n \cos(\omega_n t + \phi_n) \sin \frac{n\pi x}{l}$$

where $\omega_n = v n\pi/l$. Using the result from the previous problem, show that

$$T_n = \frac{\pi^2 v^2 n^2 \rho}{4l} A_n^2 \sin^2(\omega_n t + \phi_n)$$

and

$$V_n = \frac{\pi^2 n^2 \rho}{4l} A_n^2 \cos^2(\omega_n t + \phi_n)$$

where T_n and V_n are the kinetic energy and potential energy of the nth normal mode. Now use the fact that $v = (\tau/\rho)^{1/2}$ to show that

$$E_n = \frac{\pi^2 v^2 n^2 \rho}{4l} A_n^2$$

Note that the total energy, or intensity, is proportional to the square of the amplitude. Although we have shown this proportionality only for the case of a vibrating string, it is a general result and shows that the intensity of a wave is proportional to the square of its amplitude. If we had carried everything through in complex notation instead of sines and cosines, then we would have found that E_n is proportional to $|A_n|^2$ instead of just A_n^2.

15–15. Use a CAS to animate the wave motion pictured in Figure 15.2.

15–16. Use a CAS to construct your own version of how two standing waves combine to give a traveling wave, as in Figure 15.3.

15–17. Use a CAS to animate the wave motion pictured in Figure 15.4.

CHAPTER 16

The Schrödinger Equation

The Schrödinger equation is a fundamental equation of quantum mechanics and is one of the most famous equations in the physical sciences. It is often displayed as an eigenvalue equation

$$\hat{H}\psi = E\psi \tag{16.1}$$

where $\hat{H}$ is the Hamiltonian operator, E is the energy, and ψ is called the *wave function* of the system. The wave function has the physical interpretation that the magnitude of its square is related to certain probabilities. As we mentioned in Chapter 11, physical quantities are represented by operators in quantum mechanics, and the Schrödinger equation reflects the fact that the energy of a system is represented by its Hamiltonian operator. The Hamiltonian operator describing a single particle can be written as

$$\hat{H} = -\frac{\hbar^2}{2m}\nabla^2 + V(x, y, z) \tag{16.2}$$

where m is the mass of the particle, ∇^2 is the Laplacian operator

$$\nabla^2 = \frac{\partial^2}{\partial x^2} + \frac{\partial^2}{\partial y^2} + \frac{\partial^2}{\partial z^2}$$

and $V(x, y, z)$ is the potential that the particle experiences. We shall solve the Schrödinger equation for a particle in a box, a rigid rotator, and the electron in a hydrogen atom. Each of these systems is defined by the form of the potential $V(x, y, z)$.

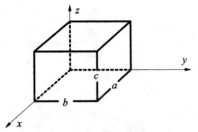

Figure 16.1. A rectangular parallelepiped of sides a, b, and c. In the quantum-mechanical problem of a particle in a box, the particle is restricted to lie within a potential-free paralellepiped.

16.1   A Particle in a Box

The simplest quantum-mechanical problem is a "particle in a box." In this problem, the potential is such that the particle is restricted to lie within a potential-free rectangular parallelepiped with sides of lengths a, b, and c (Figure 16.1). Thus, $V(x, y, z) = 0$ within the box, and the Schrödinger equation for this system is

$$-\frac{\hbar^2}{2m}\left(\frac{\partial^2\psi}{\partial x^2} + \frac{\partial^2\psi}{\partial y^2} + \frac{\partial^2\psi}{\partial z^2}\right) = E\psi(x, y, z) \qquad \begin{array}{l} 0 \le x \le a \\ 0 \le y \le b \\ 0 \le z \le c \end{array} \quad (16.3)$$

The wave function $\psi(x, y, z)$ satisfies the boundary conditions that it vanishes at the walls of the box, and so

$$\begin{array}{ll} \psi(0, y, z) = \psi(a, y, z) = 0 & \text{for all } y \text{ and } z \\ \psi(x, 0, z) = \psi(x, b, z) = 0 & \text{for all } x \text{ and } z \quad (16.4) \\ \psi(x, y, 0) = \psi(x, y, c) = 0 & \text{for all } x \text{ and } y \end{array}$$

If we apply the method of separation of variables to Equation 16.3, we obtain (Problem 16–1)

$$\psi_{n_x,n_y,n_z}(x, y, z) = A_x A_y A_z \sin\frac{n_x\pi x}{a} \sin\frac{n_y\pi y}{b} \sin\frac{n_z\pi z}{c} \quad (16.5)$$

with n_x, n_y, and n_z independently assuming the values $1, 2, 3, \ldots$. The normalization constant $A_x A_y A_z$ is found from the equation

$$\int_0^a dx \int_0^b dy \int_0^c dz\, \psi^*(x, y, z)\psi(x, y, z) = 1 \quad (16.6)$$

Problem 16–2 shows that $A_x A_y A_z = (8/abc)^{1/2}$ and so the normalized wave functions of a particle in a three-dimensional box are

$$\psi_{n_x,n_y,n_z}(x, y, z) = \left(\frac{8}{abc}\right)^{1/2} \sin\frac{n_x\pi x}{a} \sin\frac{n_y\pi y}{b} \sin\frac{n_z\pi z}{c} \qquad \begin{array}{l} n_x = 1, 2, 3, \ldots \\ n_y = 1, 2, 3, \ldots \\ n_z = 1, 2, 3, \ldots \end{array}$$
$$(16.7)$$

If we substitute Equation 16.7 into Equation 16.3, we obtain (Problem 16–3)

$$E_{n_x,n_y,n_z} = \frac{h^2}{8m}\left(\frac{n_x^2}{a^2} + \frac{n_y^2}{b^2} + \frac{n_z^2}{c^2}\right) \qquad \begin{array}{l} n_x = 1, 2, 3, \ldots \\ n_y = 1, 2, 3, \ldots \\ n_z = 1, 2, 3, \ldots \end{array} \quad (16.8)$$

The allowed energy levels of a particle that is restricted to a potential-free region in the shape of a cube are shown in Figure 16.2. The energy of the particle is restricted to only certain values, or is *quantized*. In addition, some of the energy levels are degenerate.

The quantization of the allowed energies given by Equation 16.8 arises naturally through the boundary conditions that the wave function must satisfy. We shall see that this is the case for the other quantum-mechanical problems that we discuss in this chapter. One of the great triumphs of the Schrödinger equation is

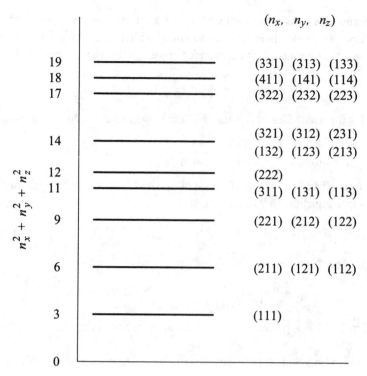

Figure 16.2. The allowed energy levels for a particle in a box with $a = b = c$.

that quantum numbers arise naturally through boundary conditions, rather than being introduced in an ad hoc manner as in the early quantum theory.

16.2 A Rigid Rotator

The next quantum-mechanical problem that we shall discuss is that of a rigid rotator, consisting of two masses, m_1 and m_2, separated by a fixed distance l. The rigid rotator serves as a simple but useful model of a rotating diatomic molecule. If we choose relative coordinates, we can consider one of the two masses to be fixed at the origin of a spherical coordinate system (Figure 16.3) with the other mass taking on the reduced mass $\mu = m_1 m_2/(m_1 + m_2)$ (Problem 16–4). The orientation of a rigid rotator is completely specified by the two angles θ and ϕ, and so rigid rotator wave functions depend upon these two variables. The rigid rotator wave functions are customarily denoted by $Y(\theta, \phi)$, and the Schrödinger equation for a rigid rotator reads

$$-\frac{\hbar^2}{2\mu}\nabla^2 Y(\theta, \phi) = EY(\theta, \phi)$$

where ∇^2 is the Laplacian operator. There is no potential energy in this case; the total energy is simply the rotational kinetic energy.

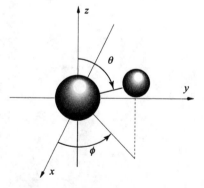

Figure 16.3. The model of a rigid rotator. We can choose our coordinates so that one of the masses sits at the origin and the other takes on a reduced mass $\mu = m_1 m_2/(m_1 + m_2)$.

Because one mass is fixed at the origin, it is natural to use spherical coordinates in this case. We use ∇^2 given by Equation 14.24 with no term involving derivatives with respect to r because r is a constant for a rigid rotator. Therefore, ∇^2 is given by

$$\nabla^2 = \frac{1}{l^2 \sin\theta} \frac{\partial}{\partial\theta} \left(\sin\theta \frac{\partial}{\partial\theta} \right) + \frac{1}{l^2 \sin^2\theta} \frac{\partial^2}{\partial\phi^2}$$

where l is the fixed value of r. Thus, the Schrödinger equation for a rigid rotator is

$$-\frac{\hbar^2}{2I} \left[\frac{1}{\sin\theta} \frac{\partial}{\partial\theta} \left(\sin\theta \frac{\partial}{\partial\theta} \right) + \frac{1}{\sin^2\theta} \frac{\partial^2}{\partial\phi^2} \right] Y(\theta,\phi) = E Y(\theta,\phi) \qquad (16.9)$$

where $I = \mu l^2$ is the moment of inertia of the rigid rotator (see Problem 16–7). If we multiply Equation 16.9 by $\sin^2\theta$ and let

$$\beta = \frac{2IE}{\hbar^2} \qquad (16.10)$$

Equation 16.9 becomes

$$\sin\theta \frac{\partial}{\partial\theta} \left(\sin\theta \frac{\partial Y}{\partial\theta} \right) + \frac{\partial^2 Y}{\partial\phi^2} + (\beta \sin^2\theta) Y = 0 \qquad (16.11)$$

Equation 16.11 is a partial differential equation in two independent variables, θ and ϕ.

To solve Equation 16.11, we use the method of separation of variables and let

$$Y(\theta,\phi) = \Theta(\theta)\Phi(\phi) \qquad (16.12)$$

If we substitute Equation 16.12 into Equation 16.11 and divide by $\Theta(\theta)\Phi(\phi)$, we find that

$$\frac{\sin\theta}{\Theta(\theta)} \frac{d}{d\theta} \left(\sin\theta \frac{d\Theta}{d\theta} \right) + \beta \sin^2\theta + \frac{1}{\Phi(\phi)} \frac{d^2\Phi}{d\phi^2} = 0$$

Because θ and ϕ are independent variables, we must have

$$\frac{\sin\theta}{\Theta(\theta)} \frac{d}{d\theta} \left(\sin\theta \frac{d\Theta}{d\theta} \right) + \beta \sin^2\theta = m^2 \qquad (16.13)$$

and

$$\frac{1}{\Phi(\phi)} \frac{d^2\Phi}{d\phi^2} = -m^2 \qquad (16.14)$$

where m^2 is a separation constant. We use m^2 as a separation constant in anticipation of using its square in later equations.

Equation 16.14 can be written as

$$\frac{d^2\Phi}{d\phi^2} + m^2\Phi(\phi) = 0$$

This is a second-order differential equation with constant coefficients, and its two solutions are

$$\Phi(\phi) = A_m e^{im\phi} \quad \text{and} \quad \Phi(\phi) = A_{-m} e^{-im\phi} \qquad (16.15)$$

The requirement that $\Phi(\phi)$ be a single-valued function of ϕ is $\Phi(\phi+2\pi) = \Phi(\phi)$, which leads to

$$A_m e^{im(\phi+2\pi)} = A_m e^{im\phi} \tag{16.16a}$$

and

$$A_{-m} e^{-im(\phi+2\pi)} = A_{-m} e^{-im\phi} \tag{16.16b}$$

Equations 16.16 imply that $e^{\pm i2\pi m} = 1$. We can write $e^{\pm i2\pi m} = 1$ in terms of sines and cosines as $\cos(2\pi m) \pm i \sin(2\pi m) = 1$, which implies that $m = 0, \pm 1, \pm 2, \ldots$, because $\cos 2\pi m = 1$ and $\sin 2\pi m = 0$ for $m = 0, \pm 1, \pm 2, \ldots$. Thus, Equations 16.15 can be written as one equation:

$$\Phi_m(\phi) = A_m e^{im\phi} \qquad m = 0, \pm 1, \pm 2, \ldots \tag{16.17}$$

We can find the value of A_m by requiring that the $\Phi_m(\phi)$ be normalized.

EXAMPLE 16–1
Determine the value of A_m in Equation 16.15 by requiring that $\Phi_m(\phi)$ be normalized.

SOLUTION: The normalization condition is

$$\int_0^{2\pi} \Phi_m^*(\phi)\Phi_m(\phi)\,d\phi = 1$$

Using Equation 16.17, we have

$$|A_m|^2 \int_0^{2\pi} d\phi = 1$$

or $A_m = (2\pi)^{1/2}$. Thus, the normalized version of Equation 16.17 is

$$\Phi_m(\phi) = \frac{1}{(2\pi)^{1/2}} e^{im\phi} \qquad m = 0, \pm 1, \pm 2, \ldots \tag{16.18}$$

The differential equation for $\Theta(\theta)$, Equation 16.13, is not as easy to solve as Equation 16.14 because it does not have constant coefficients. It is convenient to let $x = \cos\theta$ and $\Theta(\theta) = P(x)$ in Equation 16.13. (This x should not be confused with the cartesian coordinate x.) Because $0 \le \theta \le \pi$, the range of x is $-1 \le x \le +1$. Under the change of variable, $x = \cos\theta$, Equation 16.13 becomes (Problem 16–8)

$$(1-x^2)\frac{d^2 P}{dx^2} - 2x\frac{dP}{dx} + \left[\beta - \frac{m^2}{1-x^2}\right] P(x) = 0 \tag{16.19}$$

with $m = 0, \pm 1, \pm 2, \ldots$.

Equation 16.19 may not look familiar, but when $m = 0$, it becomes

$$(1 - x^2)\frac{d^2 P}{dx^2} - 2x\frac{dP}{dx} + \beta P(x) = 0$$

Table 16.1. The first few Legendre polynomials, $P_l(x)$.

$P_0(x) = 1$
$P_1(x) = x$
$P_2(x) = \frac{1}{2}(3x^2 - 1)$
$P_3(x) = \frac{1}{2}(5x^3 - 3x)$
$P_4(x) = \frac{1}{8}(35x^4 - 30x^2 + 3)$

This is Legendre's equation, which we studied in some detail in Section 7.2. We saw there that the solutions that are finite at $x = \pm1$ ($\theta = 0$ or π) are the Legendre polynomials $P_l(x)$ and that $\beta = l(l + 1)$, where $l = 0, 1, 2, \ldots$. Table 16.1 lists the first few Legendre polynomials. Recall that the Legendre polynomials satisfy the orthogonality relation

$$\int_{-1}^{1} dx\, P_l(x)P_{l'}(x) = \frac{2}{2l + 1}\delta_{ll'} \tag{16.20}$$

When we substitute $\beta = l(l + 1)$ into Equation 16.19, we get

$$(1 - x^2)\frac{d^2 P_l^{|m|}}{dx^2} - 2x\frac{dP_l^{|m|}}{dx} + \left[l(l + 1) - \frac{m^2}{1 - x^2}\right]P_l^{|m|}(x) = 0 \tag{16.21}$$

We have denoted the solutions to this equation by $P_l^{|m|}(x)$ because of the two parameters, l and m, in Equation 16.21. We've used $|m|$, the absolute value of m, to label the solutions because Equation 16.21 depends only upon m^2, and so the sign of m is not relevant. Equation 16.21 can be solved by the series method, much like Legendre's equation can. It turns out that m must equal $0, \pm1, \pm2, \ldots, \pm l$ for the solutions to be well behaved at $x = \pm1$ ($\theta = 0$ and π). The solutions, called *associated Legendre functions*, can be written in terms of the Legendre polynomials by

$$P_l^{|m|}(x) = (1 - x^2)^{|m|/2}\frac{d^{|m|}}{dx^{|m|}}P_l(x) \tag{16.22}$$

Because the leading term in $P_l(x)$ is x^l (see Table 16.1), Equation 16.22 shows that $P_l^{|m|}(x) = 0$ if $m > l$. The first few associated Legendre functions (Problem 16–9) are given in Table 16.2.

Before we discuss a few of the properties of the associated Legendre functions, let's be sure to realize that θ and not x is the variable of physical interest. Table 16.2 lists the associated Legendre functions in terms of $\cos\theta$ and $\sin\theta$ as well as x. The factors $(1 - x^2)^{1/2}$ in Table 16.2 become $\sin\theta$ when the associated Legendre functions are expressed in the variable θ. Remember that $P_l^{|m|}(x) = 0$ when $|m| > l$.

Because $x = \cos\theta$, then $dx = -\sin\theta\, d\theta$ and Equation 16.20 can be written as

$$\int_{-1}^{1} P_l(x)P_{l'}(x)dx = \int_{0}^{\pi} P_l(\cos\theta)P_{l'}(\cos\theta)\sin\theta\, d\theta = \frac{2\delta_{ll'}}{2l + 1} \tag{16.23}$$

Because the differential volume element in spherical coordinates is $d\tau = r^2\sin\theta\, dr\, d\theta\, d\phi$, we see that the factor $\sin\theta\, d\theta$ in Equation 16.23 is the "θ" part of $d\tau$ in spherical coordinates.

Table 16.2. The first few associated Legendre functions, $P_l^{|m|}(x)$.

$P_0^0(x) = 1$

$P_1^0(x) = x = \cos\theta$
$P_1^1(x) = (1 - x^2)^{1/2} = \sin\theta$

$P_2^0(x) = \frac{1}{2}(3x^2 - 1) = \frac{1}{2}(3\cos^2\theta - 1)$
$P_2^1(x) = 3x(1 - x^2)^{1/2} = 3\cos\theta\sin\theta$
$P_2^2(x) = 3(1 - x^2) = 3\sin^2\theta$

$P_3^0(x) = \frac{1}{2}(5x^3 - 3x) = \frac{1}{2}(5\cos^3\theta - 3\cos\theta)$
$P_3^1(x) = \frac{3}{2}(5x^2 - 1)(1 - x^2)^{1/2} = \frac{3}{2}(5\cos^2\theta - 1)\sin\theta$
$P_3^2(x) = 15x(1 - x^2) = 15\cos\theta\sin^2\theta$
$P_3^3(x) = 15(1 - x^2)^{3/2} = 15\sin^3\theta$

The associated Legendre functions satisfy the relation

$$\int_{-1}^{1} P_l^{|m|}(x) P_{l'}^{|m|}(x) dx = \int_0^\pi P_l^{|m|}(\cos\theta) P_{l'}^{|m|}(\cos\theta) \sin\theta\, d\theta$$

$$= \frac{2}{(2l+1)} \frac{(l+|m|)!}{(l-|m|)!} \delta_{ll'} \qquad (16.24)$$

(Remember that $0! = 1$ (Section 4.1).) Equation 16.24 can be used to show that the normalization constant of the associated Legendre functions is

$$N_{lm} = \left[\frac{(2l+1)}{2} \frac{(l-|m|)!}{(l+|m|)!} \right]^{1/2}$$

Thus, the (normalized) $\Theta(\theta)$ part of Equation 16.12 is given by

$$\Theta_l^m(\theta) = \left[\frac{2l+1}{2} \frac{(l-|m|)!}{(l+|m|)!} \right]^{1/2} P_l^{|m|}(\cos\theta) \qquad (16.25)$$

EXAMPLE 16–2
Use Equation 16.24 in both the x and θ variables and Table 16.2 to prove that $P_1^1(x)$ and $P_2^1(x)$ are orthogonal.

SOLUTION: According to Equation 16.24, we must prove that

$$\int_{-1}^{1} P_1^1(x) P_2^1(x)\, dx = 0$$

From Table 16.2, we have

$$\int_{-1}^{1} [(1 - x^2)^{1/2}][3x(1 - x^2)^{1/2}]\, dx = 3\int_{-1}^{1} x(1 - x^2)\, dx = 0$$

In terms of θ, we have (from Equation 16.24 and Table 16.2)

$$\int_0^\pi (\sin\theta)(3\cos\theta\sin\theta)\sin\theta\,d\theta = 3\int_0^\pi \sin^3\theta\cos\theta\,d\theta = 0$$

Returning to the original problem now, the solutions to Equation 16.9, which are not only the rigid rotator wave functions but, as we shall see, also the angular parts of the hydrogen atomic orbitals, are given by $\Theta_l^m(\cos\theta)\Phi_m(\phi)$ (Equation 16.12). Using Equations 16.17 and 16.25, we see that the normalized functions

$$Y_l^m(\theta,\phi) = i^{m+|m|}\left[\frac{(2l+1)}{4\pi}\frac{(l-|m|)!}{(l+|m|)!}\right]^{1/2} P_l^{|m|}(\cos\theta)e^{im\phi} \qquad (16.26)$$

with $l = 0,\ 1,\ 2,\ldots$ and $m = 0,\ \pm 1,\ \pm 2,\ldots,\ \pm l$ satisfy Equation 16.9. The peculiar-looking factor of $i^{m+|m|}$ in Equation 16.26 is simply a convention that is used by most authors. This factor is equal to 1 when m is odd and negative and is equal to -1 when m is odd and positive (Problem 16–11). The spherical harmonics given in Table 16.3 display this convention. The $Y_l^m(\theta,\phi)$ form an orthonormal set

$$\int_0^\pi d\theta\sin\theta\int_0^{2\pi} d\phi\, Y_l^m(\theta,\phi)^* Y_{l'}^k(\theta,\phi) = \delta_{ll'}\delta_{mk} \qquad (16.27)$$

Note that the $Y_l^m(\theta,\phi)$ are orthonormal with respect to $\sin\theta\,d\theta d\phi$ and not just $d\theta d\phi$. Chapter 14 shows that the factor $\sin\theta\,d\theta d\phi$ is a differential area element on the surface of a sphere of unit radius. According to Equation 16.27, the $Y_l^m(\theta,\phi)$ are orthonormal over a spherical surface and so are called *spherical harmonics*.

Table 16.3. The first few spherical harmonics, $Y_l^m(\theta,\phi)$.[a]

$$Y_0^0 = \frac{1}{(4\pi)^{1/2}} \qquad\qquad Y_1^0 = \left(\frac{3}{4\pi}\right)^{1/2}\cos\theta$$

$$Y_1^1 = -\left(\frac{3}{8\pi}\right)^{1/2}\sin\theta\, e^{i\phi} \qquad\qquad Y_1^{-1} = \left(\frac{3}{8\pi}\right)^{1/2}\sin\theta\, e^{-i\phi}$$

$$Y_2^0 = \left(\frac{5}{16\pi}\right)^{1/2}(3\cos^2\theta - 1) \qquad\qquad Y_2^1 = -\left(\frac{15}{8\pi}\right)^{1/2}\sin\theta\cos\theta\, e^{i\phi}$$

$$Y_2^{-1} = \left(\frac{15}{8\pi}\right)^{1/2}\sin\theta\cos\theta\, e^{-i\phi} \qquad\qquad Y_2^2 = \left(\frac{15}{32\pi}\right)^{1/2}\sin^2\theta\, e^{2i\phi}$$

$$Y_2^{-2} = \left(\frac{15}{32\pi}\right)^{1/2}\sin^2\theta\, e^{-2i\phi}$$

[a] The negative signs in $Y_1^1(\theta,\phi)$ and $Y_2^1(\theta,\phi)$ are simply a convention (see Equation 16.26).

EXAMPLE 16–3

Show that $Y_1^{-1}(\theta, \phi)$ is normalized and that it is orthogonal to $Y_2^1(\theta, \phi)$.

SOLUTION: Using $Y_1^{-1}(\theta, \phi)$ from Table 16.3, the normalization condition, Equation 16.27 gives

$$I = \int_0^\pi d\theta \sin\theta \int_0^{2\pi} d\phi \, Y_1^{-1}(\theta, \phi)^* Y_1^{-1}(\theta, \phi)$$

$$= \frac{3}{8\pi} \int_0^\pi d\theta \sin\theta \sin^2\theta \int_0^{2\pi} d\phi \stackrel{?}{=} 1$$

Letting $x = \cos\theta$, we have

$$I = \frac{3}{8\pi} \cdot 2\pi \int_{-1}^1 (1 - x^2) \, dx = \frac{3}{4}\left(2 - \frac{2}{3}\right) = 1$$

The orthogonality condition is

$$\int_0^\pi d\theta \sin\theta \int_0^{2\pi} d\phi \, Y_2^1(\theta, \phi)^* Y_1^{-1}(\theta, \phi)$$

$$= -\left(\frac{15}{8\pi}\right)^{1/2}\left(\frac{3}{8\pi}\right)^{1/2} \int_0^\pi d\theta \sin\theta \int_0^{2\pi} d\phi \, (e^{-i\phi} \sin\theta \cos\theta)(e^{-i\phi} \sin\theta)$$

$$= -\left(\frac{45}{64\pi^2}\right)^{1/2} \int_0^\pi d\theta \sin^3\theta \cos\theta \int_0^{2\pi} d\phi \, e^{-2i\phi}$$

The integral over ϕ is zero because it is the integral of $\cos 2\phi$ and $\sin 2\phi$ over complete cycles. Thus, we see that $Y_1^{-1}(\theta, \phi)$ and $Y_2^1(\theta, \phi)$ are orthogonal.

In summary, the Schrödinger equation for a rigid rotator is given by Equation 16.9 and the energy is given by Equation 16.10, $E = \hbar^2\beta/2I$. Using the fact that $\beta = l(l + 1)$ with $l = 0, 1, 2, \ldots$, we can write Equation 16.9 as

$$\hat{H}Y_l^m(\theta, \phi) = E_l Y_l^m(\theta, \phi) = \frac{\hbar^2 l(l + 1)}{2I} Y_l^m(\theta, \phi) \qquad l = 0, 1, 2, \ldots \quad (16.28)$$

with $\hat{H}$ given by

$$\hat{H} = -\frac{\hbar^2}{2I}\left[\frac{1}{\sin\theta}\frac{\partial}{\partial\theta}\left(\sin\theta\frac{\partial}{\partial\theta}\right) + \frac{1}{\sin^2\theta}\frac{\partial^2}{\partial\phi^2}\right]$$

and the $Y_l^m(\theta, \phi)$ are the spherical harmonics (Equation 16.26). Once again, we obtain a set of discrete energy levels, $E_l = \hbar^2 l(l + 1)/2I$ with $l = 0, 1, 2, \ldots$. In addition to the allowed energies given in Equation 16.28, we also find that each energy level has a degeneracy g_l given by $g_l = 2l + 1$. This degeneracy is due to the fact that l can take on the $2l + 1$ values $-m$ through $+m$ and that the energy doesn't depend on m.

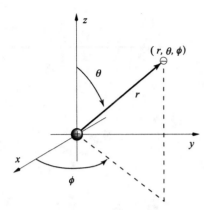

Figure 16.4. The spherical coordinates used to describe a hydrogen atom. The hydrogen nucleus is situated at the origin and the position of the electron is given by the three spherical coordinates, r, θ, and ϕ.

16.3 The Electron in a Hydrogen Atom

The final problem that we shall discuss in this chapter is the Schrödinger equation for a hydrogen atom, one of the great triumphs of quantum mechanics. As our model, we shall picture the hydrogen atom as a proton fixed at the origin and an electron of mass m interacting with the proton through a coulombic potential,

$$V(r) = -\frac{e^2}{4\pi\epsilon_0 r} \tag{16.29}$$

where e is the charge on the proton, ϵ_0 is the permittivity of free space, and r is the distance between the electron and the proton. The model suggests that we use a spherical coordinate system with the proton at the origin (Figure 16.4). Therefore, the Schrödinger equation for a hydrogen atom can be written as

$$-\frac{\hbar^2}{2m}\nabla^2\psi(r,\theta,\phi) - \frac{e^2}{4\pi\epsilon_0 r}\psi(r,\theta,\phi) = E\psi(r,\theta,\phi)$$

or

$$-\frac{\hbar^2}{2m}\left[\frac{1}{r^2}\frac{\partial}{\partial r}\left(r^2\frac{\partial\psi}{\partial r}\right) + \frac{1}{r^2\sin\theta}\frac{\partial}{\partial\theta}\left(\sin\theta\frac{\partial\psi}{\partial\theta}\right) + \frac{1}{r^2\sin^2\theta}\frac{\partial^2\psi}{\partial\phi^2}\right]$$
$$-\frac{e^2}{4\pi\epsilon_0 r}\psi(r,\theta,\phi) = E\psi(r,\theta,\phi) \tag{16.30}$$

At first sight, Equation 16.30 looks daunting, but it is straightforward to show that the substitution $\psi(r,\theta,\phi) = R(r)Y_l^m(\theta,\phi)$ yields an ordinary differential equation for $R(r)$ (Problem 16–18):

$$-\frac{\hbar^2}{2mr^2}\frac{d}{dr}\left(r^2\frac{dR}{dr}\right) + \left[\frac{\hbar^2 l(l+1)}{2mr^2} - \frac{e^2}{4\pi\epsilon_0 r} - E\right]R(r) = 0 \tag{16.31}$$

Equation 16.31 is called the *radial equation* for the hydrogen atom and is the only new equation that we have to study in order to have a complete solution to the hydrogen atom.

Equation 16.31 is an ordinary differential equation in r and can be solved by the series method. One finds that in order for solutions to be acceptable (in other words, continuous, finite, and normalizable), the energy must be quantized according to

$$E_n = -\frac{me^4}{8\epsilon_0^2 h^2 n^2} \qquad n = 1, 2, \ldots \tag{16.32}$$

Thus, we obtain one of the most important atomic properties: the energy of an electron in an atom is quantized. It is this property that leads directly to atomic spectra being line spectra.

In the course of solving Equation 16.31, one finds not only that an integer n occurs naturally but that n must satisfy the condition $n \geq l + 1$, which is usually written as

$$0 \leq l \leq n - 1 \qquad n = 1, 2, \ldots \tag{16.33}$$

because the smallest possible value of l is zero. (Equation 16.33 might be familiar from general chemistry.) The solutions to Equation 16.31 depend upon two quantum numbers, n and l, and are given by

$$R_{nl}(r) = - \left\{ \frac{(n-l-1)!}{2n[(n+l)!]^3} \right\}^{1/2} \left(\frac{2}{na_0} \right)^{l+3/2} r^l e^{-r/na_0} L_{n+l}^{2l+1} \left(\frac{2r}{na_0} \right) \quad (16.34)$$

where $a_0 = 4\pi\varepsilon_0\hbar^2/me^2$ is a distance, called the *Bohr radius*, and where the $L_{n+l}^{2l+1}(2r/na_0)$ are the *associated Laguerre polynomials*.

We discussed the Laguerre polynomials and the associated Laguerre polynomials in Chapter 8. We derived the first few Laguerre polynomials in Example 8–5 and discussed some of their properties in Section 8.2. The associated Laguerre polynomials are related to the Laguerre polynomials, $L_n(x)$, by the relation

$$L_n^\alpha(x) = \frac{d^\alpha}{dx^\alpha} L_n(x) \quad (16.35)$$

(Problem 16–19). The first few associated Laguerre polynomials are given in Table 16.4. The functions given by Equation 16.34 may look complicated, but notice that each one is just a polynomial multiplied by an exponential.

Table 16.4. The first few associated Laguerre polynomials.

$n = 1$; $l = 0$	$L_1^1(x) = -1$
$n = 2$; $l = 0$	$L_2^1(x) = -2!(2-x)$
$l = 1$	$L_3^3(x) = -3!$
$n = 3$; $l = 0$	$L_3^1(x) = -3!(3 - 3x + \frac{1}{2}x^2)$
$l = 1$	$L_4^3(x) = -4!(4-x)$
$l = 2$	$L_5^5(x) = -5!$

The complete hydrogen atomic wave functions are

$$\psi_{nlm}(r, \theta, \phi) = R_{nl}(r)Y_l^m(\theta, \phi) \quad (16.36)$$

Because $\hat{H}$ is a Hermitian operator, the ψ_{nlm} must be orthogonal. The orthonormality condition is given by

$$\int_0^{2\pi} d\phi \int_0^\pi d\theta \sin\theta \int_0^\infty dr\, r^2 \psi_{nlm}^*(r, \theta, \phi)\psi_{n'l'm'}(r, \theta, \phi) = \delta_{nn'}\delta_{ll'}\delta_{mm'}$$
$$(16.37)$$

EXAMPLE 16–4
Show that the hydrogen-like atomic wave function ψ_{210} is normalized and that it is orthogonal to ψ_{200}.

SOLUTION: We first have to construct ψ_{200} and ψ_{210} from Equation 16.36, Table 16.3, and Table 16.4:

$$\psi_{200}(r, \theta, \phi) = R_{20}(r)Y_0^0(\theta, \phi)$$

$$= -\left\{\frac{1!}{4(2!)^3}\right\}^{1/2}\left(\frac{2}{2a_0}\right)^{3/2} e^{-r/2a_0} L_2^1\left(\frac{2r}{2a_0}\right)\frac{1}{(4\pi)^{1/2}}$$

$$= \frac{a_0^{-3/2}}{(32\pi)^{1/2}}(2 - \rho)e^{-\rho/2}$$

$$\psi_{210}(r, \theta, \phi) = R_{21}(r)Y_1^0(\theta, \phi)$$

$$= -\left\{\frac{0!}{4(3!)^3}\right\}^{1/2}\left(\frac{2}{2a_0}\right)^{5/2} re^{-r/2a_0} L_3^3\left(\frac{2r}{2a_0}\right)\left(\frac{3}{4\pi}\right)^{1/2}\cos\theta$$

$$= \frac{a_0^{-3/2}}{(32\pi)^{1/2}}\rho e^{-\rho/2}\cos\theta$$

where $\rho = r/a_0$. The normalization condition is given by Equation 16.37:

$$\int_0^{2\pi} d\phi \int_0^\pi d\theta \sin\theta \int_0^\infty dr\, r^2 \psi_{210}^*(r, \theta, \phi)\psi_{210}(r, \theta, \phi)$$

$$= \int_0^{2\pi} d\phi \int_0^\pi d\theta \sin\theta \int_0^\infty dr\, r^2 \frac{\rho^2 e^{-\rho}\cos^2\theta}{32\pi a_0^3}$$

$$= \frac{2\pi}{32\pi}\int_0^\infty d\theta \sin\theta \cos^2\theta \int_0^\infty d\rho\, \rho\rho^2 e^{-\rho}$$

$$= \frac{1}{16}\cdot\frac{2}{3}\cdot 24 = 1$$

The orthogonality condition is given by

$$\int_0^{2\pi} d\phi \int_0^\pi d\theta \sin\theta \int_0^\infty dr\, r^2 \psi_{200}^*(r, \theta, \phi)\psi_{210}(r, \theta, \phi)$$

$$= \int_0^{2\pi} d\phi \int_0^\pi d\theta \sin\theta \int_0^\infty dr\, r^2 \left(\frac{\rho(2 - \rho)e^{-\rho}\cos\theta}{32\pi a_0^3}\right)$$

$$= \frac{2\pi}{32\pi}\int_0^\infty d\theta \sin\theta \cos\theta \int_0^\infty d\rho\, \rho^3(2 - \rho)e^{-\rho} = 0$$

because of the integral over θ.

We pointed out at the end of Chapter 8 that there are several definitions of the Laguerre polynomials in the literature. The various definitions are internally self-consistent, but you must be careful in jumping from source to source. The definitions that we are using here go back to Pauling and Wilson (see References) and are used by most physical chemists and many physicists.

Problems

16–1. Derive Equation 16.5.

16–2. Show that the normalization constant of $\psi(x, y, z)$ in Equation 16.5 is $A_x A_y A_z = (8/abc)^{1/2}$.

16–3. Substitute Equation 16.7 into Equation 16.3 to obtain Equation 16.8.

16–4. Consider two masses m_1 and m_2 in one dimension, interacting through a potential that depends only upon their relative separation $(x_1 - x_2)$, so that $V(x_1, x_2) = V(x_1 - x_2)$. Given that the force acting upon the jth particle is $f_j = -(\partial V / \partial x_j)$, show that $f_1 = -f_2$. What law is this?

Newton's equations for m_1 and m_2 are

$$m_1 \frac{d^2 x_1}{dt^2} = -\frac{\partial V}{\partial x_1} \quad \text{and} \quad m_2 \frac{d^2 x_2}{dt^2} = -\frac{\partial V}{\partial x_2}$$

Now introduce center-of-mass and relative coordinates by

$$X = \frac{m_1 x_1 + m_2 x_2}{M} \qquad x = x_1 - x_2$$

where $M = m_1 + m_2$, and solve for x_1 and x_2 to obtain

$$x_1 = X + \frac{m_2}{M} x \quad \text{and} \quad x_2 = X - \frac{m_1}{M} x$$

Show that Newton's equations in these coordinates are

$$m_1 \frac{d^2 X}{dt^2} + \frac{m_1 m_2}{M} \frac{d^2 x}{dt^2} = -\frac{\partial V}{\partial x} \tag{1}$$

and

$$m_2 \frac{d^2 X}{dt^2} - \frac{m_1 m_2}{M} \frac{d^2 x}{dt^2} = +\frac{\partial V}{\partial x} \tag{2}$$

Now add these two equations to find

$$M \frac{d^2 X}{dt^2} = 0$$

Interpret this result. Now divide equation 1 by m_1 and equation 2 by m_2 and subtract to obtain

$$\frac{d^2 x}{dt^2} = -\left(\frac{1}{m_1} + \frac{1}{m_2} \right) \frac{\partial V}{\partial x}$$

or

$$\mu \frac{d^2 x}{dt^2} = -\frac{\partial V}{\partial x}$$

where $\mu = m_1 m_2 / (m_1 + m_2)$ is the reduced mass. Interpret this result, and discuss how the original two-body problem has been reduced to two one-body problems.

16–5. Extend the results of Problem 16–4 to three dimensions. Realize that in three dimensions the relative separation is given by

$$r_{12} = [(x_1 - x_2)^2 + (y_1 - y_2)^2 + (z_1 - z_2)^2]^{1/2}$$

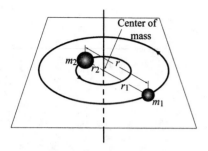

Figure 16.5. Two masses m_1 and m_2 rotating about their center of mass.

16–6.   Show that the reduced mass of two equal masses, m, is $m/2$.

16–7.   Consider two masses revolving about each other as in Figure 16.5. If there are no external forces acting on this system, then the center of mass is fixed and each mass will rotate about that point. The center of mass lies along the line joining their centers and is defined through $m_1 r_1 = m_2 r_2$. The total kinetic energy is

$$T = \frac{1}{2}m_1 v_1^2 + \frac{1}{2}m_2 v_2^2$$

Now if ω is the angular velocity of the two masses about the fixed center of mass, then $v_1 = r_1 \omega$ and $v_2 = r_2 \omega$. Now show that

$$T = \frac{1}{2} I \omega^2$$

where $r = r_1 + r_2$ and $I = m_1 r_1^2 + m_2 r_2^2 = \mu r^2$, where $\mu = m_1 m_2 / (m_1 + m_2)$ is the reduced mass. The quantity I is called the *moment of inertia* of the rotating system.

16–8.   Show that the substitution $x = \cos\theta$ converts Equation 16.13 to Equation 16.19.

16–9.   Use Equation 16.22 to generate the associated Legendre functions in Table 16.2.

16–10.   Show that the first few associated Legendre functions in Table 16.2 satisfy Equation 16.21.

16–11.   Show that the factor $i^{m+|m|} = 1$ when m is odd and negative and that it equals -1 when m is odd and positive.

16–12.   Show that the first few associated Legendre functions in Table 16.2 satisfy Equation 16.24.

16–13.   Show that the associated Legendre function, $P_3^1(x)$, in Table 16.2 satisfies the recursion formula $(2l + 1)x P_l^{|m|}(x) - (l - |m| + 1)P_{l+1}^{|m|}(x) - (l + |m|)P_{l-1}^{|m|}(x) = 0$.

16–14.   Show that the first few spherical harmonics in Table 16.3 satsify the equation

$$\sin\theta \frac{\partial}{\partial\theta}\left(\sin\theta \frac{\partial Y_l^m}{\partial\theta}\right) + \frac{\partial^2 Y_l^m}{\partial\phi^2} + l(l+1)\sin^2\theta\, Y_l^m = 0$$

16–15.   Show explicitly that the first few spherical harmonics in Table 16.3 satisfy Equation 16.9 with $E = \hbar^2 l(l+1)/2I$.

16–16.   Show that the first few spherical harmonics in Table 16.3 are orthonormal.

16–17.   Using Table 16.3, show that $|Y_1^1(\theta,\phi)|^2 + |Y_1^0(\theta,\phi)|^2 + |Y_1^{-1}(\theta,\phi)|^2 = 3/4\pi$. This is a special case of a general theorem, $\sum_{m=-l}^{l}|Y_l^m(\theta,\phi)|^2 = $ constant, known as *Unsöld's theorem*. What is the constant equal to for $l = 2$? Can you give a physical interpretation of Unsöld's theorem?

16–18.   Derive Equation 16.31.

16–19.   The associated Laguerre functions in Table 16.4 can be generated through the relation $L_n^a(x) = \dfrac{d^a}{dx^a} L_n(x)$, where the $L_n(x)$ are the Laguerre polynomials, which in turn can be generated by $L_n(x) = e^x \dfrac{d^n}{dx^n}(x^n e^{-x})$. Derivative formulas such as these are available for all the orthogonal polynomials that we discussed in Chapter 8 and are called *Rodrigues formulas*. Use the Rodrigues formulas above to generate the associated Laguerre functions in Table 16.4.

16–20. Construct $\psi_{100}(r, \theta, \phi)$ from the equations in the chapter and show that it is normalized.

16–21. Construct $\psi_{310}(r, \theta, \phi)$ from the equations in the chapter and show that it is normalized.

16–22. Show explicitly that $\psi_{210}(r, \theta, \phi)$ satisfies Equation 16.30 with E given by Equation 16.32.

16–23. Show that if $V = V(r)$ in the Schrödinger equation, then $\psi(r, \theta, \phi) = R(r)Y_l^m(\theta, \phi)$, where $R(r)$ satisfies Equation 16.31 with $-e^2/4\pi\epsilon_0 r$ replaced with $V(r)$.

16–24. The equation $\nabla^2 u + k^2 u = 0$ is called the Helmholtz equation. Show that $u(r, \theta, \phi) = f(r)Y_l^m(\theta, \phi)$ is a solution to the Helmholtz equation in spherical coordinates.

16–25. In this problem, we shall solve the Schrödinger equation for the ground-state wave function and energy of a particle confined to a sphere of radius a. The Schrödinger equation is given by Equation 16.31 with $l = 0$ (ground state) and without the $e^2/4\pi\epsilon_0 r$ term:

$$-\frac{\hbar^2}{2mr^2}\frac{d}{dr}\left(r^2\frac{d\psi}{dr}\right) = E\psi$$

Substitute $u = r\psi$ into this equation to get

$$\frac{d^2u}{dr^2} + \frac{2mE}{\hbar^2}u = 0$$

Show that the general solution to this equation is

$$u(r) = A\cos\alpha r + B\sin\alpha r$$

or

$$\psi(r) = \frac{A\cos\alpha r}{r} + \frac{B\sin\alpha r}{r}$$

where $\alpha = (2mE/\hbar^2)^{1/2}$. Which of these terms is finite at $r = 0$? Now use the fact that $\psi(a) = 0$ to prove that

$$\alpha a = \pi$$

for the ground state, or that the ground-state energy is

$$E = \frac{\pi^2\hbar^2}{2ma^2}$$

Show that the normalized ground-state wave function is

$$\psi(r) = (2\pi a)^{-1/2}\frac{\sin\pi r/a}{r}$$

CHAPTER 17

Determinants

You probably learned about determinants in high school, where you used them to solve n linear algebraic equations in n unknowns. Although this application of determinants is useful if n, the number of equations, is not very large, it is not a very useful method if n is large. If this were the only use of determinants, they would not be discussed in all applied mathematics books. It turns out that determinants arise naturally in the study of matrices, which have a multitude of applications in physical problems. To this end, then, we discuss some basic properties of determinants in this chapter.

17.1 Definition of a Determinant

Many problems in physical chemistry involve n linear algebraic equations in n unknowns. Such equations can be solved by means of *determinants*, which we discuss in this chapter. Consider the pair of linear algebraic equations

$$a_{11}x + a_{12}y = d_1$$
$$a_{21}x + a_{22}y = d_2$$

(17.1)

If we multiply the first of these equations by a_{22} and the second by a_{12} and then subtract, we obtain

$$(a_{11}a_{22} - a_{12}a_{21})x = d_1 a_{22} - d_2 a_{12}$$

or

$$x = \frac{a_{22}d_1 - a_{12}d_2}{a_{11}a_{22} - a_{12}a_{21}}$$

(17.2)

Similarly, if we multiply the first by a_{21} and the second by a_{11} and then subtract, we get

$$y = \frac{a_{11}d_2 - a_{21}d_1}{a_{11}a_{22} - a_{12}a_{21}} \tag{17.3}$$

Notice that the denominators in both Equations 17.2 and 17.3 are the same. We represent $a_{11}a_{22} - a_{12}a_{21}$ by the quantity $\begin{vmatrix} a_{11} & a_{12} \\ a_{21} & a_{22} \end{vmatrix}$, which is called a 2×2 determinant. The reason for introducing this notation is that it readily generalizes to the treatment of n linear algebraic equations in n unknowns. Generally, an $n \times n$ determinant is a square array of n^2 elements arranged in n rows and n columns. A 3×3 determinant is given by

$$\begin{vmatrix} a_{11} & a_{12} & a_{13} \\ a_{21} & a_{22} & a_{23} \\ a_{31} & a_{32} & a_{33} \end{vmatrix} = \begin{matrix} a_{11}a_{22}a_{33} + a_{21}a_{32}a_{13} + a_{12}a_{23}a_{31} \\ -a_{31}a_{22}a_{13} - a_{21}a_{12}a_{33} - a_{11}a_{23}a_{32} \end{matrix} \tag{17.4}$$

(We shall prove this soon.) Notice that the element a_{ij} occurs at the intersection of the ith row and the jth column.

Equation 17.4 and the corresponding equations for evaluating higher-order determinants can be obtained in a systematic manner. First we define a cofactor. The *cofactor*, A_{ij}, of an element a_{ij} is an $(n-1) \times (n-1)$ determinant obtained by deleting the ith row and the jth column, multiplied by $(-1)^{i+j}$. For example, A_{12}, the cofactor of element a_{12} of

$$D = \begin{vmatrix} a_{11} & a_{12} & a_{13} \\ a_{21} & a_{22} & a_{23} \\ a_{31} & a_{32} & a_{33} \end{vmatrix}$$

is

$$A_{12} = (-1)^{1+2} \begin{vmatrix} a_{21} & a_{23} \\ a_{31} & a_{33} \end{vmatrix}$$

EXAMPLE 17–1

Evaluate the cofactor of each of the first-row elements in

$$D = \begin{vmatrix} 2 & -1 & 1 \\ 0 & 3 & -1 \\ 2 & -2 & 1 \end{vmatrix}$$

SOLUTION: The cofactor of a_{11} is

$$A_{11} = (-1)^{1+1} \begin{vmatrix} 3 & -1 \\ -2 & 1 \end{vmatrix} = 3 - 2 = 1$$

The cofactor of a_{12} is

$$A_{12} = (-1)^{1+2} \begin{vmatrix} 0 & -1 \\ 2 & 1 \end{vmatrix} = -2$$

and the cofactor of a_{13} is

$$A_{13} = (-1)^{1+3} \begin{vmatrix} 0 & 3 \\ 2 & -2 \end{vmatrix} = -6$$

We can use cofactors to evaluate determinants. The value of the 3×3 determinant in Equation 17.4 can be obtained from the formula

$$\begin{vmatrix} a_{11} & a_{12} & a_{13} \\ a_{21} & a_{22} & a_{23} \\ a_{31} & a_{32} & a_{33} \end{vmatrix} = a_{11}A_{11} + a_{12}A_{12} + a_{13}A_{13} \qquad (17.5)$$

Thus, the value of D in Example 17–1 is

$$D = (2)(1) + (-1)(-2) + (1)(-6) = -2$$

Equation 17.5 is an expansion in cofactors about the first row.

EXAMPLE 17–2

Evaluate D in Example 17–1 by expanding in terms of the first column of elements instead of the first row.

SOLUTION: We shall use the formula

$$D = a_{11}A_{11} + a_{21}A_{21} + a_{31}A_{31}$$

The various cofactors are

$$A_{11} = (-1)^2 \begin{vmatrix} 3 & -1 \\ -2 & 1 \end{vmatrix} = 1$$

$$A_{21} = (-1)^3 \begin{vmatrix} -1 & 1 \\ -2 & 1 \end{vmatrix} = -1$$

and

$$A_{31} = (-1)^4 \begin{vmatrix} -1 & 1 \\ 3 & -1 \end{vmatrix} = -2$$

and so

$$D = (2)(1) + (0)(-1) + (2)(-2) = -2$$

Notice that we obtained the same answer for D as we did in Example 17–1. This result illustrates the general fact that a determinant may be evaluated by expanding in terms of the cofactors of the elements of any row or any column. If we choose the second row of D, then we obtain

$$D = (0)(-1)^3 \begin{vmatrix} -1 & 1 \\ -2 & 1 \end{vmatrix} + (3)(-1)^4 \begin{vmatrix} 2 & 1 \\ 2 & 1 \end{vmatrix} + (-1)(-1)^5 \begin{vmatrix} 2 & -1 \\ 2 & -2 \end{vmatrix} = -2$$

Although we have discussed only 3×3 determinants, the procedure is readily extended to determinants of any order.

EXAMPLE 17–3

The following *determinantal equation* arises in the Hückel molecular orbital theory of butadiene:

$$\begin{vmatrix} x & 1 & 0 & 0 \\ 1 & x & 1 & 0 \\ 0 & 1 & x & 1 \\ 0 & 0 & 1 & x \end{vmatrix} = 0$$

Expand this determinantal equation into a quartic equation for x.

SOLUTION: Expand about the first row of elements to obtain

$$x \begin{vmatrix} x & 1 & 0 \\ 1 & x & 1 \\ 0 & 1 & x \end{vmatrix} - \begin{vmatrix} 1 & 1 & 0 \\ 0 & x & 1 \\ 0 & 1 & x \end{vmatrix} = 0$$

Now expand about the first column of each of the 3×3 determinants to obtain

$$(x)(x) \begin{vmatrix} x & 1 \\ 1 & x \end{vmatrix} - (x)(1) \begin{vmatrix} 1 & 0 \\ 1 & x \end{vmatrix} - (1) \begin{vmatrix} x & 1 \\ 1 & x \end{vmatrix} = 0$$

or

$$x^2(x^2 - 1) - x(x) - (1)(x^2 - 1) = 0$$

or

$$x^4 - 3x^2 + 1 = 0$$

Because we can choose any row or column to expand the determinant, it is easiest to take the one with the most zeros.

17.2 Some Properties of Determinants

A number of properties of determinants are useful to know:

1. The value of a determinant is unchanged if the rows are made into columns in the same order; in other words, first row becomes first column,

second row becomes second column, and so on. For example,

$$\begin{vmatrix} 1 & 2 & 5 \\ -1 & 0 & -1 \\ 3 & 1 & 2 \end{vmatrix} = \begin{vmatrix} 1 & -1 & 3 \\ 2 & 0 & 1 \\ 5 & -1 & 2 \end{vmatrix}$$

Notice that these two determinants are related by flipping across the diagonal.

2. If any two rows or columns are the same, the value of the determinant is zero. For example,

$$\begin{vmatrix} 4 & 2 & 4 \\ -1 & 0 & -1 \\ 3 & 1 & 3 \end{vmatrix} = 0$$

3. If any two rows or columns are interchanged, the sign of the determinant changes. For example,

$$\begin{vmatrix} 3 & 1 & -1 \\ -6 & 4 & 5 \\ 1 & 2 & 2 \end{vmatrix} = - \begin{vmatrix} 1 & 3 & -1 \\ 4 & -6 & 5 \\ 2 & 1 & 2 \end{vmatrix}$$

4. If every element in a row or column is multiplied by a factor k, the value of the determinant is multiplied by k. For example,

$$\begin{vmatrix} 6 & 8 \\ -1 & 2 \end{vmatrix} = 2 \begin{vmatrix} 3 & 4 \\ -1 & 2 \end{vmatrix}$$

5. If any row or column is written as the sum or difference of two or more terms, the determinant can be written as the sum or difference of two or more determinants according to

$$\begin{vmatrix} a_{11} \pm a'_{11} & a_{12} & a_{13} \\ a_{21} \pm a'_{21} & a_{22} & a_{23} \\ a_{31} \pm a'_{31} & a_{32} & a_{33} \end{vmatrix} = \begin{vmatrix} a_{11} & a_{12} & a_{13} \\ a_{21} & a_{22} & a_{23} \\ a_{31} & a_{32} & a_{33} \end{vmatrix} \pm \begin{vmatrix} a'_{11} & a_{12} & a_{13} \\ a'_{21} & a_{22} & a_{23} \\ a'_{31} & a_{32} & a_{33} \end{vmatrix}$$

For example,

$$\begin{vmatrix} 3 & 3 \\ 2 & 6 \end{vmatrix} = \begin{vmatrix} 2+1 & 3 \\ -2+4 & 6 \end{vmatrix} = \begin{vmatrix} 2 & 3 \\ -2 & 6 \end{vmatrix} + \begin{vmatrix} 1 & 3 \\ 4 & 6 \end{vmatrix}$$

6. The value of a determinant is unchanged if one row or column is added or subtracted to another, as in

$$\begin{vmatrix} a_{11} & a_{12} & a_{13} \\ a_{21} & a_{22} & a_{23} \\ a_{31} & a_{32} & a_{33} \end{vmatrix} = \begin{vmatrix} a_{11}+a_{12} & a_{12} & a_{13} \\ a_{21}+a_{22} & a_{22} & a_{23} \\ a_{31}+a_{32} & a_{32} & a_{33} \end{vmatrix}$$

For example,

$$\begin{vmatrix} 1 & -1 & 3 \\ 4 & 0 & 2 \\ 1 & 2 & 1 \end{vmatrix} = \begin{vmatrix} 0 & -1 & 3 \\ 4 & 0 & 2 \\ 3 & 2 & 1 \end{vmatrix} = \begin{vmatrix} 0 & -1 & 3 \\ 4 & 0 & 2 \\ 7 & 2 & 3 \end{vmatrix}$$

In the first case we added column 2 to column 1, and in the second case we added row 2 to row 3. This procedure may be repeated n times to obtain

$$\begin{vmatrix} a_{11} & a_{12} & a_{13} \\ a_{21} & a_{22} & a_{23} \\ a_{31} & a_{32} & a_{33} \end{vmatrix} = \begin{vmatrix} a_{11} + na_{12} & a_{12} & a_{13} \\ a_{21} + na_{22} & a_{22} & a_{23} \\ a_{31} + na_{32} & a_{32} & a_{33} \end{vmatrix} \qquad (17.6)$$

This result is easy to prove:

$$\begin{vmatrix} a_{11} + na_{12} & a_{12} & a_{13} \\ a_{21} + na_{22} & a_{22} & a_{23} \\ a_{31} + na_{32} & a_{32} & a_{33} \end{vmatrix} = \begin{vmatrix} a_{11} & a_{12} & a_{13} \\ a_{21} & a_{22} & a_{23} \\ a_{31} & a_{32} & a_{33} \end{vmatrix} + n \begin{vmatrix} a_{12} & a_{12} & a_{13} \\ a_{22} & a_{22} & a_{23} \\ a_{32} & a_{32} & a_{33} \end{vmatrix}$$

$$= \begin{vmatrix} a_{11} & a_{12} & a_{13} \\ a_{21} & a_{22} & a_{23} \\ a_{31} & a_{32} & a_{33} \end{vmatrix} + 0$$

where we used rule 5 to write the first line. The second determinant on the right side equals zero because two columns are the same.

17.3 Cramer's Rule

We provided the above rules because simultaneous linear algebraic equations can be solved in terms of determinants. For simplicity, we shall consider only a pair of equations, but the final result is easy to generalize. Consider the two equations

$$a_{11}x + a_{12}y = d_1$$
$$a_{21}x + a_{22}y = d_2 \qquad (17.7)$$

If $d_1 = d_2 = 0$, the equations are said to be *homogeneous*. Otherwise, they are called *nonhomogeneous*. Let's assume at first that they are nonhomogeneous. The determinant of the coefficients of x and y is

$$D = \begin{vmatrix} a_{11} & a_{12} \\ a_{21} & a_{22} \end{vmatrix}$$

According to rule 4,

$$\begin{vmatrix} a_{11}x & a_{12} \\ a_{21}x & a_{22} \end{vmatrix} = xD$$

Furthermore, according to rule 6,

$$\begin{vmatrix} a_{11}x + a_{12}y & a_{12} \\ a_{21}x + a_{22}y & a_{22} \end{vmatrix} = xD \qquad (17.8)$$

If we substitute Equation 17.7 into Equation 17.8, then we have

$$\begin{vmatrix} d_1 & a_{12} \\ d_2 & a_{22} \end{vmatrix} = xD$$

Solving for x gives

$$x = \frac{\begin{vmatrix} d_1 & a_{12} \\ d_2 & a_{22} \end{vmatrix}}{\begin{vmatrix} a_{11} & a_{12} \\ a_{21} & a_{22} \end{vmatrix}} \tag{17.9}$$

Similarly, we get

$$y = \frac{\begin{vmatrix} a_{11} & d_1 \\ a_{21} & d_2 \end{vmatrix}}{\begin{vmatrix} a_{11} & a_{12} \\ a_{21} & a_{22} \end{vmatrix}} \tag{17.10}$$

Notice that Equations 17.9 and 17.10 are identical to Equations 17.2 and 17.3. The solution for x and y in terms of determinants is called *Cramer's rule*. Note that the determinant in the numerator is obtained by replacing the column in D that is associated with the unknown quantity with the column associated with the right sides of Equations 17.7.

EXAMPLE 17–4
Solve the equations

$$3x - y = 6$$
$$x + 2y = 5$$

SOLUTION:

$$x = \frac{\begin{vmatrix} 6 & -1 \\ 5 & 2 \end{vmatrix}}{\begin{vmatrix} 3 & -1 \\ 1 & 2 \end{vmatrix}} = \frac{17}{7}$$

$$y = \frac{\begin{vmatrix} 3 & 6 \\ 1 & 5 \end{vmatrix}}{\begin{vmatrix} 3 & -1 \\ 1 & 2 \end{vmatrix}} = \frac{9}{7}$$

This result is readily extended to more than two simultaneous equations.

EXAMPLE 17–5
Solve the equations

$$x + y + z = 2$$
$$2x - y - z = 1$$
$$x + 2y - z = -3$$

SOLUTION: The extension of Equations 17.9 and 17.10 is

$$x = \frac{\begin{vmatrix} 2 & 1 & 1 \\ 1 & -1 & -1 \\ -3 & 2 & -1 \end{vmatrix}}{\begin{vmatrix} 1 & 1 & 1 \\ 2 & -1 & -1 \\ 1 & 2 & -1 \end{vmatrix}} = \frac{9}{9} = 1$$

Similarly,

$$y = \frac{\begin{vmatrix} 1 & 2 & 1 \\ 2 & 1 & -1 \\ 1 & -3 & -1 \end{vmatrix}}{\begin{vmatrix} 1 & 1 & 1 \\ 2 & -1 & -1 \\ 1 & 2 & -1 \end{vmatrix}} = \frac{-9}{9} = -1$$

and

$$z = \frac{\begin{vmatrix} 1 & 1 & 2 \\ 2 & -1 & 1 \\ 1 & 2 & -3 \end{vmatrix}}{\begin{vmatrix} 1 & 1 & 1 \\ 2 & -1 & -1 \\ 1 & 2 & -1 \end{vmatrix}} = \frac{18}{9} = 2$$

Although Cramer's rule is easy to use for 2×2 and 3×3 systems of equations, it gets increasingly more difficult to implement as the system of equations gets larger. We shall present an alternative method in Chapter 23 that is easier to use for large systems of equations.

What happens if $d_1 = d_2 = 0$ in Equation 17.7? In that case, we find that $x = y = 0$, which is an obvious solution, called a *trivial solution*. The only way that we could obtain a nontrivial solution for a set of homogeneous equations is for the denominator in Equations 17.9 and 17.10 to be zero, or for

$$D = \begin{vmatrix} a_{11} & a_{12} \\ a_{21} & a_{22} \end{vmatrix} = 0 \tag{17.11}$$

This is an important enough result that we should emphasize it:

A homogenous system of linear algebraic equations has a nontrivial solution only if the determinant of the coefficients is equal to zero.

In discussing Hückel molecular orbital theory, we shall meet equations such as

$$c_1(H_{11} - E S_{11}) + c_2(H_{12} - E S_{12}) = 0$$

and

$$c_1(H_{12} - E S_{12}) + c_2(H_{22} - E S_{22}) = 0$$

where the H_{ij} and S_{ij} are known quantities and c_1, c_2, and E are to be determined. We can appeal to Equation 17.11, which says that for a nontrivial solution (in other words, one for which both c_1 and c_2 are not equal to zero) to exist, we must have

$$\begin{vmatrix} H_{11} - ES_{11} & H_{12} - ES_{12} \\ H_{12} - ES_{12} & H_{22} - ES_{22} \end{vmatrix} = 0 \qquad (17.12)$$

When this determinant is expanded, we obtain a quadratic equation in E, yielding two roots. The determinant in Equation 17.12 is called a *secular determinant*, and Equation 17.12 itself constitutes a *secular determinantal equation*.

EXAMPLE 17–6

Find the roots of the determinantal equation

$$\begin{vmatrix} 2 - \lambda & 3 \\ 3 & 4 - \lambda \end{vmatrix} = 0$$

SOLUTION: Expand the determinant to obtain $(2 - \lambda)(4 - \lambda) - 9 = 0$ or $\lambda^2 - 6\lambda - 1 = 0$. The two roots are

$$\lambda = \frac{6}{2} \pm \frac{\sqrt{40}}{2} = 3 \pm \sqrt{10}$$

Although we considered only two simultaneous homogeneous algebraic equations, Equation 17.11 is readily extended to any number.

Problems

17–1. Evaluate the determinant

$$D = \begin{vmatrix} 2 & 1 & 1 \\ -1 & 3 & 2 \\ 2 & 0 & 1 \end{vmatrix}$$

Add column 2 to column 1 to get

$$\begin{vmatrix} 3 & 1 & 1 \\ 2 & 3 & 2 \\ 2 & 0 & 1 \end{vmatrix}$$

and evaluate it. Compare your result with the value of D. Now add row 2 to row 1 of D to get

$$\begin{vmatrix} 1 & 4 & 3 \\ -1 & 3 & 2 \\ 2 & 0 & 1 \end{vmatrix}$$

and evaluate it. Compare your result with the value of D above.

17–2. Interchange columns 1 and 3 in D in Problem 17–1 and evaluate the resulting determinant. Compare your result with the value of D. Interchange rows 1 and 2 and do the same.

17–3. Evaluate

$$D = \begin{vmatrix} 2 & -1 & 1 \\ 0 & 3 & -1 \\ 2 & -2 & 1 \end{vmatrix}$$

17–4. Evaluate the determinant

$$D = \begin{vmatrix} 1 & 6 & 1 \\ -2 & 4 & -2 \\ 1 & -3 & 1 \end{vmatrix}$$

Can you determine its value by inspection? What about

$$D = \begin{vmatrix} 2 & 6 & 1 \\ -4 & 4 & -2 \\ 2 & -3 & 1 \end{vmatrix}$$

17–5. Starting with D in Problem 17–1, add two times the third row to the second row and evaluate the resulting determinant.

17–6. Evaluate

$$D = \begin{vmatrix} 1 & \sin x & \cos x \\ 0 & \cos x & -\sin x \\ 0 & -\sin x & -\cos x \end{vmatrix}$$

17–7. The following determinantal equation arises in the Hückel molecular orbital theory of a trimethylenemethane molecule:

$$\begin{vmatrix} x & 1 & 1 & 1 \\ 1 & x & 0 & 0 \\ 1 & 0 & x & 0 \\ 1 & 0 & 0 & x \end{vmatrix} = 0$$

Determine the values of x that satisfy this equation.

17–8. The following determinantal equation arises in the Hückel molecular orbital theory of a cyclobutadiene molecule:

$$\begin{vmatrix} x & 1 & 0 & 1 \\ 1 & x & 1 & 0 \\ 0 & 1 & x & 1 \\ 1 & 0 & 1 & x \end{vmatrix} = 0$$

Determine the values of x that satisfy this equation.

17–9. The following determinant arises in the Hückel molecular orbital theory of a benzene molecule:

$$D = \begin{vmatrix} x & 1 & 0 & 0 & 0 & 1 \\ 1 & x & 1 & 0 & 0 & 0 \\ 0 & 1 & x & 1 & 0 & 0 \\ 0 & 0 & 1 & x & 1 & 0 \\ 0 & 0 & 0 & 1 & x & 1 \\ 1 & 0 & 0 & 0 & 1 & x \end{vmatrix}$$

Determine the values of x that satisfy this equation.

17–10. Show that

$$\begin{vmatrix} \cos\theta & -\sin\theta & 0 \\ \sin\theta & \cos\theta & 0 \\ 0 & 0 & 1 \end{vmatrix} = 1$$

17–11. Find the three roots of the determinantal equation

$$\begin{vmatrix} 1-\lambda & 1 & 0 \\ 1 & 1-\lambda & 1 \\ 0 & 1 & 1-\lambda \end{vmatrix} = 0$$

17–12. Solve the following set of equations using Cramer's rule:

$$x + y = 2$$
$$3x - 2y = 5$$

17–13. Solve the following set of equations using Cramer's rule:

$$x + 2y + 3z = -5$$
$$-x - 3y + z = -14$$
$$2x + y + z = 1$$

17–14. Determine the values of x for which the following equations have a nontrivial solution.

$$xc_1 + c_2 + c_4 = 0$$
$$c_1 + xc_2 + c_3 = 0$$
$$c_2 + xc_3 + c_4 = 0$$
$$c_1 + c_3 + xc_4 = 0$$

17–15. Use Cramer's rule to solve

$$x + 2y = 3$$
$$2x + 4y = 1$$

What went wrong here? Why?

17–16. For what values of λ do the following equations have a nontrivial solution?

$$x + y = \lambda x$$
$$-x + y = \lambda y$$

CHAPTER 18

Matrices

When Heisenberg formulated quantum mechanics in the 1920s, he worked in terms of concrete experimental quantities such as spectroscopic transition probabilities. This approach led to the use of quantities with two subscripts, corresponding to the initial and final states of the system being described. In manipulating these quantities, he found to his surprise that they satisfied a set of rules that were not commutative; that is, the result of the product of two quantities depended upon the order of multiplication. Heisenberg did not know about matrices at the time, but he soon learned that his noncommuting quantities were actually matrices, which had been investigated by mathematicians almost a century earlier. Nowadays, matrices are used extensively in physics and chemistry. Modern computational chemistry is formulated almost exclusively in terms of matrices.

18.1 Matrix Algebra

Many physical operations such as magnification, rotation, and reflection through a plane can be represented mathematically by quantities called matrices. A matrix is a two-dimensional array that obeys a certain set of rules called *matrix algebra*.

Consider the lower of the two vectors shown in Figure 18.1. The x and y components of the vector are given by $x_1 = r\cos\alpha$ and $y_1 = r\sin\alpha$, where r is the length of $\mathbf{r}_1$. Now let's rotate the vector counterclockwise through an angle θ, so that $x_2 = r\cos(\alpha + \theta)$ and $y_2 = r\sin(\alpha + \theta)$ (see Figure 18.1). Using trigonometric formulas, we can write

$$x_2 = r\cos(\alpha + \theta) = r\cos\alpha\cos\theta - r\sin\alpha\sin\theta$$

$$y_2 = r\sin(\alpha + \theta) = r\cos\alpha\sin\theta + r\sin\alpha\cos\theta$$

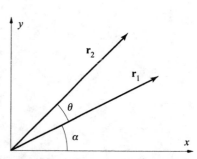

Figure 18.1. An illustration of the rotation of a vector $\mathbf{r}$ through an angle θ.

231

or

$$x_2 = x_1 \cos\theta - y_1 \sin\theta$$
$$y_2 = x_1 \sin\theta + y_1 \cos\theta \tag{18.1}$$

We can display the set of coefficients of x_1 and y_1 in Equation 18.1 in the form

$$R = \begin{pmatrix} \cos\theta & -\sin\theta \\ \sin\theta & \cos\theta \end{pmatrix} \tag{18.2}$$

We have expressed R in the form of a *matrix*, which is an array of numbers (or functions in this case) that obeys the rules of matrix algebra. We shall denote a matrix by a sans serif symbol (e.g., A, B, etc.). Unlike determinants (Chapter 17), matrices do not have to be square arrays. Furthermore, unlike determinants, matrices cannot be reduced to a single number. The matrix R in Equation 18.2 corresponds to a rotation through an angle θ.

We can write Equation 18.1 symbolically as

$$\mathbf{r}_2 = R\,\mathbf{r}_1 \tag{18.3}$$

In other words, the matrix R operates on the vector $\mathbf{r}_1$ to give a new vector $\mathbf{r}_2$ (see Figure 18.1). Equation 18.3 is equivalent to Equation 18.1. It's easy to show (see Problem 18–1) that the transformation in Equation 18.3 is a linear transformation, in the sense that

$$R(\alpha\,\mathbf{u} + \beta\,\mathbf{v}) = \alpha\,R\,\mathbf{u} + \beta\,R\,\mathbf{v} \tag{18.4}$$

where $\mathbf{u}$ and $\mathbf{v}$ are two vectors and α and β are two scalars. A matrix operation is a linear operation.

The entries in a matrix A are called its *matrix elements* and are denoted by a_{ij}, where, as in the case of determinants, i designates the row and j designates the column. Two matrices, A and B, are equal if and only if they are of the same dimension and $a_{ij} = b_{ij}$ for all i and j. In other words, equal matrices are identical. Matrices can be added or subtracted only if they have the same number of rows and columns, in which case the elements of the resultant matrix are given by $a_{ij} + b_{ij}$. Thus, if

$$A = \begin{pmatrix} -3 & 6 & 4 \\ 1 & 0 & 2 \end{pmatrix} \quad \text{and} \quad B = \begin{pmatrix} 2 & 1 & 1 \\ -6 & 4 & 3 \end{pmatrix}$$

then

$$C = A + B = \begin{pmatrix} -1 & 7 & 5 \\ -5 & 4 & 5 \end{pmatrix}$$

If we write

$$A + A = 2A = \begin{pmatrix} -6 & 12 & 8 \\ 2 & 0 & 4 \end{pmatrix}$$

we see that scalar multiplication of a matrix means that each element is multiplied by the scalar. Thus,

$$cA = \begin{pmatrix} c\,a_{11} & c\,a_{12} \\ c\,a_{21} & c\,a_{22} \end{pmatrix} \tag{18.5}$$

EXAMPLE 18–1

Using the matrices A and B above, form the matrix $D = 3A - 2B$.

SOLUTION:

$$D = 3 \begin{pmatrix} -3 & 6 & 4 \\ 1 & 0 & 2 \end{pmatrix} - 2 \begin{pmatrix} 2 & 1 & 1 \\ -6 & 4 & 3 \end{pmatrix}$$

$$= \begin{pmatrix} -9 & 18 & 12 \\ 3 & 0 & 6 \end{pmatrix} - \begin{pmatrix} 4 & 2 & 2 \\ -12 & 8 & 6 \end{pmatrix} = \begin{pmatrix} -13 & 16 & 10 \\ 15 & -8 & 0 \end{pmatrix}$$

One of the most important aspects of matrices is matrix multiplication. For simplicity, we shall discuss the multiplication of square matrices first. Consider some linear transformation of (x_1, y_1) into (x_2, y_2),

$$x_2 = a_{11}x_1 + a_{12}y_1$$
$$y_2 = a_{21}x_1 + a_{22}y_1$$
(18.6)

represented by the matrix equation $\mathbf{r}_2 = A\mathbf{r}_1$, where

$$A = \begin{pmatrix} a_{11} & a_{12} \\ a_{21} & a_{22} \end{pmatrix}$$
(18.7)

Now let's transform (x_2, y_2) into (x_3, y_3),

$$x_3 = b_{11}x_2 + b_{12}y_2$$
$$y_3 = b_{21}x_2 + b_{22}y_2$$
(18.8)

represented by the matrix equation $\mathbf{r}_3 = B\mathbf{r}_2$, where

$$B = \begin{pmatrix} b_{11} & b_{12} \\ b_{21} & b_{22} \end{pmatrix}$$
(18.9)

Let the transformation of (x_1, y_1) directly into (x_3, y_3) be given by

$$x_3 = c_{11}x_1 + c_{12}y_1$$
$$y_3 = c_{21}x_1 + c_{22}y_1$$
(18.10)

represented by the matrix equation $\mathbf{r}_3 = C\mathbf{r}_1$, where

$$C = \begin{pmatrix} c_{11} & c_{12} \\ c_{21} & c_{22} \end{pmatrix}$$
(18.11)

Symbolically, we can write that $\mathbf{r}_3 = B\mathbf{r}_2 = BA\mathbf{r}_1$, or

$$C = BA$$

In other words, C results from transforming from (x_1, y_1) to (x_2, y_2) by means of A, followed by transforming (x_2, y_2) to (x_3, y_3) by means of B. Let's find the relation between the elements of C and those of A and B. Substitute Equations 18.6 into 18.8 to obtain

$$x_3 = b_{11}(a_{11}x_1 + a_{12}y_1) + b_{12}(a_{21}x_1 + a_{22}y_1)$$
$$y_3 = b_{21}(a_{11}x_1 + a_{12}y_1) + b_{22}(a_{21}x_1 + a_{22}y_1)$$

(18.12)

or

$$x_3 = (b_{11}a_{11} + b_{12}a_{21})x_1 + (b_{11}a_{12} + b_{12}a_{22})y_1$$
$$y_3 = (b_{21}a_{11} + b_{22}a_{21})x_1 + (b_{21}a_{12} + b_{22}a_{22})y_1$$

Thus, we see that

$$C = BA = \begin{pmatrix} b_{11} & b_{12} \\ b_{21} & b_{22} \end{pmatrix}\begin{pmatrix} a_{11} & a_{12} \\ a_{21} & a_{22} \end{pmatrix} = \begin{pmatrix} b_{11}a_{11} + b_{12}a_{21} & b_{11}a_{12} + b_{12}a_{22} \\ b_{21}a_{11} + b_{22}a_{21} & b_{21}a_{12} + b_{22}a_{22} \end{pmatrix}$$

(18.13)

This result may look complicated, but it has a nice pattern that we shall illustrate two ways. Mathematically, the ijth element of C is given by the formula

$$c_{ij} = \sum_k b_{ik}a_{kj}$$

(18.14)

Notice that we sum over the middle index. For example,

$$c_{11} = \sum_{k=1}^{2} b_{1k}a_{k1} = b_{11}a_{11} + b_{12}a_{21}$$

as in Equation 18.13. A more pictorial way is to notice that any element in C can be obtained by multiplying elements in any row in B by the corresponding elements in any column in A, adding them, and then placing them in C where the row and column intersect. In terms of vectors, we take the dot product of the row of B and the column of A and place the result at their intersection. For example, c_{11} is obtained by multiplying the elements of row 1 of B with the elements of column 1 of A, or by the scheme

$$\rightarrow \begin{pmatrix} b_{11} & b_{12} \\ b_{21} & b_{22} \end{pmatrix}\begin{pmatrix} a_{11} & a_{12} \\ a_{21} & a_{22} \end{pmatrix} = \begin{pmatrix} b_{11}a_{11} + b_{12}a_{21} & \cdot \\ \cdot & \cdot \end{pmatrix}$$

and c_{12} is obtained by

$$\rightarrow \begin{pmatrix} b_{11} & b_{12} \\ b_{21} & b_{22} \end{pmatrix}\begin{pmatrix} a_{11} & a_{12} \\ a_{21} & a_{22} \end{pmatrix} = \begin{pmatrix} \cdot & b_{11}a_{12} + b_{12}a_{22} \\ \cdot & \cdot \end{pmatrix}$$

EXAMPLE 18–2

Find C = BA if

$$B = \begin{pmatrix} 1 & 2 & 1 \\ 3 & 0 & -1 \\ -1 & -1 & 2 \end{pmatrix} \quad \text{and} \quad A = \begin{pmatrix} -3 & 0 & -1 \\ 1 & 4 & 0 \\ 1 & 1 & 1 \end{pmatrix}$$

SOLUTION:

$$C = \begin{pmatrix} 1 & 2 & 1 \\ 3 & 0 & -1 \\ -1 & -1 & 2 \end{pmatrix} \begin{pmatrix} -3 & 0 & -1 \\ 1 & 4 & 0 \\ 1 & 1 & 1 \end{pmatrix}$$

$$= \begin{pmatrix} -3+2+1 & 0+8+1 & -1+0+1 \\ -9+0-1 & 0+0-1 & -3+0-1 \\ 3-1+2 & 0-4+2 & 1+0+2 \end{pmatrix}$$

$$= \begin{pmatrix} 0 & 9 & 0 \\ -10 & -1 & -4 \\ 4 & -2 & 3 \end{pmatrix}$$

EXAMPLE 18–3

The matrix R given by Equation 18.2 represents a rotation through the angle θ. Show that R^2 represents a rotation through an angle 2θ. In other words, show that R^2 represents two sequential applications of R.

SOLUTION:

$$R^2 = \begin{pmatrix} \cos\theta & -\sin\theta \\ \sin\theta & \cos\theta \end{pmatrix} \begin{pmatrix} \cos\theta & -\sin\theta \\ \sin\theta & \cos\theta \end{pmatrix}$$

$$= \begin{pmatrix} \cos^2\theta - \sin^2\theta & -2\sin\theta\cos\theta \\ 2\sin\theta\cos\theta & \cos^2\theta - \sin^2\theta \end{pmatrix}$$

Using standard trigonometric identities, we get

$$R^2 = \begin{pmatrix} \cos 2\theta & -\sin 2\theta \\ \sin 2\theta & \cos 2\theta \end{pmatrix}$$

which represents rotation through an angle 2θ.

An important aspect of matrix multiplication is that BA does not necessarily equal AB. For example, if

$$A = \begin{pmatrix} 0 & 2 \\ 1 & 0 \end{pmatrix} \quad \text{and} \quad B = \begin{pmatrix} 3 & 0 \\ 0 & -1 \end{pmatrix}$$

then

$$AB = \begin{pmatrix} 0 & 2 \\ 1 & 0 \end{pmatrix} \begin{pmatrix} 3 & 0 \\ 0 & -1 \end{pmatrix} = \begin{pmatrix} 0 & -2 \\ 3 & 0 \end{pmatrix}$$

and

$$BA = \begin{pmatrix} 3 & 0 \\ 0 & -1 \end{pmatrix} \begin{pmatrix} 0 & 2 \\ 1 & 0 \end{pmatrix} = \begin{pmatrix} 0 & 6 \\ -1 & 0 \end{pmatrix}$$

and so $AB \neq BA$. If it does happen that $AB = BA$, then A and B are said to *commute*.

EXAMPLE 18–4

Do the matrices A and B commute if

$$A = \begin{pmatrix} 2 & 1 \\ 0 & 1 \end{pmatrix} \quad \text{and} \quad B = \begin{pmatrix} 1 & 1 \\ 0 & 1 \end{pmatrix}$$

SOLUTION:

$$AB = \begin{pmatrix} 2 & 3 \\ 0 & 1 \end{pmatrix}$$

and

$$BA = \begin{pmatrix} 2 & 2 \\ 0 & 1 \end{pmatrix}$$

so they do not commute.

Another property of matrix multiplication that differs from ordinary scalar multiplication is that the equation

$$AB = O$$

where O is the zero matrix (all elements equal to zero) does not imply that A or B necessarily is a zero matrix. For example,

$$\begin{pmatrix} 1 & 1 \\ 2 & 2 \end{pmatrix} \begin{pmatrix} -1 & 1 \\ 1 & -1 \end{pmatrix} = \begin{pmatrix} 0 & 0 \\ 0 & 0 \end{pmatrix} = O$$

We emphasize here that determinants and matrices are entirely different entities, even though they are both two-dimensional arrays of numbers. A determinant is a number, whereas a matrix is an array of numbers and cannot be reduced to a single number. Nevertheless, we can associate a determinant with a square matrix by writing

$$\det A = |A| = \begin{vmatrix} a_{11} & a_{12} & \cdots & a_{1n} \\ a_{21} & a_{22} & \cdots & a_{2n} \\ \vdots & \vdots & \vdots & \vdots \\ a_{n1} & a_{n2} & \cdots & a_{nn} \end{vmatrix}$$

Thus, the determinant of R given by Equation 18.2 is

$$|R| = \begin{vmatrix} \cos\theta & -\sin\theta \\ \sin\theta & \cos\theta \end{vmatrix} = \cos^2\theta + \sin^2\theta = 1$$

If $\det A = 0$, then A is said to be a *singular matrix*.

Matrices do not have to be square to be multiplied together, but either Equation 18.14 or the pictorial method illustrated above suggests that the number of columns of B must be equal to the number of rows of A. When this is so, A and B are said to be *compatible*. For example, Equations 18.6 can be written in matrix form as

$$\begin{pmatrix} x_2 \\ y_2 \end{pmatrix} = \begin{pmatrix} a_{11} & a_{12} \\ a_{21} & a_{22} \end{pmatrix} \begin{pmatrix} x_1 \\ y_1 \end{pmatrix} \tag{18.15}$$

18.2 Inverse of a Matrix

A transformation that leaves a vector unaltered is called the *identity transformation*, and the corresponding matrix is called the *identity matrix* or the *unit matrix*. All the elements of the unit matrix are equal to zero, except those along the diagonal, which equal one:

$$\mathsf{I} = \begin{pmatrix} 1 & 0 & 0 & \cdots & 0 \\ 0 & 1 & 0 & \cdots & 0 \\ 0 & 0 & 1 & \cdots & 0 \\ \vdots & \vdots & \vdots & \vdots & \vdots \\ 0 & 0 & 0 & \cdots & 1 \end{pmatrix}$$

The elements of I are δ_{ij}, the Kronecker delta. The unit matrix has the property that

$$\mathsf{IA} = \mathsf{AI} \tag{18.16}$$

for any matrix A. The unit matrix is an example of a *diagonal matrix*. The only nonzero elements of a diagonal matrix are along its diagonal. Diagonal matrices are necessarily square matrices.

If $\mathsf{BA} = \mathsf{AB} = \mathsf{I}$, then B is said to be the *inverse* of A and is denoted by A^{-1}. Thus, A^{-1} has the property that

$$\mathsf{AA}^{-1} = \mathsf{A}^{-1}\mathsf{A} = \mathsf{I} \tag{18.17}$$

If A represents some transformation, then A^{-1} undoes that transformation and restores the original state. It should be clear on physical grounds that the inverse of R in Equation 18.2 is

$$\mathsf{R}^{-1}(\theta) = \mathsf{R}(-\theta) = \begin{pmatrix} \cos(-\theta) & -\sin(-\theta) \\ \sin(-\theta) & \cos(-\theta) \end{pmatrix} = \begin{pmatrix} \cos\theta & \sin\theta \\ -\sin\theta & \cos\theta \end{pmatrix} \tag{18.18}$$

In other words, if $\mathsf{R}(\theta)$ represents a rotation through an angle θ, then $\mathsf{R}^{-1}(\theta) = \mathsf{R}(-\theta)$ and represents the reverse rotation. It is easy to show that R and R^{-1} satisfy Equation 18.17. Using Equations 18.2 and 18.18, we have

$$\begin{aligned} \mathsf{R}^{-1}\mathsf{R} &= \begin{pmatrix} \cos\theta & \sin\theta \\ -\sin\theta & \cos\theta \end{pmatrix} \begin{pmatrix} \cos\theta & -\sin\theta \\ \sin\theta & \cos\theta \end{pmatrix} \\ &= \begin{pmatrix} \cos^2\theta + \sin^2\theta & 0 \\ 0 & \cos^2\theta + \sin^2\theta \end{pmatrix} \\ &= \begin{pmatrix} 1 & 0 \\ 0 & 1 \end{pmatrix} \end{aligned}$$

and

$$RR^{-1} = \begin{pmatrix} \cos\theta & -\sin\theta \\ \sin\theta & \cos\theta \end{pmatrix} \begin{pmatrix} \cos\theta & \sin\theta \\ -\sin\theta & \cos\theta \end{pmatrix}$$

$$= \begin{pmatrix} \cos^2\theta + \sin^2\theta & 0 \\ 0 & \cos^2\theta + \sin^2\theta \end{pmatrix}$$

$$= \begin{pmatrix} 1 & 0 \\ 0 & 1 \end{pmatrix}$$

Before we show how to find the inverse of a given matrix, we must first define the transpose of A, A^T, as the matrix that we obtain from A by interchanging rows and columns. For example, the transpose of

$$A = \begin{pmatrix} 1 & 8 & -4 \\ 4 & -4 & -7 \\ 8 & 1 & 4 \end{pmatrix}$$

is

$$A^T = \begin{pmatrix} 1 & 4 & 8 \\ 8 & -4 & 1 \\ -4 & -7 & 4 \end{pmatrix}$$

We obtain A^T from A by simply flipping it about its diagonal. In terms of the elements of A, we have $a_{ij}^T = a_{ji}$. If $A^T = A$, then the matrix is said to be *symmetric*. In terms of the elements of A, we have $a_{ij} = a_{ji}$. Most matrices in quantum mechanics are symmetric.

A standard procedure for finding the inverse of a matrix is the following:

1. Replace each element of A by its cofactor in the corresponding determinant (see Chapter 17 for a definition of a cofactor). Denote this matrix by A_{cof}.

2. Take the transpose of the cofactor matrix and denote it by A_{cof}^T.

3. Divide each element of A_{cof}^T by the determinant of A. The result is A^{-1}.

In an equation, we have

$$A^{-1} = \frac{A_{cof}^T}{|A|} \tag{18.19}$$

For example, if

$$A = \begin{pmatrix} 1 & 2 \\ 3 & 4 \end{pmatrix}$$

the cofactors of A are $A_{cof,11} = 4$, $A_{cof,12} = -3$, $A_{cof,21} = -2$, and $A_{cof,22} = 1$. The matrix of cofactors is $A_{cof} = \begin{pmatrix} 4 & -3 \\ -2 & 1 \end{pmatrix}$ and its transpose is $A_{cof}^T = \begin{pmatrix} 4 & -2 \\ -3 & 1 \end{pmatrix}$. The determinant of $A = -2$, and so according to Equation 18.19

$$A^{-1} = \frac{A_{cof}^T}{|A|} = -\frac{1}{2}\begin{pmatrix} 4 & -2 \\ -3 & 1 \end{pmatrix} = \begin{pmatrix} -2 & 1 \\ \frac{3}{2} & -\frac{1}{2} \end{pmatrix}$$

which you can readily verify by showing that $AA^{-1} = A^{-1}A = I$.

It should be evident from our procedure to find the inverse of a matrix that singular matrices (those whose determinants are equal to zero) do not have inverses.

EXAMPLE 18–5

Find the inverse of

$$A = \begin{pmatrix} 1 & -1 & 0 \\ 0 & 1 & 1 \\ 2 & 2 & 0 \end{pmatrix}$$

SOLUTION: The determinant of A is equal to -4 and the matrix of cofactors is

$$A_{cof} = \begin{pmatrix} -2 & 2 & -2 \\ 0 & 0 & -4 \\ -1 & -1 & 1 \end{pmatrix}$$

Using Equation 18.19, we have

$$A^{-1} = \frac{A_{cof}^{T}}{|A|} = -\frac{1}{4} \begin{pmatrix} -2 & 0 & -1 \\ 2 & 0 & -1 \\ -2 & -4 & 1 \end{pmatrix} = \begin{pmatrix} \frac{1}{2} & 0 & \frac{1}{4} \\ -\frac{1}{2} & 0 & \frac{1}{4} \\ \frac{1}{2} & 1 & \frac{1}{4} \end{pmatrix}$$

You can readily verify that $AA^{-1} = A^{-1}A = I$.

The inverse operation satisfies the relation (Problem 18–11)

$$(AB)^{-1} = B^{-1}A^{-1} \tag{18.20}$$

(Does this make sense to you physically? If it does not, recall that AB represents doing the operation B first, followed by the operation A. Think what sequence you must follow to reverse the result of AB.)

18.3 Orthogonal Matrices

It is useful to look upon an n-dimensional vector as an $n \times 1$ matrix, which consists of 1 column and n rows. Thus, the elements of v are $(v_{11}, v_{21}, \ldots, v_{n1})$ (the second subscript is essentially redundant here). Such a matrix is called a *column matrix*, which we write as (see Equation 18.15)

$$v = \begin{pmatrix} v_{11} \\ v_{21} \\ \vdots \\ v_{n1} \end{pmatrix}$$

To save space, we usually write v as $v = (v_{11}, v_{21}, \ldots, v_{n1})$, keeping in mind that it is a column matrix. The transpose of v is a row matrix

$$v^{T} = (v_{11} \ v_{21} \ \cdots \ v_{n1})$$

where, once again, the second subscript is redundant. We can write the square of the length of a vector v in matrix notation as

$$v^\mathsf{T}v = \sum_{j=1}^{n} v_{1j}^\mathsf{T} v_{j1} = \sum_{j=1}^{n} v_{j1}^2 \tag{18.21}$$

in terms of an equation, or pictorially as

$$(v_{11}\ v_{21}\ldots v_{n1}) \begin{pmatrix} v_{11} \\ v_{21} \\ \vdots \\ v_{n1} \end{pmatrix} = v_{11}^2 + v_{21}^2 + \cdots + v_{n1}^2$$

Equation 18.21 is the matrix equivalent of a dot product. We write $\mathbf{u} \cdot \mathbf{v}$ in vector notation and $u^\mathsf{T}v$ in matrix notation.

The matrix described by Equation 18.2 corresponds to a rotation of a vector through an angle θ in a counterclockwise direction. Since Ru corresponds to a rotation of u, we should expect that the length of u does not get altered under R (Problem 18–3). Let $v = Au$ be a linear transformation of u that preserves its length. Then

$$v^\mathsf{T}v = (Au)^\mathsf{T}(Au)$$

Now, just as $(AB)^{-1} = B^{-1}A^{-1}$ (Equation 18.20), it turns out that $(AB)^\mathsf{T} = B^\mathsf{T}A^\mathsf{T}$ (Problem 18–10). Using this result, we write

$$v^\mathsf{T}v = (Au)^\mathsf{T}(Au) = u^\mathsf{T}A^\mathsf{T}Au$$

If v and u are to have the same length, then we must have $A^\mathsf{T}A = I$, or that $A^\mathsf{T} = A^{-1}$. A matrix with this property is said to be *orthogonal*, and $v = Au$ is said to be an *orthogonal transformation*. Orthogonal matrices preserve the length of vectors. The relation $A^{-1} = A^\mathsf{T}$ is important enough to display as a separate equation:

$$A^{-1} = A^\mathsf{T} \qquad \text{(orthogonal matrix)} \tag{18.22}$$

EXAMPLE 18–6
Show that R given by Equation 18.2 is an orthogonal matrix.

SOLUTION: We need to show that $R^\mathsf{T} = R^{-1}$ or that $RR^\mathsf{T} = I$.

$$R^\mathsf{T}R = \begin{pmatrix} \cos\theta & \sin\theta \\ -\sin\theta & \cos\theta \end{pmatrix} \begin{pmatrix} \cos\theta & -\sin\theta \\ \sin\theta & \cos\theta \end{pmatrix}$$
$$= \begin{pmatrix} \cos^2\theta + \sin^2\theta & 0 \\ 0 & \sin^2\theta + \cos^2\theta \end{pmatrix} = \begin{pmatrix} 1 & 0 \\ 0 & 1 \end{pmatrix}$$

It is also true that $RR^\mathsf{T} = I$.

The orthogonality expression

$$A^TA = AA^T = I \qquad (18.23)$$

implies that the rows of A^T and the columns of A are orthonormal. The rows of A^T, however, are the columns of A, so Equation 18.23 implies that the column vectors (also the row vectors) of an orthogonal matrix are orthonormal. To see this analytically, insert the fact that $a_{ij}^T = a_{ji}$ into $A^TA = I$ to get

$$\sum_{j=1}^{n} a_{ij}^T a_{jk} = \delta_{ik} = \sum_{j=1}^{n} a_{ji} a_{jk} \qquad (18.24)$$

Equation 18.24 says that the ith and kth columns of A are orthonormal vectors. If we use $A A^T = I$ instead of $A^TA = I$, we find that the ith and kth rows of A are orthonormal vectors (Problem 18–12).

EXAMPLE 18–7
First show that

$$A = \frac{1}{9}\begin{pmatrix} 1 & 8 & -4 \\ 4 & -4 & -7 \\ 8 & 1 & 4 \end{pmatrix}$$

is orthogonal and then show explicitly that its rows (columns) are ortho-normal vectors.

SOLUTION:

$$A^TA = \frac{1}{81}\begin{pmatrix} 1 & 4 & 8 \\ 8 & -4 & 1 \\ -4 & -7 & 4 \end{pmatrix}\begin{pmatrix} 1 & 8 & -4 \\ 4 & -4 & -7 \\ 8 & 1 & 4 \end{pmatrix} = \frac{1}{81}\begin{pmatrix} 81 & 0 & 0 \\ 0 & 81 & 0 \\ 0 & 0 & 81 \end{pmatrix}$$

$$AA^T = \frac{1}{81}\begin{pmatrix} 1 & 8 & -4 \\ 4 & -4 & -7 \\ 8 & 1 & 4 \end{pmatrix}\begin{pmatrix} 1 & 4 & 8 \\ 8 & -4 & 1 \\ -4 & -7 & 4 \end{pmatrix} = \frac{1}{81}\begin{pmatrix} 81 & 0 & 0 \\ 0 & 81 & 0 \\ 0 & 0 & 81 \end{pmatrix}$$

The columns are normalized because

$$\frac{1}{81}(1 + 16 + 64) = 1, \quad \frac{1}{81}(64 + 16 + 1) = 1, \text{ and } \frac{1}{81}(16 + 49 + 16) = 1$$

Column 1 and column 2 are orthogonal vectors because

$$(1 \quad 4 \quad 8)\begin{pmatrix} 8 \\ -4 \\ 1 \end{pmatrix} = 8 - 16 + 8 = 0$$

Similarly, columns 1 and 3 and columns 2 and 3 are orthogonal. For rows 2 and 3,

$$(4 \quad -4 \quad -7) \begin{pmatrix} 8 \\ 1 \\ 4 \end{pmatrix} = 32 - 4 - 28 = 0$$

Rows 1 and 2 and rows 1 and 3 are also orthogonal.

Using the fact that $\det(AB) = \det(A)\det(B)$, you can show that the determinant of an orthogonal matrix is equal to ± 1 (Problem 18–16). If $\det(A) = 1$, then A corresponds to a pure rotation, and if $\det(A) = -1$, A corresponds to a rotation *and* an inversion through the origin.

18.4 Unitary Matrices

Up to this point, we have considered only real vectors and real matrices. We frequently deal with complex vectors and matrices in quantum mechanics, and so we shall end this chapter with an extension of our results to complex vectors and matrices. We'll refer to this as a complex space. We define a scalar product in a complex space by

$$u^{*T}v = u_1^* v_1 + u_2^* v_2 + \cdots + u_n^* v_n \tag{18.25}$$

The scalar product, $u^{*T}u$, is called the *norm* of u and is given by

$$u^{*T}u = (u_1^* u_1 + u_2^* u_2 + \cdots + u_n^* u_n)^{1/2} \tag{18.26}$$

According to Equation 18.26, the norm is guaranteed to be a real quantity. The norm of a vector with real components is its length. Thus, the norm is the complex space analog of length.

EXAMPLE 18–8

Determine the norm of the vector $u = (2i, 4, 1 - 3i, 2 + i)$.

SOLUTION: Using Equation 18.26, we have

$$u^{*T}u = [(-2i)(2i) + (4)(4) + (1 + 3i)(1 - 3i) + (2 - i)(2 + i)]$$
$$= [4 + 16 + 10 + 5] = 35$$

The norm of u is $\sqrt{35}$.

The analogy of a symmetric matrix ($A = A^T$) in a complex space is

$$A^\dagger = (A^*)^T = A \qquad \text{(Hermitian matrix)} \tag{18.27}$$

where $A^\dagger = (A^*)^T$ is called the *Hermitian conjugate* of A. Using the dagger notation, Equation 18.25 becomes

$$u^\dagger v = u_1^* v_1 + u_2^* v_2 + \cdots + u_n^* v_n \tag{18.28}$$

A linear transformation, $v = A\,u$, that preserves lengths in a complex space must satisfy

$$v^\dagger v = (A^* u^*)^T (Au) = u^{*T}(A^*)^T Au = u^\dagger A^\dagger A u = u^\dagger u$$

If the norm of v is to equal the norm of u, then $A^\dagger A = I$, or

$$A^\dagger = A^{-1} \qquad \text{(unitary matrix)} \tag{18.29}$$

A matrix that satisfies Equation 18.29 is said to be *unitary*. Equation 18.29 also says that

$$A^\dagger A = AA^\dagger = I \tag{18.30}$$

A unitary matrix is the analog of an orthogonal matrix in a complex space. Unitary transformations preserve the norms of vectors.

Equation 18.30 shows that the rows (columns) of a unitary matrix are orthonormal in the sense that $v_i^\dagger v_j = \delta_{ij}$. Using the fact that $a_{ij}^\dagger = a_{ji}^*$, Equation 18.30 becomes

$$\sum_{j=1}^{n} a_{ij}^\dagger a_{jk} = \sum_{j=1}^{n} a_{ji}^* a_{jk} = \delta_{ik}$$

which shows that the rows of $A^\dagger$ are orthonormal. The second expression in Equation 18.30 shows that the columns are orthonormal. Problem 18–22 has you show that the determinant of a unitary matrix is of absolute value 1.

EXAMPLE 18–9
Show that

$$A = \frac{1}{5}\begin{pmatrix} -1 + 2i & -4 - 2i \\ 2 - 4i & -2 - i \end{pmatrix}$$

is unitary.

SOLUTION:

$$A^\dagger = (A^*)^T = \frac{1}{5}\begin{pmatrix} -1 - 2i & 2 + 4i \\ -4 + 2i & -2 + i \end{pmatrix}$$

and

$$A^\dagger A = \frac{1}{25}\begin{pmatrix} -1 - 2i & 2 + 4i \\ -4 + 2i & -2 + i \end{pmatrix}\begin{pmatrix} -1 + 2i & -4 - 2i \\ 2 - 4i & -2 - i \end{pmatrix} = \frac{1}{25}\begin{pmatrix} 25 & 0 \\ 0 & 25 \end{pmatrix}$$

$$AA^\dagger = \frac{1}{25}\begin{pmatrix} -1 + 2i & -4 - 2i \\ 2 - 4i & -2 - i \end{pmatrix}\begin{pmatrix} -1 - 2i & 2 + 4i \\ -4 + 2i & -2 + i \end{pmatrix} = \frac{1}{25}\begin{pmatrix} 25 & 0 \\ 0 & 25 \end{pmatrix}$$

Note also that the rows and columns of A are orthonormal and that det A $=$ $(20 - 15i)/25$, which is of unit magnitude.

Problems

18–1. Show that the matrix transformation given by Equations 18.1 through 18.3 is a linear transformation.

18–2. Determine the geometric result of the following matrices acting on a two-dimensional vector u:

(a) $\begin{pmatrix} -1 & 0 \\ 0 & -1 \end{pmatrix}$ (b) $\begin{pmatrix} 1 & 0 \\ 0 & -1 \end{pmatrix}$ (c) $\begin{pmatrix} -1 & 0 \\ 0 & 1 \end{pmatrix}$

18–3. Show that u and Ru have the same length if R is given by Equation 18.2.

18–4. Given the two matrices

$$A = \begin{pmatrix} 1 & 0 & -1 \\ -1 & 2 & 0 \\ 0 & 1 & 1 \end{pmatrix} \quad \text{and} \quad B = \begin{pmatrix} -1 & 1 & 0 \\ 3 & 0 & 2 \\ 1 & 1 & 1 \end{pmatrix}$$

form the matrices $C = 2A - 3B$ and $D = 6B - A$. Evaluate the determinant of each matrix.

18–5. Given the three matrices

$$A = \frac{1}{2}\begin{pmatrix} 0 & 1 \\ 1 & 0 \end{pmatrix} \quad B = \frac{1}{2}\begin{pmatrix} 0 & -i \\ i & 0 \end{pmatrix} \quad C = \frac{1}{2}\begin{pmatrix} 1 & 0 \\ 0 & -1 \end{pmatrix}$$

show that $A^2 + B^2 + C^2 = \frac{3}{4}I$, where I is a unit matrix. Also show that $AB - BA = iC$, $BC - CB = iA$, and $CA - AC = iB$.

18–6. Given the matrices

$$A = \frac{1}{\sqrt{2}}\begin{pmatrix} 0 & 1 & 0 \\ 1 & 0 & 1 \\ 0 & 1 & 0 \end{pmatrix} \quad B = \frac{1}{\sqrt{2}}\begin{pmatrix} 0 & -i & 0 \\ i & 0 & -i \\ 0 & i & 0 \end{pmatrix} \quad C = \begin{pmatrix} 1 & 0 & 0 \\ 0 & 0 & 0 \\ 0 & 0 & -1 \end{pmatrix}$$

show that $AB - BA = iC$, $BC - CB = iA$, $CA - AC = iB$, and $A^2 + B^2 + C^2 = 2I$, where I is a unit matrix.

18–7. A three-dimensional rotation about the z axis can be represented by the matrix

$$R(\theta) = \begin{pmatrix} \cos\theta & -\sin\theta & 0 \\ \sin\theta & \cos\theta & 0 \\ 0 & 0 & 1 \end{pmatrix}$$

Show that det R $= |R| = 1$. Also show that

$$R^{-1}(\theta) = R(-\theta) = \begin{pmatrix} \cos\theta & \sin\theta & 0 \\ -\sin\theta & \cos\theta & 0 \\ 0 & 0 & 1 \end{pmatrix}$$

18–8. Show that the matrix R in Problem 18–7 is orthogonal.

18–9. Consider the matrices A and S:

$$A = \begin{pmatrix} 1 & 0 & 1 \\ 0 & 1 & 0 \\ 1 & 0 & 1 \end{pmatrix} \qquad S = \begin{pmatrix} \frac{1}{\sqrt{2}} & 0 & \frac{1}{\sqrt{2}} \\ 0 & 1 & 0 \\ \frac{1}{\sqrt{2}} & 0 & -\frac{1}{\sqrt{2}} \end{pmatrix}$$

First show that S is orthogonal. Then evaluate the matrix $D = S^{-1}AS = S^{T}AS$. What form does D have?

18–10. Show that $(AB)^{T} = B^{T}A^{T}$.

18–11. Show that $(AB)^{-1} = B^{-1}A^{-1}$.

18–12. Prove that the rows of an orthogonal matrix are orthonormal.

18–13. Prove that the product of two orthogonal matrices is orthogonal.

18–14. Starting with

$$A = \begin{vmatrix} a_{11} & a_{12} \\ a_{21} & a_{22} \end{vmatrix} \qquad \text{and} \qquad B = \begin{vmatrix} b_{11} & b_{12} \\ b_{21} & b_{22} \end{vmatrix}$$

show that $\det AB = \det BA = \det A \cdot \det B$. Although you have shown that this relation is true only for 2×2 determinants, it is true for $n \times n$ determinants. (The proof is a little long.)

18–15. Starting with

$$A = \begin{vmatrix} 1 & 2 & -1 \\ 0 & 1 & 2 \\ 1 & 3 & -1 \end{vmatrix} \qquad \text{and} \qquad B = \begin{vmatrix} 1 & 2 & -1 \\ 0 & 3 & 1 \\ 2 & 1 & 3 \end{vmatrix}$$

show that $\det AB = \det BA = \det A \cdot \det B$. (See the previous problem.)

18–16. Using the fact that $\det(AB) = \det(A)\det(B)$ (see the previous two problems), show that the determinant of an orthogonal matrix is equal to ± 1. You need to use the fact that $\det(A^{T}) = \det(A)$ (see Chapter 17).

18–17. Using the fact that $\det AB = \det A \cdot \det B$ (see Problem 18–14), show that $\det A^{-1} = 1/\det A$.

18–18. Show that $A = \dfrac{1}{3}\begin{pmatrix} 1 & 2 & 2 \\ 2 & 1 & -2 \\ -2 & 2 & -1 \end{pmatrix}$ is orthogonal.

18–19. Find the norm of the vector $v = (i, 3, -2i)$.

18–20. Show that $v = (i, 1, i)$ and $u = (-i, 2, -i)$ are orthogonal.

18–21. Show that $A = \begin{pmatrix} 1 & i & 1-i \\ -i & 0 & -1+i \\ 1+i & -1-i & 3 \end{pmatrix}$ is Hermitian.

18–22. Show that the determinant of a unitary matrix is of absolute value unity. Verify this result for the unitary matrices in Example 18–9 and Problems 18–23 and 18–25. *Hint*: You need the relation $\det(AB) = \det(A)\det(B)$ (Problem 18–14).

18–23. Show that $A = \dfrac{1}{6}\begin{pmatrix} 2-4i & 4i \\ -4i & -2-4i \end{pmatrix}$ is unitary.

18–24. Show that the rows and columns of the unitary matrix in the previous problem are orthonormal.

18–25. Show that $A = \dfrac{1}{\sqrt{3}} \begin{pmatrix} 1 & 1 & 1 & 0 \\ 1 & 0 & -1 & -i \\ 1 & -1 & 0 & i \\ 0 & -i & i & 1 \end{pmatrix}$ is unitary.

18–26. Consider the simultaneous algebraic equations

$$x_1 + x_2 = 3$$

$$4x_1 - 3x_2 = 5$$

Show that this pair of equations can be written in the matrix form

$$A x = c \tag{1}$$

where

$$x = \begin{pmatrix} x_1 \\ x_2 \end{pmatrix} \qquad c = \begin{pmatrix} 3 \\ 5 \end{pmatrix} \quad \text{and} \quad A = \begin{pmatrix} 1 & 1 \\ 4 & -3 \end{pmatrix}$$

Now multiply equation 1 from the left by A^{-1} to obtain

$$x = A^{-1} c \tag{2}$$

Now show that

$$A^{-1} = -\frac{1}{7} \begin{pmatrix} -3 & -1 \\ -4 & 1 \end{pmatrix}$$

and that

$$x = \begin{pmatrix} x_1 \\ x_2 \end{pmatrix} = -\frac{1}{7} \begin{pmatrix} -3 & -1 \\ -4 & 1 \end{pmatrix} \begin{pmatrix} 3 \\ 5 \end{pmatrix} = \begin{pmatrix} 2 \\ 1 \end{pmatrix}$$

or that $x_1 = 2$ and $x_2 = 1$. Do you see how this procedure generalizes to any number of simultaneous equations?

18–27. Solve the following simultaneous algebraic equations by the matrix inverse method developed in the previous problem:

$$x_1 + x_2 - x_3 = 1$$

$$2x_1 - 2x_2 + x_3 = 6$$

$$x_1 + 3x_3 = 0$$

First show that

$$A^{-1} = \frac{1}{13} \begin{pmatrix} 6 & 3 & 1 \\ 5 & -4 & 3 \\ -2 & -1 & 4 \end{pmatrix}$$

and evaluate $x = A^{-1} c$.

CHAPTER 19

Matrix Eigenvalue Problems

The Schrödinger equation

$$\hat{H}\psi = E\psi \tag{19.1}$$

is an eigenvalue problem; ψ is the eigenfunction and E is the corresponding eigenvalue. We've seen in Chapter 18 that operators can be represented by matrices, and so the matrix equation

$$\mathsf{A}\mathsf{c} = \lambda\,\mathsf{c} \tag{19.2}$$

which is analogous to Equation 19.1, is called a *matrix eigenvalue problem*, where c is an *eigenvector* of the matrix A and λ is the corresponding *eigenvalue*. Equations 19.1 and 19.2 suggest that there is a strong relationship between the Schrödinger equation and a matrix eigenvalue problem. Most research in computational quantum chemistry is formulated in terms of matrices and matrix eigenvalue problems.

To see explicitly the relation between the Schrödinger equation and a matrix eigenvalue problem, we expand the (unknown) eigenfunction ψ in Equation 19.1 in terms of some convenient set of functions ϕ_j:

$$\psi = \sum_{j=1}^{N} c_j \phi_j \tag{19.3}$$

The set of functions $\{\phi_j\}$ is called a *basis set* and the ϕ_j are called *basis functions*. As N gets larger and larger, we expect ψ given by Equation 19.3 to become a better and better approximation to the exact ψ, if we choose the ϕ_j judiciously. The unknown nature of ψ is now represented by the set of unknown coefficients $\{c_j\}$. We substitute Equation 19.3 into Equation 19.1, multiply by ϕ_i^*, and then integrate over all the coordinates to obtain the set of algebraic equations

$$H_{11}c_1 + H_{12}c_2 + \cdots + H_{1N}c_N = E(c_1 S_{11} + c_2 S_{12} + \cdots + c_N S_{1N})$$

$$H_{21}c_1 + H_{22}c_2 + \cdots + H_{2N}c_N = E(c_1 S_{21} + c_2 S_{22} + \cdots + c_N S_{1N})$$

$$\vdots \qquad \vdots \qquad \qquad \vdots \quad = \qquad \vdots \tag{19.4}$$

$$H_{N1}c_1 + H_{N2}c_2 + \cdots + H_{NN}c_N = E(c_1 S_{N1} + c_2 S_{N2} + \cdots + c_N S_{NN})$$

where the

$$H_{ij} = \int d\tau\, \phi_i^* \hat{H} \phi_j \tag{19.5a}$$

and the

$$S_{ij} = \int d\tau\, \phi_i^* \phi_j \tag{19.5b}$$

are called *matrix elements*. All the matrix elements are known quantities because the ϕ_i are known. The $d\tau$ in Equations 19.5 denotes the integration element, be it dx in one dimension, $dx\,dy$ in two dimensions, etc.

We can write Equation 19.4 as a matrix eigenvalue problem,

$$\mathsf{H}\mathsf{c} = E\,\mathsf{S}\mathsf{c} \tag{19.6}$$

This type of equation appears often in quantum chemistry. Equation 19.6 becomes the same as Equation 19.2 (with $\mathsf{A} = \mathsf{S}^{-1}\mathsf{H}$) if we multiply Equation 19.6 from the left by S^{-1}. Thus, we see that the Schrödinger equation can be expressed as a matrix eigenvalue problem.

19.1 The Eigenvalue Problem

Let's look at Equation 19.2 more closely. Equation 19.2 represents the system of homogeneous linear equations

$$(a_{11} - \lambda)c_1 + a_{12}c_2 + \cdots + a_{1N}c_N = 0$$

$$a_{21}c_1 + (a_{22} - \lambda)c_2 + \cdots + a_{2N}c_N = 0$$

$$\vdots \qquad \vdots \qquad \qquad \vdots \quad = \vdots \tag{19.7}$$

$$a_{N1}c_1 + a_{N2}c_2 + \cdots + (a_{NN} - \lambda)c_N = 0$$

As we have seen a number of times before, the determinant of the c_j's must be equal to zero in order to have a nontrivial solution; in other words, a solution where not all the $c_j = 0$. Thus, we write

$$\det(\mathsf{A} - \lambda\,\mathsf{I}) = 0 \tag{19.8}$$

which leads to an Nth-degree polynomial equation in (the unknown) λ. The polynomial equation given by Equation 19.8 is called the *characteristic equation* or the *secular equation* of A. The roots of this equation give us N eigenvalues of λ for which Equation 19.2 is satisfied. Associated with each eigenvalue is an eigenvector. We obtain each eigenvector by substituting one of the values of λ into

Equations 19.7 and then solving for the c_j's. The following Example illustrates this procedure.

EXAMPLE 19–1

Find the eigenvalues and eigenvectors of

$$A = \begin{pmatrix} a & 1 \\ 1 & a \end{pmatrix}$$

where a is a constant.

SOLUTION: The determinant of $A - \lambda I$ is given by

$$\det(A - \lambda I) = \begin{vmatrix} a - \lambda & 1 \\ 1 & a - \lambda \end{vmatrix} = (a - \lambda)^2 - 1 = 0$$

and so the eigenvalues are given by the solution to $(a - \lambda)^2 - 1 = 0$, or $\lambda = a \pm 1$. The equations for the eigenvectors are (see Equation 19.7)

$$(a - \lambda)c_1 + c_2 = 0$$
$$c_1 + (a - \lambda)c_2 = 0 \tag{1}$$

If we substitute $\lambda = a + 1$ into these equations, we obtain

$$-c_1 + c_2 = 0$$
$$c_1 - c_2 = 0$$

or $c_1 = c_2$. Thus, the eigenvector corresponding to $\lambda = a + 1$ is (c_1, c_1), where c_1 is an arbitrary constant. We can fix the value of c_1 by requiring that the eigenvector be normalized, in which case we have

$$c_1 = \begin{pmatrix} 1/\sqrt{2} \\ 1/\sqrt{2} \end{pmatrix}$$

The other eigenvector is found by substituting $\lambda = a - 1$ into equations 1. In this case, we obtain $c_1 = -c_2$, and so the normalized eigenvector corresponding to $\lambda = a - 1$ is

$$c_2 = \begin{pmatrix} 1/\sqrt{2} \\ -1/\sqrt{2} \end{pmatrix}$$

You can readily verify that $A c_1 = \lambda_1 c_1$ and that $A c_2 = \lambda_2 c_2$.

EXAMPLE 19–2

Find the eigenvalues and eigenvectors of the matrix

$$A = \begin{pmatrix} -1 & 2 \\ 2 & 2 \end{pmatrix}$$

SOLUTION: The characteristic equation is

$$\begin{vmatrix} -1 - \lambda & 2 \\ 2 & 2 - \lambda \end{vmatrix} = \lambda^2 - \lambda - 6 = 0$$

or $\lambda = 3$ and -2. Thus, this 2×2 eigenvalue problem yields two distinct eigenvalues.

If we substitute $\lambda = 3$ into Equations 19.7, we have

$$-4c_1 + 2c_2 = 0$$
$$2c_1 - c_2 = 0$$

These two equations are the same, and yield $c_2 = 2c_1$. Thus, the eigenvalue corresponding to $\lambda = 3$ is $c_1 = (a, 2a)$, where a is any nonzero constant. For $\lambda = -2$, we find

$$c_1 + 2c_2 = 0$$
$$2c_1 + 4c_2 = 0$$

or $c_2 = -c_1/2$. Therefore, the eigenvector corresponding to $\lambda = -2$ is $c_2 = (b, -b/2)$, where b is any nonzero constant. We can fix the values of a and b by requiring that the eigenvectors be normalized, in which case we have

$$c_1 = \frac{1}{\sqrt{5}} \begin{pmatrix} 1 \\ 2 \end{pmatrix} \quad \text{and} \quad c_2 = \frac{1}{\sqrt{5}} \begin{pmatrix} 2 \\ -1 \end{pmatrix}$$

It shouldn't be surprising that both eigenvectors are determined only to within an arbitrary factor because you can see from Equation 19.2 that if c is a solution, so is any multiple of c. Thus, Equation 19.2 can determine c only to within a multiplicative constant. We usually fix the multiplicative constant by normalizing the eigenvectors.

In Examples 19–1 and 19–2, we solved a 2×2 eigenvalue problem. The algebra was simple because we had to solve only a quadratic equation to find the two eigenvalues. The algebra increases drastically as we go on to problems of dimension greater than two; even a 3×3 system leads to a cubic equation for λ, which is usually quite tedious to solve, and the tedium grows rapidly with the size of the matrix. As we've mentioned a number of times, computer algebra systems, such as *MathCad*, *Maple*, and *Mathematica*, can solve for all the eigenvalues and corresponding eigenvectors of a sizable matrix in seconds.

Note that the two eigenvectors in Example 19–1 are orthogonal because

$$c_1 \cdot c_2 = \left[\frac{1}{\sqrt{2}} \frac{1}{\sqrt{2}} + \frac{1}{\sqrt{2}} \left(-\frac{1}{\sqrt{2}} \right) \right] = 0$$

The eigenvectors in Example 19–2 are also orthogonal. The reason that the eigenvectors of the matrices in Examples 19–1 and 19–2 are orthogonal is that they are eigenvectors corresponding to discrete eigenvalues of symmetric matrices; in other words, matrices such that $A = A^T$, or $a_{ij} = a_{ji}$. It turns out that the eigenvectors corresponding to discrete eigenvalues of Hermitian matrices will also be orthogonal. Recall from the previous chapter that a Hermitian matrix has the property $A = (A^*)^T = A^\dagger$, or $a_{ij} = a_{ji}^*$. It is the complex space analog of a symmetric matrix.

This result is completely analogous to the fact that the nondegenerate eigenfunctions of Hermitian operators are orthogonal. Recall from Section 11.3 that the definition of a Hermitian operator $\hat{A}$ is

$$\int d\tau\, \psi_i^* \hat{A}\psi_j = \int d\tau\, (\hat{A}\psi_i)^* \psi_j = \int d\tau\, \psi_j(\hat{A}\psi_i)^* \qquad (19.9)$$

If we let $a_{ij} = \int d\tau\, \psi_i^* \hat{A}\psi_j$, then Equation 19.9 says that

$$a_{ij} = a_{ji}^* \qquad \text{(Hermitian matrix)} \qquad (19.10)$$

and so we see that there is a close relation between a Hermitian operator and a Hermitian matrix. Because matrices are representations of operators, all matrices in quantum mechanics must be Hermitian because, as we shall show below, the eigenvalues of a Hermitian matrix are real, just as the eigenvalues of a Hermitian operator are real (see Section 11.3).

19.2 The Eigenvalues and Eigenvectors of Hermitian Matrices

Let's prove that the eigenvalues of a Hermitian matrix are real. Since a symmetric matrix is a special case of a Hermitian matrix if its elements happen to be real, the following proof is valid for symmetric matrices as well. Let A be a Hermitian matrix with eigenvalue λ and corresponding (nontrivial) eigenvector c. Then

$$A c = \lambda c \qquad \text{and} \qquad A^* c^* = \lambda^* c^*$$

Multiply the first equation from the left by $(c^*)^T = c^\dagger$ to get

$$c^\dagger A c = \lambda c^\dagger c \qquad (19.11)$$

Now transpose the second equation and use the relation $(AB)^T = B^T A^T$ to write

$$(A^* c^*)^T = (c^*)^T (A^*)^T = c^\dagger A^\dagger = \lambda^* c^\dagger$$

Multiply this result from the right by c to obtain

$$c^\dagger A^\dagger c = \lambda^* c^\dagger c \qquad (19.12)$$

Now subtract Equation 19.12 from 19.11 to get

$$c^\dagger (A - A^\dagger) c = c^\dagger (\lambda - \lambda^*) c = (\lambda - \lambda^*) c^\dagger c$$

But $A = A^\dagger$ because A is Hermitian, and $c^\dagger c > 0$ since $c \neq 0$, so $\lambda = \lambda^*$, or, in other words, λ is real. In quantum mechanics, measurable quantities correspond to eigenvalues of matrices, and so these matrices are required to be Hermitian because measurable quantities must be real.

The proof that we just presented that the eigenvalues of a Hermitian matrix are real can be slightly modified to prove that the eigenvectors corresponding to distinct eigenvalues of a Hermitian matrix are orthogonal. To prove this, let A be a Hermitian matrix with distinct eigenvalues λ_1 and λ_2 and corresponding eigenvectors c_1 and c_2, respectively, so that

$$A c_1 = \lambda_1 c_1 \qquad \text{and} \qquad A c_2 = \lambda_2 c_2 \tag{19.13}$$

Multiply the first of these equations from the left by $c_2^{*T} = c_2^\dagger$ to obtain

$$c_2^\dagger A c_1 = \lambda_1 c_2^\dagger c_1 \tag{19.14}$$

Now take the complex conjugate and then the transpose the second of Equations 19.13 to obtain

$$(A^* c_2^*)^T = c_2^{*T} A^{*T} = c_2^\dagger A^\dagger = (\lambda_2^* c_2^*)^T = \lambda_2 c_2^\dagger$$

where in the last step, we recognized that λ_2 is real. Now multiply this result from the right by c_1 to obtain

$$c_2^\dagger A^\dagger c_1 = \lambda_2 c_2^\dagger c_1 \tag{19.15}$$

Subtract Equation 19.15 from 19.14 to obtain

$$c_2^\dagger (A - A^\dagger) c_1 = (\lambda_1 - \lambda_2) c_2^\dagger c_1$$

The left-hand side here equals zero because A is Hermitian ($A = A^\dagger$), and so we have

$$(\lambda_1 - \lambda_2) c_2^\dagger c_1 = 0$$

Because $\lambda_1 \neq \lambda_2$, we have $c_2^\dagger c_1 = 0$, meaning that the eigenvectors of distinct eigenvalues are orthogonal.

EXAMPLE 19–3
Find the eigenvalues and eigenvectors of

$$A = \begin{pmatrix} 1 & 0 & 1 \\ 0 & 1 & 0 \\ 1 & 0 & 1 \end{pmatrix}$$

and show that the eigenvectors are mutually orthogonal.

SOLUTION: The characteristic equation of A is

$$\begin{vmatrix} 1-\lambda & 0 & 1 \\ 0 & 1-\lambda & 0 \\ 1 & 0 & 1-\lambda \end{vmatrix} = (1-\lambda)^3 - (1-\lambda) = (1-\lambda)(\lambda^2 - 2\lambda) = 0$$

or $\lambda = 0, 2$, and 1. The corresponding equations for the coefficients of the secular equation are

$$(c_1 - \lambda) + 0 + c_3 = 0$$
$$0 + (c_2 - \lambda) + 0 = 0$$
$$c_1 + 0 + (c_3 - \lambda) = 0$$

For $\lambda = 0$, we have

$$c_1 \quad\;\; + c_3 = 0$$
$$\qquad c_2 \qquad = 0$$
$$c_1 \quad\;\; + c_3 = 0$$

Therefore, $c_3 = -c_1$ and $c_2 = 0$, or $c_0 = (a, 0, -a)$, where a is an arbitrary constant. For $\lambda = 1$ and 2, we get $c_1 = (0, b, 0)$ and $c_2 = (c, 0, c)$, respectively, where b and c are arbitrary constants. To show that these three eigenvectors are orthogonal, we form the scalar product of the eigenvectors

$$c_0^{\mathsf{T}} c_1 = \mathbf{c}_0 \cdot \mathbf{c}_1 = (a \; 0 \; -a) \begin{pmatrix} 0 \\ b \\ 0 \end{pmatrix} = 0$$

$$c_0^{\mathsf{T}} c_2 = \mathbf{c}_0 \cdot \mathbf{c}_2 = (a \; 0 \; -a) \begin{pmatrix} c \\ 0 \\ c \end{pmatrix} = 0$$

and

$$c_1^{\mathsf{T}} c_2 = \mathbf{c}_1 \cdot \mathbf{c}_2 = (0 \; b \; 0) \begin{pmatrix} c \\ 0 \\ c \end{pmatrix} = 0$$

We have expressed these equations both in matrix notation $(c_i^{\mathsf{T}} c_j)$ and vector notation $(\mathbf{c}_i \cdot \mathbf{c}_j)$.

Even if the eigenvalues of a Hermitian matrix are not distinct, we can construct an orthogonal set of eigenvectors. Suppose that the eigenvalue λ_1 is k-fold degenerate. In this case, we have

$$A c_i = \lambda_1 c_i \qquad \text{for } i = 1, 2, \ldots, k \qquad (19.16)$$

We can use the Gram–Schmidt orthogonalization procedure that we introduced in Section 8.2 to construct k orthogonal eigenvectors from the k linearly independent eigenvectors in Equation 19.16. Consequently, we can say that an $n \times n$ Hermitian matrix has n mutually orthogonal eigenvectors, just as we can say that a Hermitian operator has n mutually orthogonal eigenfunctions.

19.3 Some Applied Eigenvalue Problems

In this section, we shall present a few examples of eigenvalue problems that arise for physical systems. A wide variety of applications of eigenvalue problems involve systems of simultaneous first-order linear differential equations with constant coefficients. For example, a sequence of radioactive decays from a parent isotope through subsequent generations A $\longrightarrow$ B $\longrightarrow$ C $\longrightarrow$ D, or the time dependence of the occupation of energy levels of the molecules in a lasing material, or a study of the stability of equilibrium systems, or calculations of currents in electrical networks lead to sets of differential equations of the form

$$\frac{dx_1}{dt} = a_{11}x_1 + a_{12}x_2 + \cdots + a_{1n}x_n$$

$$\frac{dx_2}{dt} = a_{21}x_1 + a_{22}x_2 + \cdots + a_{2n}x_n$$

$$\vdots$$

$$\frac{dx_n}{dt} = a_{n1}x_1 + a_{n2}x_2 + \cdots + a_{nn}x_n$$

where the a_{ij}s are constants. We can write this set of equations in matrix form:

$$\frac{d\mathsf{x}}{dt} = \dot{\mathsf{x}} = \mathsf{A}\mathsf{x} \tag{19.17}$$

where the overdot is standard notation for time derivatives.

Let's consider a 2×2 case first. The rate equations for the decay scheme A $\to$ B $\to$ C are

$$\dot{A} = -k_1 A \qquad \text{and} \qquad \dot{B} = k_1 A - k_2 B \tag{19.18}$$

We don't need to consider the equation for $\dot{C}$ because the law of conservation of mass says that $A + B + C = $ constant. To keep the notation simple, let's write Equations 19.18 as

$$\dot{x}_1 = -k_1 x_1 \qquad \text{and} \qquad \dot{x}_2 = k_1 x_1 - k_2 x_2 \tag{19.19}$$

In the one-dimensional case, $\dot{x} = ax$, the solution is simply $x(t) = x(0)e^{at}$. Therefore, let's try a solution to Equations 19.19 of the form $x_j(t) = u_j e^{\lambda t}$, where the u_j and λ are to be determined. If we substitute this into Equations 19.19, we obtain

$$u_1(\lambda + k_1) = 0 \qquad \text{and} \qquad k_1 u_1 - u_2(\lambda + k_2) = 0 \tag{19.20}$$

To have a nontrivial solution for u_1 and u_2, the determinant of the coefficients must equal zero, so that we have

$$\begin{vmatrix} \lambda + k_1 & 0 \\ k_1 & -(\lambda + k_2) \end{vmatrix} = -(\lambda + k_1)(\lambda + k_2) = 0$$

This is just the determinantal characteristic equation of the coefficient matrix in Equations 19.19. The eigenvalues are $\lambda = -k_1$ and $\lambda = -k_2$. To determine u_1

and u_2 associated with each value of λ, we set $\lambda = -k_1$ in Equations 19.20 to find that $0u_1 = 0$ and $k_1u_1 = u_2(k_2 - k_1)$. The first equation doesn't tell us anything, but the second says that $u_1 = (u_1, u_2) = (1, k_1/(k_2 - k_1))$, where we set $u_1 = 1$ arbitrarily at this point. Similarly, for $\lambda = -k_2$, we find that $u_1(k_1 - k_2) = 0$ and $k_1u_1 = 0$, which says that $u_1 = 0$ and u_2 is arbitrary. Thus, we find that $u_2 = (u_1, u_2) = (0, 1)$, where we set $u_2 = 1$ for now. Therefore,

$$x_1 = \begin{pmatrix} 1 \\ \dfrac{k_1}{k_2 - k_1} \end{pmatrix} e^{-k_1 t} \quad \text{and} \quad x_2 = \begin{pmatrix} 0 \\ 1 \end{pmatrix} e^{-k_2 t}$$

are both solutions to Equations 19.19. The general solution is a linear combination of $x_1(t)$ and $x_2(t)$, or

$$x(t) = \begin{pmatrix} x_1(t) \\ x_2(t) \end{pmatrix} = c_1 \begin{pmatrix} 1 \\ \dfrac{k_1}{k_2 - k_1} \end{pmatrix} e^{-k_1 t} + c_2 \begin{pmatrix} 0 \\ 1 \end{pmatrix} e^{-k_2 t} \tag{19.21}$$

We can determine the values of c_1 and c_2 by employing the initial values of $x_1(t)$ and $x_2(t)$. Suppose that $x_1(0) = A_0$ and $x_2(0) = 0$. Then Equation 19.21 reads

$$x_1(0) = c_1 = A_0$$

$$x_2(0) = \frac{c_1 k_1}{k_2 - k_1} + c_2 = 0$$

or $c_2 = -c_1 k_1/(k_2 - k_1)$. Thus, Equation 19.21 becomes

$$x(t) = A_0 \begin{pmatrix} 1 \\ \dfrac{k_1}{k_2 - k_1} \end{pmatrix} e^{-k_1 t} - \frac{A_0 k_1}{k_2 - k_1} \begin{pmatrix} 0 \\ 1 \end{pmatrix} e^{-k_2 t}$$

or, in terms of $x_1(t)$ and $x_2(t)$,

$$x_1(t) = A_0 e^{-k_1 t}$$

$$x_2(t) = \frac{A_0 k_1}{k_2 - k_1} e^{-k_1 t} - \frac{A_0 k_1}{k_2 - k_1} e^{-k_2 t}$$

in agreement with Equation 6.7. The advantage of the method that we are using here is that it is readily extended to n simultaneous equations.

In general, we substitute $x(t) = ue^{\lambda t}$ into Equation 19.17 to obtain

$$(A - \lambda I)u = 0 \tag{19.22}$$

Let the eigenvalues and the corresponding eigenvectors of A be $\lambda_1, \ldots, \lambda_n$ and $u_1, \ldots, u_n$, so that

$$Au_j = \lambda_j u_j \qquad j = 1, 2, \ldots, n \tag{19.23}$$

Substituting $x(t) = u_j e^{\lambda_j t}$ into Equation 19.17 shows that this is a solution to Equation 19.17. The general solution to Equation 19.17 can be written as a linear combination of the individual solutions $u_j e^{\lambda_j t}$:

$$x(t) = c_1 u_1 e^{\lambda_1 t} + c_2 u_2 e^{\lambda_2 t} + \cdots + c_n u_n e^{\lambda_n t} \tag{19.24}$$

To prove that this is indeed a solution to Equation 19.17, we substitute it into Equation 19.17 and show that the two sides are the same (Problem 19–10). Equation 19.24 expresses the solution to Equation 19.17 in terms of the eigenvalues and eigenvectors of A. The c_j in Equation 19.24 depend upon the initial conditions $x_1(0), x_2(0), \ldots, x_n(0)$.

EXAMPLE 19–4

The radioactive decay of a certain nucleus satisfies the equations

$$\frac{dA}{dt} = -k_1 A$$

$$\frac{dB}{dt} = k_1 A - k_2 B$$

$$\frac{dC}{dt} = k_2 B$$

Solve this set of equations for $k_1 = 2$ and $k_2 = 1$ with the initial conditions $A(0) = A_0$ and $B(0) = C(0) = 0$.

SOLUTION: We write these equations in matrix notation:

$$\frac{d\mathsf{x}}{dt} = \mathsf{M}\mathsf{x} = \begin{pmatrix} -2 & 0 & 0 \\ 2 & -1 & 0 \\ 0 & 1 & 0 \end{pmatrix} \mathsf{x}$$

where the components of x are (A, B, C). The eigenvalues and corresponding eigenvectors of M are $\lambda_1 = -2$, $\lambda_2 = -1$, $\lambda_3 = 0$, and $\mathsf{u}_1 = (1, -2, 1)$, $\mathsf{u}_2 = (0, -1, 1)$, $\mathsf{u}_3 = (0, 0, 1)$. Thus, the general solution is

$$\begin{pmatrix} A \\ B \\ C \end{pmatrix} = c_1 \begin{pmatrix} 1 \\ -2 \\ 1 \end{pmatrix} e^{-2t} + c_2 \begin{pmatrix} 0 \\ -1 \\ 1 \end{pmatrix} e^{-t} + c_3 \begin{pmatrix} 0 \\ 0 \\ 1 \end{pmatrix}$$

Using $A(0) = A_0$ and $B(0) = C(0) = 0$ gives $c_1 = A_0$, $c_2 = -2A_0$, and $c_3 = A_0$, so that

$$A(t) = A_0 e^{-2t}$$

$$B(t) = 2A_0(e^{-t} - e^{-2t})$$

$$C(t) = A_0(1 - 2e^{-t} + e^{-2t})$$

Figure 19.1 shows $A(t)$, $B(t)$, and $C(t)$ plotted against t. Can you account for the shape of the curve $B(t)$?

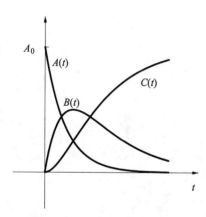

Figure 19.1. The solutions to the radioactive decay rate equations in Example 19–4.

Although we illustrated this approach with 2×2 and 3×3 systems, the size of the system is really irrelevant, particularly with the CAS that are currently available.

Vibrating or oscillating mechanical systems can be formulated as eigenvalue problems. Our first example consists of two particles of mass m connected by three identical springs of relaxed length l and constrained to move horizontally, as shown in Figure 19.2. We shall displace the masses horizontally from their equilibrium positions, then let them go, and investigate their subsequent motion. If x_1 and x_2 denote small displacements of the two particles from their equilibrium positions, then the potential energy of the system is given by

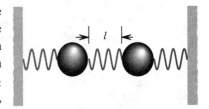

Figure 19.2. Two particles of mass m connected by three identical springs of relaxed length l and constrained to move longitudinally.

$$V = \frac{k}{2}x_1^2 + \frac{k}{2}(x_2 - x_1)^2 + \frac{k}{2}x_2^2 \tag{19.25}$$

where k is the force constant of each spring. Newton's equation for each particle is

$$m\frac{d^2x_1}{dt^2} = -\frac{\partial V}{\partial x_1} = k(x_2 - 2x_1)$$

$$m\frac{d^2x_2}{dt^2} = -\frac{\partial V}{\partial x_2} = k(x_1 - 2x_2) \tag{19.26}$$

Write Equations 19.26 in matrix notation:

$$m\frac{d^2\mathsf{x}}{dt^2} = \mathsf{A}\,\mathsf{x} \tag{19.27}$$

where

$$\mathsf{A} = \begin{pmatrix} -2k & k \\ k & -2k \end{pmatrix} \tag{19.28}$$

We know that this system oscillates in time, so we'll try a solution of the form $\mathsf{x}(t) = \mathsf{u}\,e^{i\omega t}$. Of course, $\mathsf{x}_1(t)$ and $\mathsf{x}_2(t)$ are real quantities, so we anticipate taking the real parts of $\mathsf{x}_1(t)$ and $\mathsf{x}_2(t)$ in the final result. Substituting $\mathsf{x}(t) = \mathsf{u}\,e^{i\omega t}$ into Equation 19.27 gives

$$-m\omega^2 \begin{pmatrix} u_1 \\ u_2 \end{pmatrix} = \begin{pmatrix} -2k & k \\ k & -2k \end{pmatrix} \begin{pmatrix} u_1 \\ u_2 \end{pmatrix}$$

Thus, ω^2 turns out to be given by

$$\begin{vmatrix} 2k - m\omega^2 & -k \\ -k & 2k - m\omega^2 \end{vmatrix} = 0 \tag{19.29}$$

and so we find two *characteristic frequencies*

$$\omega_1 = \left(\frac{k}{m}\right)^{1/2} \quad \text{and} \quad \omega_2 = \left(\frac{3k}{m}\right)^{1/2} \tag{19.30}$$

with corresponding eigenvectors

$$\mathsf{u}_1 = \begin{pmatrix} a \\ a \end{pmatrix} \quad \text{and} \quad \mathsf{u}_2 = \begin{pmatrix} a \\ -a \end{pmatrix} \tag{19.31}$$

where a is an arbitrary (possibly complex) constant. The motion of the two masses is given by

$$\mathsf{x}(t) = c_1 \begin{pmatrix} a \\ a \end{pmatrix} e^{i\omega_1 t} + c_2 \begin{pmatrix} a \\ -a \end{pmatrix} e^{i\omega_2 t} \tag{19.32}$$

or

$$x_1(t) = b_1 e^{i\omega_1 t} + b_2 e^{i\omega_2 t}$$
$$x_2(t) = b_1 e^{i\omega_1 t} - b_2 e^{i\omega_2 t}$$
(19.33)

where $b_1 = ac_1$ and $b_2 = ac_2$.

Equations 19.33 say that the masses will vibrate in phase if $b_2 = 0$ ($b_1 \neq 0$) and $180°$ out of phase if $b_1 = 0$ ($b_2 \neq 0$). To prove these last statements, write $b_1 = A_1 e^{i\phi_1}$ and $b_2 = 0$, so that the real parts of $x_1(t)$ and $x_2(t)$ are

$$x_1(t) = A_1 \cos(\omega_1 t + \phi_1) \quad \text{and} \quad x_2(t) = A_1 \cos(\omega_1 t + \phi_1) \quad (19.34)$$

Equation 19.34 shows that the two masses vibrate back and forth in unison. Similarly, write $b_2 = A_2 e^{i\phi_2}$ and $b_1 = 0$ and find that the real parts of $x_1(t)$ and $x_2(t)$ are

$$x_1(t) = A_2 \cos(\omega_2 t + \phi_2)$$
(19.35)
$$x_2(t) = -A_2 \cos(\omega_2 t + \phi_2) = A_2 \cos(\omega_2 t + \phi_2 + \pi)$$

Thus, in this case, the two masses vibrate in opposite directions.

Equations 19.34 and 19.35 are called the *normal modes* of vibration of the two-mass system. If the initial conditions are just so, the system will vibrate either in phase with frequency ω_1 or $180°$ out of phase with frequency ω_2 (Figure 19.3). Usually, however, the motion of the system will be described by a superposition of the two normal modes according to

$$x_1(t) = A_1 \cos(\omega_1 t + \phi_1) + A_2 \cos(\omega_2 t + \phi_2)$$

and

$$x_2(t) = A_1 \cos(\omega_1 t + \phi_1) - A_2 \cos(\omega_2 t + \phi_2)$$

The four constants, A_1, A_2, ϕ_1, and ϕ_2, can be determined from the values of $x_1(0)$, $x_2(0)$, $\dot{x}_1(0)$, and $\dot{x}_2(0)$.

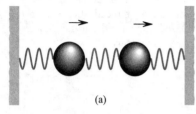

(a)

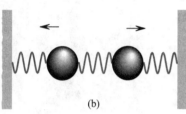

(b)

Figure 19.3. An illustration of the normal modes of the system shown in Figure 19.2. In part (a), the masses vibrate in phase, and in part (b), the masses vibrate out of phase.

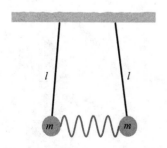

Figure 19.4. Two pendula of length l coupled by a harmonic spring and constrained to move in a single plane.

EXAMPLE 19–5

Figure 19.4 shows two pendula of length l constrained to move in a single plane coupled by a harmonic spring with force constant k. Problem 19–17 has you show that the equations of motion are

$$m \frac{d^2 s_1}{dt^2} = -\frac{\partial V}{\partial s_1} = -\frac{mg s_1}{l} + k(s_2 - s_1)$$

and

$$m \frac{d^2 s_2}{dt^2} = -\frac{\partial V}{\partial s_2} = -\frac{mg s_2}{l} - k(s_2 - s_1)$$

where s_1 and s_2 measure the distances of the masses along their arc of motion from their vertical positions. Solve for the characteristic frequencies and the normal modes of this system.

SOLUTION: The equations of motion in matrix form are

$$m\frac{d^2 \mathsf{s}}{dt^2} = \mathsf{A}\,\mathsf{s} = \begin{pmatrix} -\left(\dfrac{mg}{l}+k\right) & k \\ k & -\left(\dfrac{mg}{l}+k\right) \end{pmatrix} \mathsf{s}$$

Substitute $\mathsf{s} = \mathsf{u}e^{i\omega t}$ into these equations to get the eigenvalue equation

$$\begin{vmatrix} \dfrac{mg}{l}+k-m\omega^2 & -k \\ -k & \dfrac{mg}{l}+k-m\omega^2 \end{vmatrix} = 0$$

which gives

$$\omega_1 = (g/l)^{1/2} \quad \text{and} \quad \omega_2 = (g/l)^{1/2}(1+2kl/mg)^{1/2}$$

For $\omega_1^2 = g/l$, the eigenvector is given by

$$\begin{pmatrix} k & -k \\ -k & k \end{pmatrix}\begin{pmatrix} u_1 \\ u_2 \end{pmatrix} = 0$$

or $u_1 = u_2$. For $\omega_2^2 = g/l(1+2kl/mg)$, we get $u_1 = -u_2$. Thus, the two eigenvectors can be written as $\mathsf{u}_1 = (a, a)$ and $\mathsf{u}_2 = (b, -b)$, where a and b are arbitrary constants. The motion of the two masses is given by

$$\begin{pmatrix} s_1 \\ s_2 \end{pmatrix} = c_1 \begin{pmatrix} a \\ a \end{pmatrix} e^{i\omega_1 t} + c_2 \begin{pmatrix} b \\ -b \end{pmatrix} e^{i\omega_2 t}$$

or

$$s_1(t) = \alpha_1 e^{i\omega_1 t} + \alpha_2 e^{i\omega_2 t}$$
$$s_2(t) = \alpha_1 e^{i\omega_1 t} - \alpha_2 e^{i\omega_2 t}$$

where $\alpha_1 = c_1 a$ and $\alpha_2 = c_2 b$. Just as for the motion described by Equations 19.33, the motion corresponding to ω_1 is in phase and that corresponding to ω_2 is $180°$ out of phase (Figure 19.5).

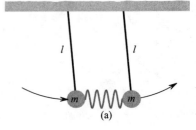

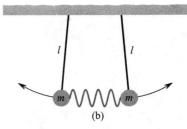

Figure 19.5. An illustration of the two normal modes of the coupled pendula shown in Figure 19.4. In part (a), the masses oscillate in phase, and in part (b), the masses oscillate out of phase.

19.4 Diagonalization of a Matrix

Let's go back to Equation 19.2, which we will write in the form

$$\mathsf{A}\,\mathsf{c}_k = \lambda_k\,\mathsf{c}_k \qquad k = 1,\,2,\,\ldots,\,N \tag{19.36}$$

There are N eigenvalues λ_k and N corresponding eigenvectors c_k. Now let's normalize the c_k and form a matrix

$$\mathsf{S} = (\mathsf{c}_1,\,\mathsf{c}_2,\,\ldots,\,\mathsf{c}_N) \tag{19.37}$$

where the notation means that the columns of S are the normalized eigenvectors of A. We call S the *modal matrix* of A. Because the columns of S consist of the normalized eigenvectors of A, and because these eigenvectors form an orthonormal set if A is symmetric (which it usually is), S is an orthogonal matrix. In other words, $S^{-1} = S^T$. Furthermore, the matrix S has a remarkable property that we can discover by operating on S with A to obtain (Problem 19–19)

$$AS = (A c_1, A c_2, \ldots, A c_N)$$
$$= (\lambda_1 c_1, \lambda_2 c_2, \ldots, \lambda_N c_N)$$
$$= SD \tag{19.38}$$

where

$$D = \begin{pmatrix} \lambda_1 & 0 & 0 & \cdots & 0 \\ 0 & \lambda_2 & 0 & \cdots & 0 \\ \vdots & \vdots & \vdots & \ddots & \vdots \\ 0 & 0 & 0 & \cdots & \lambda_N \end{pmatrix} \tag{19.39}$$

is a diagonal matrix whose elements are the eigenvalues of A.

If we multiply Equation 19.38 from the left by S^{-1}, then we obtain

$$D = S^{-1}AS = S^TAS \tag{19.40}$$

where we used the fact that S is orthogonal to write the last equality. Equation 19.40 is called a *similarity transformation*. We say that the matrix A has been *diagonalized* by the similarity transformation in Equation 19.40. Physically, A and D represent the same operation (such as a rotation or a reflection through a plane). Their different forms result from the fact that D is expressed in an optimum, or natural, coordinate system.

As an example, consider the potential energy given by Equation 19.25:

$$V(x_1, x_2) = kx_1^2 - kx_1x_2 + kx_2^2 \tag{19.41}$$

(We'll set $k = 1$ just for convenience, or consider $V(x_1, x_2)/k$, which amounts to the same thing.) If we define a vector $x = (x_1, x_2)$, then $V(x_1, x_2)$ in Equation 19.41 is given by

$$V = x^TVx = (x_1\ x_2) \begin{pmatrix} 1 & -\frac{1}{2} \\ -\frac{1}{2} & 1 \end{pmatrix} \begin{pmatrix} x_1 \\ x_2 \end{pmatrix} = x_1^2 - x_1x_2 + x_2^2$$

where

$$V = \begin{pmatrix} 1 & -\frac{1}{2} \\ -\frac{1}{2} & 1 \end{pmatrix} \tag{19.42}$$

We now wish to express x_1 and x_2 as a linear combination of new coordinates, η_1 and η_2, such that $V(\eta_1, \eta_2)$ has no cross terms in η_1 and η_2. If we let

$$x = S\eta \tag{19.43}$$

where S is the modal matrix of V, then we have

$$V = x^TVx = (S\eta)^TVS\eta = \eta^TS^TVS\eta = \eta^TD\eta \tag{19.44}$$

where D is the diagonal matrix $S^T V S$. Because D is a diagonal matrix, the last term in Equation 19.44 says that $V(\eta_1, \eta_2)$ consists of only squared terms in η_1 and η_2.

The eigenvalues and corresponding normalized eigenvectors of V are (Problem 19–21)

$$\lambda = \frac{1}{2}, \quad c_1 = \begin{pmatrix} \dfrac{1}{\sqrt{2}} \\ \dfrac{1}{\sqrt{2}} \end{pmatrix} \quad \text{and} \quad \lambda = \frac{3}{2}, \quad c_2 = \begin{pmatrix} \dfrac{1}{\sqrt{2}} \\ -\dfrac{1}{\sqrt{2}} \end{pmatrix} \quad (19.45)$$

and so the modal matrix is

$$S = \begin{pmatrix} \dfrac{1}{\sqrt{2}} & \dfrac{1}{\sqrt{2}} \\ \dfrac{1}{\sqrt{2}} & -\dfrac{1}{\sqrt{2}} \end{pmatrix}$$

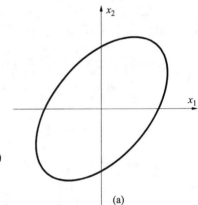

x_2

x_1

(a)

According to Equation 19.43, $x = S\eta$, or

$$\eta = S^{-1}x = S^T x = \begin{pmatrix} \dfrac{1}{\sqrt{2}} & \dfrac{1}{\sqrt{2}} \\ \dfrac{1}{\sqrt{2}} & -\dfrac{1}{\sqrt{2}} \end{pmatrix} \begin{pmatrix} x_1 \\ x_2 \end{pmatrix}$$

$$= \begin{pmatrix} \dfrac{1}{\sqrt{2}}(x_1 + x_2) \\ \dfrac{1}{\sqrt{2}}(x_1 - x_2) \end{pmatrix} = \begin{pmatrix} \eta_1 \\ \eta_2 \end{pmatrix} \quad (19.46)$$

where we used the fact that S is an orthogonal matrix ($S^{-1} = S^T$). Now, using Equation 19.44, we have

$$V(\eta_1, \eta_2) = \eta^T D \eta = \frac{1}{2}\eta_1^2 + \frac{3}{2}\eta_2^2$$

Therefore, $V(\eta_1, \eta_2)$ has no cross terms, and it turns out that the equations of motion expressed in these coordinates are uncoupled. These are the optimal coordinates to use to determine the motion of a particle under the potential V.

This result has a nice geometric interpretation. Figure 19.6a shows $V(x_1, x_2)$ plotted in an x_1, x_2 coordinate system and Figure 19.6b shows $V(\eta_1, \eta_2)$ plotted in an η_1, η_2 coordinate system. Notice that diagonalizing V corresponds to choosing a coordinate system where the ellipse lies along the coordinate axes rather than being tilted.

Diagonalizing a matrix A is *completely equivalent* to solving the eigenvalue problem in Equation 19.36, or, because Equations 19.1 and 19.2 are equivalent, diagonalizing the Hamiltonian matrix is completely equivalent to solving the Schrödinger equation. Because of the central importance of matrix diagonalization in quantum mechanics, there are many sophisticated and efficient algorithms for matrix diagonalization in the numerical analysis literature.

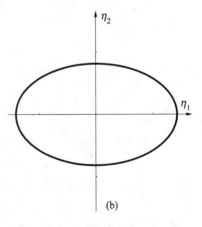

η_2

η_1

(b)

Figure 19.6. (a) The function $V(x_1, x_2) = x_1^2 - x_1 x_2 + x_2^2 = 1$ plotted in an x_1, x_2 coordinate system and (b) the function $V(\eta_1, \eta_2) = \frac{1}{2}\eta_1^2 + \frac{3}{2}\eta_2^2 = 1$ plotted in an η_1, η_2 coordinate system.

Problems

19–1. Determine the eigenvalues and eigenvectors of $A = \begin{pmatrix} 1 & 1 \\ 1 & 1 \end{pmatrix}$.

19–2. Determine the eigenvalues and eigenvectors of $A = \begin{pmatrix} 1 & -2 \\ -2 & 1 \end{pmatrix}$.

19–3. Determine the eigenvalues and eigenvectors of $A = \begin{pmatrix} 1 & 0 & 1 \\ 0 & 1 & 0 \\ 1 & 0 & 0 \end{pmatrix}$.

19–4. Determine the eigenvalues and eigenvectors of $A = \begin{pmatrix} 1 & 0 & -1 \\ 0 & 1 & 0 \\ -1 & 0 & 1 \end{pmatrix}$.

19–5. Show that the eigenvalues of A^T are the same as those of A.

19–6. Show that the eigenvalues of a unitary matrix are of absolute value 1.

19–7. Show that the eigenvalues of A^{-1} are λ_j^{-1}.

19–8. Show that $A^n x = \lambda^n x$ if $Ax = \lambda x$.

19–9. The *Cayley–Hamilton theorem* says that every square matrix satisfies its own characteristic equation. Verify this theorem for the matrices given in the first three problems.

19–10. Show that Equation 19.24 is a solution to Equation 19.17.

19–11. Solve the simultaneous equations

$$\dot{x}_1 = -x_1 + x_2$$
$$\dot{x}_2 = -3x_2$$

by trying a solution of the form $x_j(t) = u_j e^{\lambda t}$.

19–12. Solve the simultaneous equations

$$\dot{x}_1 = x_1 + x_2$$
$$\dot{x}_2 = 4x_1 + x_2$$

by trying a solution of the form $x_j(t) = u_j e^{\lambda t}$.

19–13. Solve the simultaneous equations $\dfrac{dx}{dt} = Ax = \begin{pmatrix} 2 & 5 \\ 1 & 6 \end{pmatrix} x$ if $x(0) = (3, -2)$.

19–14. Solve the simultaneous equations $\dfrac{dx}{dt} = Ax = \begin{pmatrix} 2 & -2 & 0 \\ 1 & -2 & -1 \\ -2 & 1 & -2 \end{pmatrix} x$ if $x(0) = (1, 1, 0)$.

19–15. Show that the two-mass vibrating system shown in Figure 19.2 will vibrate solely in one of its two normal modes only if the initial displacement is $x_1(0) = x_2(0)$ or $x_1(0) = -x_2(0)$.

19–16. Show that the potential energy in Equation 19.25 has no cross terms if it is expressed in terms of the two normal modes $y_1 = (x_1 + x_2)/\sqrt{2}$ and $y_2 = (x_1 - x_2)/\sqrt{2}$. Can you interpret this result?

19–17. Show that the equations of motion of the masses in Figure 19.4 are those given in Example 19–5.

19–18. Consider the three-mass, four-spring system shown in Figure 19.7. Show that the potential energy of this system is given by $V = 3x_1^2 + (x_2 - x_1)^2 + \frac{1}{2}(x_3 - x_2)^2 + \frac{1}{2}x_3^2$, and that Newton's equations for this system are

$$8\frac{d^2x_1}{dt^2} = -8x_1 + 2x_2$$

$$2\frac{d^2x_2}{dt^2} = -3x_2 + 2x_1 + x_3$$

$$2\frac{d^2x_3}{dt^2} = -2x_3 + x_2$$

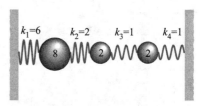

Figure 19.7. The three-mass, four-spring system that is analyzed in Problem 19–18. The system vibrates only longitudinally.

Determine the three fundamental frequencies associated with this system.

19–19. Verify that $(\lambda_1 c_1, \lambda_2 c_2, \ldots, \lambda_N c_N) = S D$ in Equation 19.38.

19–20. The three eigenvectors of A in Problem 19–4 are $c_1(1, 0, -1)$, $c_2(0, 1, 0)$, and $c_3(1, 0, 1)$, where c_1, c_2, and c_3 are arbitrary. Choose them so that the three eigenvectors are normalized. Now form the modal matrix S, whose columns consist of the three normalized eigenvectors. Find the inverse of S and then show explicitly that $S^{-1} = S^T$, or that S is indeed orthogonal.

19–21. Verify that the eigenvalues and normalized eigenvectors of V defined in Equation 19.42 are given by Equation 19.45.

19–22. Diagonalize the matrix in Problem 19–1.

19–23. Diagonalize the matrix in Problem 19–2.

19–24. Diagonalize the matrix in Problem 19–4.

19–25. The *trace* of a matrix A, Tr A, is the sum of the diagonal elements of A. First show that Tr AB = Tr BA. Now show that Tr D = Tr A, where $D = S^{-1}AS$.

19–26. Programs such as *MathCad*, *Maple*, and *Mathematica* can find the eigenvalues and corresponding eigenvectors of large matrices in seconds. Use one of these programs to find the eigenvalues and corresponding eigenvectors of

$$A = \begin{pmatrix} a & 1 & 0 & 0 & 0 & 1 \\ 1 & a & 1 & 0 & 0 & 0 \\ 0 & 1 & a & 1 & 0 & 0 \\ 0 & 0 & 1 & a & 1 & 0 \\ 0 & 0 & 0 & 1 & a & 1 \\ 1 & 0 & 0 & 0 & 1 & a \end{pmatrix}$$

This matrix occurs in the Hückel molecular orbital theory of benzene.

CHAPTER 20

Vector Spaces

You may have noticed a similarity between complex numbers and two-dimensional vectors. Each can be represented by a line from the origin of a coordinate system to some point, and each obeys the same parallelogram law of addition. You also may have noticed a similarity between vectors and orthogonal polynomials; we have sets of orthogonal unit vectors and sets of orthogonal polynomials, and we can expand one vector in terms of orthogonal unit vectors (such as **i**, **j**, and **k**), and we can expand functions in terms of orthogonal polynomials. Thus, we see an underlying similarity between complex numbers, vectors, and orthogonal polynomials. There is a mathematical formalism that treats all these quantities in a unified manner and allows us to exploit the similarities between them. This formalism is called a *vector space*—a set of quantities (called *vectors*) that obey a certain set of algebraic rules. Although we use the terms *vectors* and *vector spaces*, ordinary vectors are but one example of a great many quantities that form a vector space.

This chapter may seem to be the most abstract of all the chapters in the book, but the idea of a vector space plays a central role in quantum mechanics. In fact, at one time, instructors of physical chemistry recommended that students take a course in differential equations to be better prepared for quantum chemistry; but nowadays many instructors recommend a course in linear algebra, where vector spaces are taught.

20.1 The Axioms of a Vector Space

We first shall present the axioms that define a vector space and then give a number of examples. We define a *vector space* V to be a set of objects (which we call *vectors*) that satisfy the following requirements:

A1. Two vectors, $\mathbf{x}$ and $\mathbf{y}$, may be added to give another vector, $\mathbf{x} + \mathbf{y}$, which is also in V; in other words, if $\mathbf{x}$ and $\mathbf{y}$ are in V, so is $\mathbf{x} + \mathbf{y}$. (We say that the set is *closed* under addition.)

A2. Addition is commutative; that is, $\mathbf{x} + \mathbf{y} = \mathbf{y} + \mathbf{x}$.

A3. Addition is associative; that is, $(\mathbf{x} + \mathbf{y}) + \mathbf{z} = \mathbf{x} + (\mathbf{y} + \mathbf{z}) = \mathbf{x} + \mathbf{y} + \mathbf{z}$.

A4. There exists in V a unique *zero vector*, 0, such that $\mathbf{x} + 0 = 0 + \mathbf{x} = \mathbf{x}$ for any $\mathbf{x}$ in V.

A5. For every $\mathbf{x}$ in V, there is an additive inverse $-\mathbf{x}$ such that $\mathbf{x} + (-\mathbf{x}) = 0$.

M1. Any vector $\mathbf{x}$ may be multiplied by a scalar c, and the result, $c\,\mathbf{x}$, is in V. (We say that the set is closed under scalar multiplication.)

M2. Scalar multiplication is associative; that is, for any two numbers a and b, $a(b\,\mathbf{x}) = (ab)\mathbf{x}$.

M3. Scalar multiplication is distributive over addition; that is,
$a(\mathbf{x} + \mathbf{y}) = a\mathbf{x} + a\mathbf{y}$ and $(a + b)\mathbf{x} = a\,\mathbf{x} + b\,\mathbf{x}$.

M4. For the unit scalar, $1\mathbf{x} = \mathbf{x}$ for every $\mathbf{x}$ in V.

These nine properties are called the *axioms of a vector space*. If the scalars are real numbers, V is called a *real vector space*; if they are complex, V is called a *complex vector space*. You certainly don't need to memorize the above axioms. As you look them over, they should all be fairly reasonable.

Ordinary vectors certainly satisfy all the above properties of a vector space, and form what is called a *Euclidean vector space*. The elements of a vector space, however, need not be geometric vectors. The real numbers form a vector space, as do the complex numbers. So does the set of all $n \times n$ matrices (Problem 20–1). Another vector space consists of all ordered n-tuples of real numbers, where the sum of $(u_1, u_2, \ldots, u_n)$ and $(v_1, v_2, \ldots, v_n)$ is defined as $(u_1 + v_1, u_2 + v_2, \ldots, u_n + v_n)$ and the product of an n-tuple by a scalar is defined as $a(u_1, u_2, \ldots, u_n) = (au_1, au_2, \ldots, au_n)$ (Problem 20–3). We'll use ordered n-tuples fairly often to illustrate the properties of vector spaces, so we'll designate the space of all ordered n-tuples of real numbers by R^n and that of complex numbers by C^n.

Although a vector space is an abstract mathematical construct, and the "vectors" of a vector space can be any objects that satisfy the axioms, it is still sometimes helpful to visualize ordinary vectors as the "vectors" in order to ascribe physical meaning to some of the definitions and the results that we draw from them. The ordered n-tuples that we used above may be considered the components of an n-dimensional vector in some coordinate system.

Sets of functions can form a vector space. For example, the set of all polynomials of degree less than or equal to n with real or complex coefficients forms a vector space. The polynomials

$$P_n(x) = a_n x^n + a_{n-1} x^{n-1} + \cdots + a_1 x + a_0$$

satisfy all nine requirements of a vector space: the sum of two polynomials of degree less than or equal to n is a polynomial of degree less than or equal to n;

addition is commutative and associative; $P_n(x) = 0$ is the zero vector; $-P_n(x)$ is an additive inverse; $cP_n(x) = ca_nx^n + \cdots + ca_0$; scalar multiplication is associative and distributive over addition; and 1 is the unit multiplicative scalar. Notice that an nth-degree polynomial may be written as an ordered $n+1$–tuple $(a_n, a_{n-1}, \ldots, a_0)$.

The following Example shows that the solutions to a linear homogeneous differential equation also form a vector space.

EXAMPLE 20–1
Show that the set of functions whose first derivatives are continuous in the interval (a, b) and satisfy

$$\frac{df}{dx} + 2f(x) = 0$$

form a vector space.

SOLUTION: To show that the set is closed under addition (A1), let f and g be two elements of V (in other words, both f and g satisfy the above equation). Then

$$\frac{d}{dx}(f + g) + 2(f + g) = \frac{df}{dx} + \frac{dg}{dx} + 2f + 2g$$
$$= \left(\frac{df}{dx} + 2f\right) + \left(\frac{dg}{dx} + 2g\right) = 0 + 0 = 0$$

Addition is associative and distributive, and $f(x) = 0$ is the zero vector. To show that the set is closed under scalar multiplication (M1), consider

$$\frac{d}{dx}(af) + 2(af) = a\left(\frac{df}{dx} + 2f\right) = 0$$

Axioms M2 through M4 are also satisfied by any continuous function.

Example 20–1 suggests that the set of solutions to any linear homogeneous differential equation forms a vector space (Problem 20–6), which is indeed the case and is used extensively in the theory of differential equations.

20.2 Linear Independence

An important concept associated with vector spaces is the linear independence of vectors. We touched upon this idea in Section 6.2, but we shall study it more thoroughly here. Let $\{\mathbf{v}_j; \ j = 1, 2, \ldots, n\}$ be a set of nonzero vectors from a vector space V. (Remember, these are "vectors.") We say that the set of vectors is *linearly independent* if the only way that

$$\sum_{j=1}^{n} c_j\mathbf{v}_j = 0 \tag{20.1}$$

is for each and every $c_j = 0$. If the vectors are not linearly independent, then they are *linearly dependent*. Linear independence means that no vector can be expressed as a linear combination of the others; linear dependence means that at least one vector can be expressed as a linear combination of the others. For example, the three unit cartesian vectors, $\mathbf{i}$, $\mathbf{j}$, and $\mathbf{k}$, are linearly independent, but the three vectors $\mathbf{i}$, $\mathbf{j}$, and $2\mathbf{i} + 3\mathbf{j}$ are not because the third one is a linear combination of the first two. These three vectors lie in a plane, and intuitively you might expect that only two linearly independent vectors can lie in a plane. We can use determinants to find out whether a set of vectors is linearly independent. Let's test the three vectors $\mathbf{v}_1 = (1, 0, 0)$, $\mathbf{v}_2 = (1, -1, 1)$, and $\mathbf{v}_3 = (1, 2, -1)$ for linear independence. Is there a set of c_j not all zero such that

$$\sum_{j=1}^{3} c_j \mathbf{v}_j = (c_1, 0, 0) + (c_2, -c_2, c_2) + (c_3, 2c_3, -c_3) = (0, 0, 0)$$

or, in other words, is there a nontrivial solution to the equations

$$c_1 + c_2 + c_3 = 0$$
$$-c_2 + 2c_3 = 0$$
$$c_2 - c_3 = 0$$

The determinant of the matrix of coefficients

$$\begin{vmatrix} 1 & 1 & 1 \\ 0 & -1 & 2 \\ 0 & 1 & -1 \end{vmatrix} = -1$$

is nonzero, so the only solution is $c_1 = c_2 = c_3 = 0$, and so the three vectors are linearly independent.

EXAMPLE 20–2

Are the vectors $\mathbf{v}_1 = (1, 1, 0, 1)$, $\mathbf{v}_2 = (-1, -1, 0, -1)$, $\mathbf{v}_3 = (1, 0, 1, 1)$, and $\mathbf{v}_4 = (-1, 0, -1, -1)$ linearly independent?

SOLUTION: Equation 20.1 becomes

$$\sum_{j=1}^{4} c_j \mathbf{v}_j = (c_1, c_1, 0, c_1) + (-c_2, -c_2, 0, -c_2) + (c_3, 0, c_3, c_3)$$
$$+ (-c_4, 0, -c_4, -c_4) = 0$$

The determinant of the coefficients in these equations is

$$\begin{vmatrix} 1 & -1 & 1 & -1 \\ 1 & -1 & 0 & 0 \\ 0 & 0 & 1 & -1 \\ 1 & -1 & 1 & -1 \end{vmatrix} = 0$$

and so the vectors are not linearly independent. In fact, $\mathbf{v}_1 = -\mathbf{v}_2$ and $\mathbf{v}_3 = -\mathbf{v}_4$, and so only two of them are linearly independent.

If $\mathbf{v}_1, \mathbf{v}_2, \ldots, \mathbf{v}_n$ are vectors in a vector space V and if *any* other vector $\mathbf{u}$ in V can be expressed as a linear combination of $\mathbf{v}_1, \mathbf{v}_2, \ldots, \mathbf{v}_n$ so that

$$\mathbf{u} = c_1 \mathbf{v}_1 + c_2 \mathbf{v}_2 + \cdots + c_n \mathbf{v}_n$$

where the c_j are constants, then we say that $\mathbf{v}_1, \mathbf{v}_2, \ldots, \mathbf{v}_n$ *span* V. For example, the three unit vectors $\mathbf{i}$, $\mathbf{j}$, and $\mathbf{k}$ span the three-dimensional space R^3, as do any three non-coplanar (linearly independent) vectors.

Actually, three non-coplanar vectors plus any set of n other vectors also span R^3. You may think (correctly) that these other n vectors are not necessary to span R^3, but nevertheless the $3 + n$ vectors do span R^3 according to the definition. If the vectors $\mathbf{v}_1, \mathbf{v}_2, \ldots, \mathbf{v}_n$ in a vector space V are linearly independent and span V, then the set $\mathbf{v}_1, \mathbf{v}_2, \ldots, \mathbf{v}_n$ is called a *basis* or *basis set* for V. A basis set consists of the fewest number of vectors that are needed to span V. The unit vectors $\mathbf{i}, \mathbf{j}$, and $\mathbf{k}$, or any three non-coplanar vectors, form a basis in R^3. The number of vectors in a basis is defined as the *dimension* of the vector space. The dimension of a vector space V is equal to the maximum number of linearly independent vectors in V or the number of vectors in a basis set. When we said above that three non-coplanar vectors plus n others span R^3, but that the n other vectors are not really necessary, we were anticipating intuitively that it requires only three linearly independent vectors to span R^3 because it is a three-dimensional vector space.

If $\{\mathbf{v}_j; \ j = 1, 2, \ldots, n\}$ is a basis of V, then any vector $\mathbf{u}$ of V can be written as a linear combination of the $\mathbf{v}_j$:

$$\mathbf{u} = \sum_{j=1}^{n} u_j \mathbf{v}_j$$

We say that u_j is the jth coordinate of $\mathbf{u}$ with respect to the given basis set. Once again, remember that $\mathbf{u}$ and the $\mathbf{v}_j$ are abstract vectors, although we are using the language of ordinary vectors when we call u_j a component of $\mathbf{u}$.

EXAMPLE 20–3
Show that the vectors $(1, 0, 0)$, $(0, 1, 0)$, and $(0, 0, 1)$ form a basis in R^3. What familiar vectors do they correspond to?

SOLUTION: They form a basis in R^3 because any vector $\mathbf{u} = (x, y, z)$ in R^3 can be written as

$$\mathbf{u} = x(1, 0, 0) + y(0, 1, 0) + z(0, 0, 1)$$

These vectors correspond to the unit vectors **i**, **j**, and **k**. The components of **u** in this basis set are x, y, and z.

EXAMPLE 20–4

Show that the vector space V of ordered n-tuples is an n-dimensional space.

SOLUTION: The n vectors $(1, 0, \ldots, 0)$, $(0, 1, \ldots, 0)$, $\ldots$, $(0, 0, \ldots, 1)$ constitute a linearly independent set of vectors in V because they are mutually orthogonal. Furthermore, they span V because any other vector in V can be written as a linear combination of these vectors according to

$$\mathbf{u} = (u_1, u_2, \ldots, u_n) = u_1(1, 0, \ldots, 0) + u_2(0, 1, \ldots, 0) + \cdots + u_n(0, 0, \ldots, 1)$$

Thus, the n vectors constitute a basis and the dimension of V is n. The components of **u** in this basis set are the u_j's.

20.3 Inner Product Spaces

The idea of a vector space generalizes the spaces of two- and three-dimensional geometric vectors that we discussed in Chapter 13. In those spaces we used a dot product to define lengths of vectors and the angle between two vectors. These concepts are so useful that it is desirable to introduce them into our general vector spaces.

A vector space is called an *inner product space* if, in addition to the nine requirements that we listed in the first section, there is a rule that associates with any two vectors **u** and **v** in V a real number, written as $\langle \mathbf{u}, \mathbf{v} \rangle$ (some authors use $(\mathbf{u}, \mathbf{v})$), that satisfies the following three properties:

1. $\langle a\,\mathbf{u_1} + b\,\mathbf{u_2}, \mathbf{u_3} \rangle = a\langle \mathbf{u_1}, \mathbf{u_3} \rangle + b\langle \mathbf{u_2}, \mathbf{u_3} \rangle$ (20.2)
   where a and b are scalars. (The inner product is a linear operation.)

2. $\langle \mathbf{u}, \mathbf{v} \rangle = \langle \mathbf{v}, \mathbf{u} \rangle$ (20.3)
   (The inner product is commutative.)

3. $\langle \mathbf{u}, \mathbf{u} \rangle \geq 0$, with $\langle \mathbf{u}, \mathbf{u} \rangle = 0$ if and only if $\mathbf{u} = 0$ (20.4)
   (This property is known as *positive definiteness*.)

Problem 20–13 has you prove that the dot product that we defined in Chapter 13 for geometric vectors is an inner product, so that the Euclidean space of two- or three-dimensional geometric vectors with a defined dot product is an inner product space.

EXAMPLE 20–5

Let $\mathbf{u} = (u_1, u_2, \ldots, u_n)$ and $\mathbf{v} = (v_1, v_2, \ldots, v_n)$. Show that the product defined by

$$\langle \mathbf{u}, \mathbf{v} \rangle = u_1 v_1 + \cdots + u_n v_n$$

in the vector space R^n is an inner product.

SOLUTION: We shall verify each of the above three properties in turn:

1. $\langle a\,\mathbf{u} + b\,\mathbf{v},\ \mathbf{w} \rangle = (au_1 + bv_1)\,w_1 + (au_2 + bv_2)\,w_2 +$
$$\cdots + (au_n + bv_n)\,w_n$$
$$= au_1 w_1 + au_2 w_2 + \cdots + au_n w_n + bv_1 w_1 +$$
$$\cdots + bv_n w_n$$
$$= a\langle \mathbf{u}, \mathbf{w} \rangle + b\langle \mathbf{v}, \mathbf{w} \rangle,$$

2. $\langle \mathbf{u}, \mathbf{v} \rangle = u_1 v_1 + \cdots + u_n v_n = v_1 u_1 + \cdots + v_n u_n = \langle \mathbf{v}, \mathbf{u} \rangle$, and

3. $\langle \mathbf{u}, \mathbf{u} \rangle = u_1^2 + \cdots + u_n^2 > 0$ unless $u_1 = u_2 = \cdots = u_n = 0$.

EXAMPLE 20–6

Let V be the vector space of real-valued functions that are integrable over the interval (α, β). Show that

$$\langle f, g \rangle = \int_\alpha^\beta f(x)g(x)\,dx$$

is an inner product.

SOLUTION:

1. $\langle af_1 + bf_2, g \rangle = \int_\alpha^\beta [\,af_1(x) + bf_2(x)\,]\,g(x)\,dx$
$$= a\,\langle f_1, g \rangle + b\,\langle f_2, g \rangle$$

2. $\langle f, g \rangle = \int_\alpha^\beta f(x)g(x)\,dx = \int_\alpha^\beta g(x)f(x)\,dx = \langle g, f \rangle$

3. $\langle f, f \rangle = \int_\alpha^\beta f^2(x)\,dx \geq 0$ and is equal to zero if and only if $f(x) = 0$ throughout (α, β).

Motivated by geometric vectors, we define the length of a vector in V by

$$\|\mathbf{u}\| = \langle \mathbf{u}, \mathbf{u} \rangle^{1/2} \tag{20.5}$$

We also call $\|\mathbf{u}\|$ the *norm* of $\mathbf{u}$. For the case of R^n, the norm is given by

$$\|\mathbf{u}\| = \langle \mathbf{u}, \mathbf{u} \rangle^{1/2} = \left(u_1^2 + u_2^2 + \cdots + u_n^2\right)^{1/2} \tag{20.6}$$

This is the Pythagorean theorem in n dimensions.

The inner product satisfies an important inequality called the *Schwarz inequality* (Problem 20–19; see also Problem 13–23):

$$|\langle \mathbf{u}, \mathbf{v} \rangle| \leq \|\mathbf{u}\| \, \|\mathbf{v}\| \tag{20.7}$$

Because $-1 \leq \langle \mathbf{u}, \mathbf{v} \rangle / \|\mathbf{u}\| \, \|\mathbf{v}\| \leq +1$, we can define the angle between $\mathbf{u}$ and $\mathbf{v}$ by $\cos\theta = \langle \mathbf{u}, \mathbf{v} \rangle / \|\mathbf{u}\| \, \|\mathbf{v}\|$, where $0 \leq \theta \leq \pi$.

The norm satisfies the following properties:

$$\|\mathbf{v}\| \geq 0 \quad \text{with } \|\mathbf{v}\| = 0 \text{ if and only if } \mathbf{v} = 0 \tag{20.8}$$

$$\|c\mathbf{v}\| = |c| \, \|\mathbf{v}\| \tag{20.9}$$

$$\|\mathbf{u} + \mathbf{v}\| \leq \|\mathbf{u}\| + \|\mathbf{v}\| \tag{20.10}$$

Can you see why Equation 20.10 is called the triangle inequality?

EXAMPLE 20–7

Verify the Schwarz inequality, Equation 20.7, for $\mathbf{u} = (2, 1, -1, 2, 4)$ and $\mathbf{v} = (-1, 0, 1, -2, -3)$.

SOLUTION: First we find

$$\langle \mathbf{u}, \mathbf{v} \rangle = -2 + 0 - 1 - 4 - 12 = -19$$

$$\langle \mathbf{u}, \mathbf{u} \rangle = 4 + 1 + 1 + 4 + 16 = 26$$

$$\langle \mathbf{v}, \mathbf{v} \rangle = 1 + 0 + 1 + 4 + 9 = 15$$

$$|\langle \mathbf{u}, \mathbf{v} \rangle| = 19$$

and so $\|\mathbf{u}\| = \sqrt{\langle \mathbf{u}, \mathbf{u} \rangle} = 26^{1/2}$ and $\|\mathbf{v}\| = \sqrt{\langle \mathbf{v}, \mathbf{v} \rangle} = 15^{1/2}$. Equation 20.7 reads $19 < (26 \cdot 15)^{1/2} = 19.7$ in this case.

If $\langle \mathbf{u}, \mathbf{v} \rangle = 0$ and neither $\mathbf{u}$ nor $\mathbf{v} = 0$, then $\mathbf{u}$ and $\mathbf{v}$ are said to be *orthogonal*. Using the definition of the inner product of two functions as given in Example 20–6, we say the two functions, $f(x)$ and $g(x)$, are orthogonal over (a, b) if

$$\langle f, g \rangle = \int_a^b f(x)g(x)\,dx = 0$$

a result that we have used often in previous chapters.

For a set of vectors $\mathbf{v}_j$, $j = 1, 2, \ldots, n$, we can express orthogonality by writing $\langle \mathbf{v}_i, \mathbf{v}_j \rangle = \|\mathbf{v}_j\|^2 \delta_{ij}$, where δ_{ij} is the Kronecker delta. If the lengths of all the vectors are made to be unity by dividing each one by its norm $\|\mathbf{v}_j\|$, then the new set is *orthonormal* and we have $\langle \mathbf{v}_i, \mathbf{v}_j \rangle = \delta_{ij}$.

It's easy to show that an orthonormal set of vectors is linearly independent. Let $\mathbf{v}_1, \mathbf{v}_2, \ldots, \mathbf{v}_n$ be an orthonormal set and form the linear combination

$$c_1 \mathbf{v}_1 + c_2 \mathbf{v}_2 + \cdots + c_n \mathbf{v}_n = 0 \tag{20.11}$$

272

where the c_j's are to be determined. Now form the inner product of Equation 20.11 with each of the vectors $\mathbf{v}_1, \mathbf{v}_2, \ldots, \mathbf{v}_n$ in turn, and find that $c_j = 0$ for $j = 1, 2, \ldots, n$. Thus, the set of vectors is linearly independent.

We shall now show that every n-dimensional vector space V has an orthonormal basis by actually constructing one. Let $\mathbf{v}_j$ for $j = 1, 2, \ldots, n$ be any set of (nonzero) linearly independent vectors in V. Start with $\mathbf{v}_1$ and call it $\mathbf{u}_1$. Now take the second vector in the new set to be a linear combination of $\mathbf{v}_2$ and $\mathbf{u}_1$:

$$\mathbf{u}_2 = \mathbf{v}_2 + a_1 \mathbf{u}_1$$

such that $\langle \mathbf{u}_1, \mathbf{u}_2 \rangle = 0$. This condition gives $0 = \langle \mathbf{u}_1, \mathbf{v}_2 \rangle + a_1 \langle \mathbf{u}_1, \mathbf{u}_1 \rangle$ and so

$$\mathbf{u}_2 = \mathbf{v}_2 - \frac{\langle \mathbf{u}_1, \mathbf{v}_2 \rangle}{\langle \mathbf{u}_1, \mathbf{u}_1 \rangle} \mathbf{u}_1 \tag{20.12}$$

Now take the third member of the orthogonal set to be

$$\mathbf{u}_3 = \mathbf{v}_3 + b_1 \mathbf{u}_1 + b_2 \mathbf{u}_2$$

and require that $\langle \mathbf{u}_3, \mathbf{u}_1 \rangle = 0$ and $\langle \mathbf{u}_3, \mathbf{u}_2 \rangle = 0$. This gives (Problem 20–17)

$$\mathbf{u}_3 = \mathbf{v}_3 - \frac{\langle \mathbf{v}_3, \mathbf{u}_2 \rangle}{\langle \mathbf{u}_2, \mathbf{u}_2 \rangle} \mathbf{u}_2 - \frac{\langle \mathbf{v}_3, \mathbf{u}_1 \rangle}{\langle \mathbf{u}_1, \mathbf{u}_1 \rangle} \mathbf{u}_1 \tag{20.13}$$

and so on. This procedure is completely analogous to the one we used in Section 8.2 to derive sets of orthonormal polynomials. In fact, once you realize that orthogonal polynomials can be viewed as vectors in a vector space, then the two procedures are identical. The above procedure simply describes the Gram–Schmidt orthogonalization procedure for a vector space when the inner product is defined through Equations 20.2 through 20.4.

EXAMPLE 20–8
Construct an orthonormal basis from the three vectors, $\mathbf{v}_1 = (1, 0, 0)$, $\mathbf{v}_2 = (1, 1, 0)$, and $\mathbf{v}_3 = (1, 1, 1)$.

SOLUTION: Start with $\mathbf{v}_1 = (1, 0, 0)$, which is normalized, and call it $\mathbf{u}_1$. Now, $\langle \mathbf{u}_1, \mathbf{v}_2 \rangle = \mathbf{u}_1 \cdot \mathbf{v}_2 = 1$ and $\langle \mathbf{u}_1, \mathbf{u}_1 \rangle = \mathbf{u}_1 \cdot \mathbf{u}_1 = 1$, and so Equation 20.12 gives

$$\mathbf{u}_2 = \mathbf{v}_2 - \mathbf{u}_1 = (0, 1, 0)$$

which is normalized. Finally, Equation 20.13 with $\langle \mathbf{v}_3, \mathbf{u}_1 \rangle = \mathbf{v}_3 \cdot \mathbf{u}_1 = 1$ and $\langle \mathbf{v}_3, \mathbf{u}_2 \rangle = 1$ gives

$$\mathbf{u}_3 = \mathbf{v}_3 - \mathbf{u}_2 - \mathbf{u}_1 = (0, 0, 1)$$

Notice that $\mathbf{u}_1$, $\mathbf{u}_2$, and $\mathbf{u}_3$ are the three cartesian unit vectors, $\mathbf{i}$, $\mathbf{j}$, and $\mathbf{k}$.

20.4 Complex Inner Product Spaces

In our discussion of vector spaces so far, we have tacitly assumed that the scalars and vectors are real-valued quantities. It turns out that quantum mechanics can be formulated in terms of complex vector spaces with complex inner products, so in this section we shall extend the results of the previous sections to include complex numbers.

The definition of linear independence is not altered if we are dealing with a complex vector space instead of a real one. Nowhere in the previous sections did we specify that the vectors or the set of constants in the definition of linear independence had to be real. Let's determine whether the vectors $\mathbf{v}_1 = (1, i, -1)$, $\mathbf{v}_2 = (1 + i, 0, 1 - i)$, and $\mathbf{v}_3 = (i, -1, -i)$ are linearly independent. We look to see if there is a nontrivial solution to the equations

$$\sum_{j=1}^{3} c_j \mathbf{v}_j = 0 = (c_1, ic_1, -c_1) + ((1 + i)c_2, 0, (1 - i)c_2) + (ic_3, -c_3, -ic_3)$$

or to the equations

$$
\begin{aligned}
c_1 + (1 + i)c_2 + ic_3 &= 0 \\
ic_1 - c_3 &= 0 \\
-c_1 + (1 - i)c_2 - ic_3 &= 0
\end{aligned}
$$

The determinant of the coefficients is

$$
\begin{vmatrix}
1 & 1 + i & i \\
i & 0 & -1 \\
-1 & 1 - i & -i
\end{vmatrix} = 0
$$

and so we see that the vectors are not linearly independent. (In fact, $\mathbf{v}_3 = i\,\mathbf{v}_1$.)

The primary difference between a real and a complex inner product space is in our definition of an inner product. If we allow the scalars and vectors to be complex, Equations 20.2 through 20.4 become

1. $\langle \mathbf{u}|a\,\mathbf{v}_1 + b\,\mathbf{v}_2 \rangle = a\langle \mathbf{u}|\mathbf{v}_1 \rangle + b\langle \mathbf{u}|\mathbf{v}_2 \rangle$ (20.14)

2. $\langle \mathbf{u}|\mathbf{v} \rangle = \langle \mathbf{v}|\mathbf{u} \rangle^*$ (20.15)

3. $\langle \mathbf{u}|\mathbf{u} \rangle \geq 0$, with $\langle \mathbf{u}|\mathbf{u} \rangle = 0$ if and only if $\mathbf{u} = 0$ (20.16)

We are using a vertical line rather than a comma to separate the two vectors enclosed by the angular brackets in Equations 20.14 through 20.16 to distinguish between real and complex product spaces. (This is standard notation in quantum mechanics.)

If we take the complex conjugate of Equation 20.14,

$$\langle \mathbf{u} \,|\, a\,\mathbf{v}_1 + b\,\mathbf{v}_2 \rangle^* = a^*\langle \mathbf{u} \,|\, \mathbf{v}_1 \rangle^* + b^*\langle \mathbf{u} \,|\, \mathbf{v}_2 \rangle^*$$

and then use Equation 20.15, we obtain

$$\langle a\,\mathbf{v}_1 + b\,\mathbf{v}_2 | \mathbf{u} \rangle = a^*\langle \mathbf{v}_1|\mathbf{u} \rangle + b^*\langle \mathbf{v}_2|\mathbf{u} \rangle \qquad (20.17)$$

In particular, if we let $b = 0$ in Equation 20.14, then we have

$$\langle \mathbf{u}|a\,\mathbf{v}\rangle = a\langle \mathbf{u}|\mathbf{v}\rangle \tag{20.18}$$

and if we let $b = 0$ in Equation 20.17, then we have

$$\langle a\,\mathbf{v}|\mathbf{u}\rangle = a^*\langle \mathbf{v}|\mathbf{u}\rangle \tag{20.19}$$

These two equations say that a scalar comes out of the inner product "as is" from the second position, but as its complex conjugate from the first position.

If $\mathbf{u}$ and $\mathbf{v}$ are ordered n-tuples of complex numbers, then we can define $\langle \mathbf{u}|\mathbf{v}\rangle$ by

$$\langle \mathbf{u}|\mathbf{v}\rangle = u_1^* v_1 + u_2^* v_2 + \cdots + u_n^* v_n \tag{20.20}$$

and the *norm* of $\mathbf{u}$, $\|\mathbf{u}\|$, by

$$\|\mathbf{u}\| = \langle \mathbf{u}|\mathbf{u}\rangle = u_1^* u_1 + u_2^* u_2 + \cdots + u_n^* u_n \geq 0 \tag{20.21}$$

Equation 20.20 is sometimes called a *Hermitian inner product*. Problem 20–26 has you show that this definition satisfies Equations 20.14 through 20.16.

EXAMPLE 20–9
Given $\mathbf{u} = (1 + i, 3, 4 - i)$ and $\mathbf{v} = (3 - 4i, 1 + i, 2i)$, find $\langle \mathbf{u}|\mathbf{v}\rangle$, $\langle \mathbf{v}|\mathbf{u}\rangle$, $\|\mathbf{u}\|$, and $\|\mathbf{v}\|$.

SOLUTION:

$$\langle \mathbf{u}|\mathbf{v}\rangle = (1 - i)(3 - 4i) + 3(1 + i) + (4 + i)(2i) = 4i$$
$$\langle \mathbf{v}|\mathbf{u}\rangle = (3 + 4i)(1 + i) + 3(1 - i) - 2i(4 - i) = -4i = \langle \mathbf{u}|\mathbf{v}\rangle^*$$
$$\|\mathbf{u}\| = [(1 - i)(1 + i) + 9 + (4 + i)(4 - i)]^{1/2} = \sqrt{28}$$
$$\|\mathbf{v}\| = [(3 + 4i)(3 - 4i) + (1 - i)(1 + i) + (-2i)(2i)]^{1/2} = \sqrt{31}$$

If a set of vectors $\mathbf{v}_j$ for $j = 1, 2, \ldots, n$ satisfies the condition $\langle \mathbf{v}_i|\mathbf{v}_j\rangle = \delta_{ij}$, we say that the set is orthonormal.

EXAMPLE 20–10
Show that the three 3-tuples $\mathbf{u}_1 = (1, i, 1 + i)$, $\mathbf{u}_2 = (0, 1 - i, i)$, and $\mathbf{u}_3 = (3i - 3, 1 + i, 2)$ form an orthogonal set. How would you make them orthonormal?

SOLUTION:
$$\langle \mathbf{u}_1|\mathbf{u}_2\rangle = 0 - i(1 - i) + (1 - i)i = 0$$
$$\langle \mathbf{u}_1|\mathbf{u}_3\rangle = (3i - 3) - i(1 + i) + 2(1 - i) = 0$$
$$\langle \mathbf{u}_2|\mathbf{u}_3\rangle = 0 + (1 + i)(1 + i) - 2i = 0$$

To make them orthonormal, divide each one by its norm:

$$\|\mathbf{u}_1\| = \langle \mathbf{u}_1|\mathbf{u}_1 \rangle^{1/2} = [(1)(1) - i(i) + (1-i)(1+i)]^{1/2} = 2$$

$$\|\mathbf{u}_2\| = \langle \mathbf{u}_2|\mathbf{u}_2 \rangle^{1/2} = [(1+i)(1-i) + (-i)(i)]^{1/2} = 3^{1/2}$$

$$\|\mathbf{u}_3\| = \langle \mathbf{u}_3|\mathbf{u}_3 \rangle^{1/2} = [(-3i-3)(3i-3) + (1-i)(1+i) + 4]^{1/2} = (24)^{1/2}$$

The Schwarz inequality takes on the same form for a complex vector space:

$$|\langle \mathbf{u}|\mathbf{v} \rangle| \leq \|\mathbf{u}\|\,\|\mathbf{v}\| \tag{20.22}$$

and its proof parallels the one for a real vector space. Note that the vectors in Example 20–9 satisfy Equation 20.22 ($4 < (28 \cdot 31)^{1/2} = 29.46$).

The Gram-Schmidt procedure is also valid for complex vector spaces. Let's construct an orthonormal basis from the two vectors $\mathbf{v}_1 = (-1, 1)$ and $\mathbf{v}_2 = (i, -1)$. We start with $\mathbf{u}_1 = \mathbf{v}_1 = (-1, 1)$ and write

$$\mathbf{u}_2 = \mathbf{v}_2 + a\,\mathbf{u}_1$$

Form the inner product with $\mathbf{u}_1$ from the left to get $\langle \mathbf{u}_1|\mathbf{u}_2 \rangle = 0 = \langle \mathbf{u}_1|\mathbf{v}_2 \rangle + \langle \mathbf{u}_1|a\,\mathbf{u}_1 \rangle$, which gives

$$0 = -i - 1 + a\langle \mathbf{u}_1|\mathbf{u}_1 \rangle = -i - 1 + 2a$$

or $a = (1+i)/2$. Thus, the two orthogonal vectors are $\mathbf{u}_1 = (-1, 1)$ and

$$\mathbf{u}_2 = \mathbf{v}_2 + a\,\mathbf{u}_1 = (i, -1) + \frac{1+i}{2}(-1, 1) = \frac{1}{2}(i-1, i-1)$$

The two orthonormal vectors are obtained by dividing each by its norm:

$$\mathbf{u}_1 = \frac{1}{\sqrt{2}}(-1, 1) \quad \text{and} \quad \mathbf{u}_2 = \frac{1}{2}(i-1, i-1)$$

Before we finish this chapter, we should touch upon one last topic. It is a standard procedure in quantum mechanics to expand functions in terms of the eigenfunctions of the Hamiltonian operator of a system. The eigenfunctions may be viewed as vectors in a (usually) infinite-dimensional (often complex) vector space, and all the definitions and results of this chapter apply to this case, but with the added complication that we must deal with questions of convergence of the various expansions. The definition of the inner product of a vector space formed by a set of wave functions is

$$\langle f\,|\,g \rangle = \int_a^b f^*(x)g(x)\,dx$$

which you may have seen in your study of quantum mechanics. An infinite-dimensional inner product space is called a *Hilbert space*. The mathematical foundations of quantum mechanics are built around the idea of a Hilbert space.

Problems

20–1. Show that the set of all $n \times n$ matrices forms a vector space.

20–2. Show that the set of all two-dimensional geometric vectors forms a vector space.

20–3. Show that the set of all ordered n-tuples of real numbers forms a vector space if addition of two n-tuples $(u_1, u_2, \ldots, u_n)$ and $(v_1, v_2, \ldots, v_n)$ is defined as $(u_1 + v_1, u_2 + v_2, \ldots, u_n + v_n)$ and scalar multiplication is defined by $c(u_1, u_2, \ldots, u_n) = (cu_1, cu_2, \ldots, cu_n)$.

20–4. Show that the set of all polynomials of degree less than or equal to 3 forms a vector space. What is its dimension?

20–5. Show that the set of functions that are continuous in the interval (a, b) forms a vector space.

20–6. Show that the set of solutions to an nth-order linear homogeneous differential equation forms a vector space.

20–7. It often happens that a subset of the vectors in V forms a vector space with respect to the same addition and multiplication operations as V. In such a case, the set of vectors is said to form a *subspace* of V. A simple geometric example of a subspace is the xy plane of a three-dimensional Euclidean space. The set of all vectors that lie in the xy plane forms a two-dimensional subspace of the three-dimensional Euclidean space. Show that the set of n-tuples $(a, a, \ldots, a)$, where a is any real number, is a subspace of R^n.

20–8. Test the following vectors for linear independence: $(0, 1, 0, 0)$, $(1, 1, 0, 0)$, $(0, 1, 1, 0)$, and $(0, 0, 0, 1)$.

20–9. Test the following vectors for linear independence: $(1, 1, 1)$, $(1, -1, 1)$, and $(-1, 1, -1)$.

20–10. Is the vector $(1, 0, 2)$ in the set spanned by $(1, 1, 1)$, $(1, -1, -1)$, $(3, 1, 1)$?

20–11. Show that the set of vectors $\{(1, 1, 1, 1), (1, -1, 1, -1), (1, 2, 3, 4), (1, 0, 2, 0)\}$ is a basis for R^4.

20–12. Find the coordinates of $(1, 2, 3)$ with respect to the basis $(1, 1, 0)$, $(1, 0, 1)$, $(1, 1, 1)$ in R^3.

20–13. Show that the dot product that we defined in Chapter 13 for geometric vectors is an inner product.

20–14. Show that the two geometric vectors $\mathbf{u} = \mathbf{i} + 2\mathbf{j} - \mathbf{k}$ and $\mathbf{v} = -\mathbf{i} + \mathbf{j} + 2\mathbf{k}$ satisfy the Schwarz inequality.

20–15. Show that the two functions $f_1(x) = 1 + x$ and $f_2(x) = x$ over the interval $(0, 1)$ satisfy the Schwarz inequality given the definition of the inner product in Example 20–6.

20–16. Using the inner product defined in Example 20–5, show that the norm of an ordered n-tuple of real numbers is $\|(u_1, u_2, \ldots, u_n)\| = (u_1^2 + u_2^2 + \cdots + u_n^2)^{1/2}$.

20–17. Derive Equation 20.13.

20–18. Construct an orthonormal basis from the three vectors $(1, -1, 0)$, $(1, 1, 0)$, and $(0, 1, 1)$.

20–19. We shall prove the Schwarz inequality in this problem. Start with $\langle \mathbf{u} + \lambda\mathbf{v}, \mathbf{u} + \lambda\mathbf{v} \rangle \geq 0$, where λ is an arbitrary constant. Expand this inner product to write

$$\lambda^2 \langle \mathbf{v}, \mathbf{v} \rangle + 2\lambda \langle \mathbf{u}, \mathbf{v} \rangle + \langle \mathbf{u}, \mathbf{u} \rangle \geq 0$$

This inequality must be true for any value of λ, so choose $\lambda = -\langle \mathbf{u}, \mathbf{v} \rangle \langle \mathbf{v}, \mathbf{v} \rangle$ to derive Equation 20.7.

20–20. Show that the three vectors $(1, i, -1)$, $(1 + i, 0, 1 - i)$, and $(i, -1, -i)$ are not linearly independent by expressing one of them as a linear combination of the others.

20–21. Determine if the vectors $(1, 1, -i)$, $(0, i, i)$, and $(0, 1, -1)$ are linearly independent.

20–22. Suppose that $\langle \mathbf{u} \mid \mathbf{v} \rangle = 2 + i$. Determine (a) $\langle (1 - i)\mathbf{u} \mid \mathbf{v} \rangle$ and (b) $\langle \mathbf{u} \mid 2i\,\mathbf{v} \rangle$.

20–23. Determine if the four matrices $\mathbf{I} = \begin{pmatrix} 1 & 0 \\ 0 & i \end{pmatrix}$, $\sigma_x = \begin{pmatrix} 0 & 1 \\ 1 & 0 \end{pmatrix}$, $\sigma_y = \begin{pmatrix} 0 & -i \\ i & 0 \end{pmatrix}$, and $\sigma_z = \begin{pmatrix} 1 & 0 \\ 0 & -1 \end{pmatrix}$ are linearly independent. These matrices are called the *Pauli spin matrices*.

20–24. Verify Equation 20.15 explicitly for (a) $\mathbf{u} = (1 + i, 1)$; $\mathbf{v} = (-i, -1)$ and (b) $\mathbf{u} = (3, -i, 2i)$; $\mathbf{v} = (1, 3i, -1)$.

20–25. Let $\mathbf{u} = (1, 1)$ and $\mathbf{v} = (1, -2)$. Verify Equations 20.18 and 20.19 explicitly for $a = i$.

20–26. Show that the inner product defined by Equation 20.20 satisfies Equations 20.14 through 20.16.

CHAPTER 21

Probability

Probability theory permeates the physical sciences, from the probabilistic inter-pretation of a wave function in quantum mechanics to the kinetic theory of gases to the molecular interpretation of entropy. It would be difficult to overstate the im-portance of probability theory not only in physical chemistry but in all the physical and biological sciences as well. In this chapter we shall introduce some of the basic ideas of probability theory and discuss some of its applications.

21.1 Discrete Distributions

Consider some experiment, such as the tossing of a coin, the rolling of a die, or the measurement of the z component of the angular momentum of a nucleus, that has n possible outcomes, each with probability p_j, where $j = 1, 2, \ldots, n$. If the experiment is repeated indefinitely, we intuitively expect that

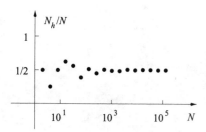

Figure 21.1. A computer simulation of the tossing of a coin. The graph shows the ratio of the number of heads obtained to the number of tosses as a function of N, the number of tosses.

$$p_j = \lim_{N \to \infty} \frac{N_j}{N} \qquad j = 1, 2, \ldots, n \qquad (21.1)$$

where N_j is the number of times that the event j occurs and N is the total number of repetitions of the experiment (see Figure 21.1). Because $0 \leq N_j \leq N$, p_j must satisfy the condition

$$0 \leq p_j \leq 1 \qquad (21.2)$$

When $p_j = 1$, we say the event j is a certainty and when $p_j = 0$, we say it is impossible. In addition, because

$$\sum_{j=1}^{n} N_j = N$$

we have the normalization condition,

$$\sum_{j=1}^{n} p_j = 1 \tag{21.3}$$

Equation 21.3 means that the probability that *some* event occurs is a certainty.

Many events are descriptive in nature, such as the event that the next card drawn from a deck of cards will be red. In many other cases, events have a natural numerical value. Suppose that each of the possible events $E_1, E_2, \ldots, E_n$ from an experiment has a numerical value. Then we can represent the outcomes by a random variable, X. A *random variable* is a rule or a formula that assigns numerical values to each of the possible events in an experiment. We denote a random variable by a capital letter and a particular observation of the random variable by a lowercase letter. The random variable X takes on the values $x_1, x_2, \ldots, x_n$ with respective probabilities $p(x_1), p(x_2), \ldots, p(x_n)$. The set of probabilities $\{p(x_j)\}$ can be considered to be the values that the function $p(X)$ assumes at the n points $\{x_j\}$. The function $p(X)$ is called the *probability density* of the distribution, and we write

$$p(x_j) = \text{Prob}\{X = x_j\} \tag{21.4}$$

A probability density is shown in Figure 21.2.

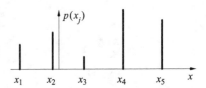

Figure 21.2. The discrete probability density, $p(x_j) = \text{Prob}\{X = x_j\}$.

It is often convenient to interpret a set of probabilities $\{p(x_j)\}$ as a unit mass distributed along the x axis such that m_j is the quantity of mass located at the point x_j. Because $\sum_j p(x_j) = 1$, we must have $\sum_j m_j = 1$. The probability distribution, then, can be pictured as the distribution of a unit mass along the x axis.

The *mean value* of X is given by

$$\langle x \rangle = \sum_{j=1}^{n} x_j p(x_j) \tag{21.5}$$

(The notation $\bar{x}$ is sometimes used in applied statistics (see the next chapter) and in the physical sciences.) By referring to our analogy of a unit mass distribution, Equation 21.5 says that the mean value is simply the center of mass. More generally, if $f(X)$ is a function of X, we define the mean value of $f(X)$ by

$$\langle f(x) \rangle = \sum_{j=1}^{n} f(x_j) p(x_j)$$

If $f(X) = X^n$, then $\langle x^n \rangle$ is called the nth *moment* of the probability distribution $\{p(x_j)\}$. Note that the second moment

$$\langle x^2 \rangle = \sum_{j=1}^{n} x_j^2 p(x_j) \tag{21.6}$$

corresponds to the moment of inertia of a mass distribution. You should be aware of the fact that $\langle x^2 \rangle$ does not necessarily equal $\langle x \rangle^2$, as the following Example shows.

EXAMPLE 21–1

Given the following probability data, calculate $\langle x \rangle$, $\langle x \rangle^2$, and $\langle x^2 \rangle$.

x	1	2	3	4	5
$p(x)$	0.10	0.15	0.05	0.50	0.20

SOLUTION: We have

$$\langle x \rangle = (1)(0.10) + (2)(0.15) + (3)(0.05) + (4)(0.50) + (5)(0.20)$$
$$= 3.55$$

$$\langle x \rangle^2 = 12.60$$

$$\langle x^2 \rangle = (1)(0.10) + (4)(0.15) + (9)(0.05) + (16)(0.50) + (25)(0.20)$$
$$= 14.14$$

Note that $\langle x^2 \rangle \neq \langle x \rangle^2$. We shall prove below that this is a general result.

A physically more interesting quantity than $\langle x^2 \rangle$ is the *second central moment*, or the *variance*, defined by

$$\sigma_x^2 = \text{Var}[X] = \langle (x - \langle x \rangle)^2 \rangle = \sum_{j=1}^{n} (x_j - \langle x \rangle)^2 p_j \qquad (21.7)$$

As the notation suggests, we denote the square root of the quantity in Equation 21.7 by σ_x, which is called the *standard deviation*. From the summation in Equation 21.7, we can see that σ_x^2 will be large if x_j is likely to differ from $\langle x \rangle$, because in that case $(x_j - \langle x \rangle)$ and so $(x_j - \langle x \rangle)^2$ will be large for the significant values of p_j. On the other hand, σ_x^2 will be small if x_j is not likely to differ from $\langle x \rangle$, or if the x_j cluster tightly around $\langle x \rangle$, because then $(x_j - \langle x \rangle)^2$ will be small for the significant values of p_j. Thus, we see that either the variance or the standard deviation is a measure of the spread of the distribution about its mean.

Equation 21.7 shows that σ_x^2 is a sum of positive terms, and so $\sigma_x^2 \geq 0$. Furthermore,

$$\sigma_x^2 = \sum_{j=1}^{n} (x_j - \langle x \rangle)^2 p_j = \sum_{j=1}^{n} (x_j^2 - 2\langle x \rangle x_j + \langle x \rangle^2) p_j$$

$$= \sum_{j=1}^{n} x_j^2 p_j - 2 \sum_{j=1}^{n} \langle x \rangle x_j p_j + \sum_{j=1}^{n} \langle x \rangle^2 p_j$$

The first term here is just $\langle x^2 \rangle$ (see Equation 21.6). To evaluate the second and third terms, we need to realize that $\langle x \rangle$, the mean of x_j, is just a number and so can be factored out of the summations, leaving a summation of the form $\sum x_j p_j$ in the second term and $\sum p_j$ in the third term. The summation $\sum x_j p_j$ is $\langle x \rangle$ by definition, and the summation $\sum p_j$ is unity because of normalization (Equation 21.3).

Putting all this together, we find that

$$\sigma_x^2 = \langle x^2 \rangle - 2\langle x \rangle^2 + \langle x \rangle^2$$

$$= \langle x^2 \rangle - \langle x \rangle^2 \geq 0 \qquad (21.8)$$

Because $\sigma_x^2 \geq 0$, we see that $\langle x^2 \rangle \geq \langle x \rangle^2$. A consideration of Equation 21.7 shows that $\sigma_x^2 = 0$ or $\langle x \rangle^2 = \langle x^2 \rangle$ only when $x_j = \langle x \rangle$ with a probability of one, a case that is not really probabilistic because the event j occurs on every trial.

We shall now discuss two well-known and important discrete probability distributions. First let us consider n successive tosses of a coin and ask what is the probability that heads comes up exactly m times. The result of any n successive tosses can be represented by a sequence of h's and t's, such as $hhhthttt \cdots tth$. There are n positions in any sequence and there are two choices (h or t) for each position. Thus, there are 2^n possible sequences, and since they are all equally likely, the probability of any particular one will be 2^{-n}. The number of such sequences with heads exactly m times can be calculated in the following way: There are n choices for the position of the first head, $n - 1$ choices for the second, and so on until the last of the m heads, in which case there are $n - m + 1$ choices. Thus, there are

$$\underbrace{n(n-1)\cdots(n-m+1)}_{m \text{ times}}$$

sequences containing heads m times. This product can be written more conveniently as

$$n(n-1)\cdots(n-m+1) = \frac{n!}{(n-m)!}$$

But all the heads are identical, so there is no first head or second head; there are simply heads. This means that we have overcounted by a factor of $m!$, which is the number of ways of permuting m heads. Thus, the number of ways of arranging m h's and $(n - m)$ t's is equal to the number of ways we can arrange m heads among n positions, or by

$$\frac{n!}{m!(n-m)!}$$

This quantity is called a *binomial coefficient*. (You can test this result by enumerating all the possible results for the tossing of a coin two or three times, as shown in Tables 21.1 and 21.2.) The probability of exactly m heads, p_m, then, is 2^{-n} times this result, or

$$p_m = \frac{n!}{m!(n-m)!}\left(\frac{1}{2}\right)^n$$

More generally, if a "successful" outcome occurs with probability p and an "unsuccessful" outcome occurs with probability q, where $p + q = 1$, then the

Table 21.1. An enumeration of the possible outcomes when a coin is tossed two times.

toss		head count
#1	#2	
H	H	2
H	T	1
T	H	1
T	T	0

Table 21.2. An enumeration of the possible outcomes when a coin is tossed three times.

toss			head count
#1	#2	#3	
H	H	H	3
H	H	T	2
H	T	H	2
T	H	H	2
H	T	T	1
T	H	T	1
T	T	H	1
T	T	T	0

probability of m "successful" outcomes is given by

$$p_m = \frac{n!}{m!(n-m)!}p^m q^{n-m} = \frac{n!}{m!(n-m)!}p^m(1-p)^{n-m} \tag{21.9}$$

This distribution is known as the *binomial distribution* and is applicable to the case of repeated independent trials such as the drawing of cards from a deck, replacing the drawn card after each draw. Equation 21.9 is called the binomial distribution because the binomial theorem says that

$$(x+y)^n = \sum_{m=0}^{n} \frac{n!}{m!(n-m)!}x^{n-m}y^m \tag{21.10}$$

For example,

$$(x+y)^2 = \sum_{m=0}^{2} \frac{2!}{m!(2-m)!}x^{2-m}y^m = x^2 + 2xy + y^2$$

and

$$(x+y)^3 = \sum_{m=0}^{3} \frac{3!}{m!(3-m)!}x^{3-m}y^m = x^3 + 3x^2 y + 3xy^2 + y^3$$

The left side of Equation 21.10 is symmetric in x and y, and so we can also write it in the form

$$(x+y)^n = \sum_{m=0}^{n} \frac{n!}{m!(n-m)!}x^m y^{n-m} \tag{21.11}$$

Equations 21.10 and 21.11 are equivalent. Equation 21.10 or 21.11 may also be written in the more symmetric form

$$(x+y)^n = \sum_{n_1=0}^{n}\sum_{n_2=0}^{n}{}^{*} \frac{n!}{n_1! n_2!}x^{n_1}y^{n_2} \tag{21.12}$$

where the asterisk on the summation indicates that only terms with $n_1 + n_2 = n$ are included.

The mean value of m for a binomial distribution is given by

$$\langle m \rangle = \sum_{m=0}^{n} m p_m = \sum_{m=0}^{\infty} \frac{n!}{m!(n-m)!}m p^m (1-p)^{n-m}$$

By comparing the right side here with Equation 21.11, we see that

$$\langle m \rangle = x\frac{\partial}{\partial x}(x+y)^n$$

where we set $x = p$ and $y = 1 - p$ *after* differentiating. Therefore, we see that (Problem 21–4)

$$\langle m \rangle = np \qquad \text{(binomial distribution)} \tag{21.13}$$

Figure 21.3 shows the binomial distribution plotted against m for several values of n for $p = 1/2$. Note that the distribution becomes more and more bell-shaped as n increases.

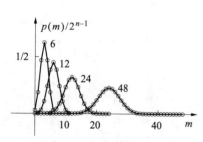

Figure 21.3. The binomial distribution plotted against m for $p = 1/2$ and for several values of n (6, 12, 24, and 48). Note that the distribution becomes increasingly bell-shaped as n increases.

EXAMPLE 21–2
Calculate the probability of getting exactly five heads in ten tosses of a coin.

SOLUTION: We use Equation 21.9 with $n = 10$, $m = 5$, and $p = 1/2$ to obtain

$$p_5 = \frac{10!}{5!5!}\left(\frac{1}{2}\right)^{10} = 0.246$$

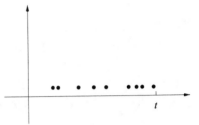

Although the binomial distribution is important and useful for its own sake, it also serves as the basis for another perhaps better-known distribution, namely, the Poisson distribution. Consider the following problem: We randomly distribute n points over some time interval $(0, t)$ and then ask what is the probability that exactly m of these points will lie in some subinterval Δt (Figure 21.4). We can think of this problem as consisting of n repeated, independent trials of placing a single particle in the interval $(0, t)$ with "success" being the probability that the particle will lie in the subinterval Δt, which is equal to $p = \Delta t/t$. Thus, the probability that m of the n particles will lie in Δt is given by

Figure 21.4. n points distributed randomly over the time interval $(0, t)$.

$$p_m = \frac{n!}{m!(n-m)!}p^m(1-p)^{n-m}$$
$$= \frac{n!}{m!(n-m)!}\left(\frac{\Delta t}{t}\right)^m\left(1 - \frac{\Delta t}{t}\right)^{n-m} \tag{21.14}$$

The mean value of m is $np = n\Delta t/t$, which we denote by $\lambda\Delta t$. We are often interested in the case where n is large and Δt is small (with $\langle m \rangle$ fixed at $\lambda\Delta t$), in which case Equation 21.14 can be cast in a much more convenient form (Problem 21–29):

$$p_m = \frac{(\lambda\Delta t)^m}{m!}e^{-\lambda\Delta t} \tag{21.15}$$

Equation 21.15 is the probability that there will be m points in an interval Δt if a large number of points n are randomly distributed over the time interval $(0, t)$ with $\lambda = n/t$. Equation 21.15 is called a *Poisson distribution*.

EXAMPLE 21–3
Determine the mean value of m for a Poisson distribution.

SOLUTION: The mean value of m is given by

$$\langle m \rangle = e^{-\lambda\Delta t}\sum_{m=0}^{\infty}\frac{m(\lambda\Delta t)^m}{m!} \tag{1}$$

We can evaluate this summation in closed form by noting that

$$\sum_{m=0}^{\infty} \frac{x^m}{m!} = e^x$$

If we differentiate both sides with respect to x and then multiply by x, we obtain

$$\sum_{m=0}^{\infty} \frac{m x^m}{m!} = x \frac{d}{dx} e^x = x e^x$$

Using this result in equation 1, then, we find that

$$\langle m \rangle = \lambda \Delta t$$

Equation 21.15 applies to radioactive decay, among many other physical phenomena. In this case, λ is the mean rate of decay, and Equation 21.15 tells us the chance that we will observe m decays in a time interval Δt. Notice that the probability of observing no decay in an interval Δt is given by $p_0 = e^{-\lambda \Delta t}$.

EXAMPLE 21–4

A radioactive sample is observed over a long period of time to emit alpha particles at a rate of 1.5 per minute. Determine the mean number of alpha particles that you would observe in a two-minute interval. Calculate the probability that you would observe 0, 1, 2, 3, 4, and equal to or greater than 5 counts.

SOLUTION: According to Example 21–3, the mean number of counts in a two-minute interval is $\lambda \Delta t = 3.0$ counts. The probability of observing m counts is given by Equation 21.15.

m	0	1	2	3	4
p_m	0.050	0.15	0.22	0.22	0.17

The probability that $m \geq 5$ is given by

$$\text{Prob}\,\{m \geq 5\} = 1 - \sum_{m=0}^{4} \text{Prob}\,\{M = m\} = 0.19$$

The Poisson distribution has an amazingly broad range of physical applications, including radioactive decay, aerial search, the arrival of electrons striking a counter, the transmission of a nervous impulse across a synapse, and the distribution of galaxies. We'll write Equation 21.15 in the general case as

$$p_m = \frac{a^m}{m!} e^{-a} \qquad (21.16)$$

where a is equal to $\langle m \rangle$ (Figure 21.5). For example, suppose it is known that there are 300 errors in a book containing 500 pages. What is the probability that a page

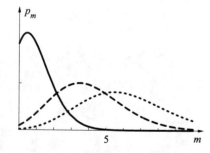

Figure 21.5. The Poisson distribution (Equation 21.16) plotted against m for $a = 1.0$ (solid), $a = 4.0$ (dashed), and $a = 6.0$ (dotted).

contains no errors? Three or more errors? Here the "time interval" is the area of a page, and we wish to calculate the frequency of errors on a page, given that there is a mean value of $a = 300/500 = 0.6$ errors per page. The probability that a page has no errors, then, is $p_0 = e^{-0.6} = 0.5488$, and the probability that there are three or more errors is given by

$$\text{Prob}\{m \geq 3\} = \sum_{m=3}^{\infty} \left(\frac{a^m}{m!}\right) e^{-a} = 1 - e^{-a} - ae^{-a} - \frac{a^2}{2}e^{-a} = 0.0231$$

See Problems 21–10 and 21–11 for some other applications of the Poisson distribution.

21.2 The Multinomial Distribution

We saw in the previous section that the number of ways of dividing N distinguishable objects into two groups, one containing N_1 objects and the other containing the $N - N_1 = N_2$ remaining objects, is given by the binomial coefficient

$$W(N_1, N_2) = \frac{N!}{(N - N_1)!N_1!} = \frac{N!}{N_1!N_2!} \tag{21.17}$$

The generalization to the distribution of N distinguishable objects into r groups, the first containing N_1, the second containing N_2, and so on, is

$$W(N_1, N_2, \ldots, N_r) = \frac{N!}{N_1!N_2! \cdots N_r!} \tag{21.18}$$

with $N_1 + N_2 + \cdots + N_r = N$. This quantity is called a multinomial coefficient because it occurs in the multinomial expansion:

$$(x_1 + x_2 + \cdots + x_r)^N = \sum_{N_1=0}^{N} \sum_{N_2=0}^{N} \cdots \sum_{N_r=0}^{N} {}^* \frac{N!}{N_1!N_2! \cdots N_r!} x_1^{N_1} x_2^{N_2} \ldots x_r^{N_r} \tag{21.19}$$

where the asterisk indicates that only terms such that $N_1 + N_2 + \cdots + N_r = N$ are included. Note how Equation 21.19 is a straightforward generalization of Equation 21.12.

EXAMPLE 21–5

A deal in bridge consists of the distribution of 52 cards into four groups of 13 cards each. Calculate the total possible number of bridge hands.

SOLUTION: We use Equation 21.18 with $N = 52$ and $N_1 = N_2 = N_3 = N_4 = 13$ to get

$$\text{number of bridge hands} = \frac{52!}{(13!)^4} = 5.36 \times 10^{28}$$

which is larger than Avogadro's number.

If we use Equation 21.18 to calculate something like the number of ways of distributing Avogadro's number of particles over their energy states, then we are forced to deal with factorials of huge numbers. Even the evaluation of 100! would be a chore, never mind 10^{23}!, unless we have a good approximation for N!. We're about to present an approximation for N! that actually improves as N gets larger. Such an approximation is called an *asymptotic approximation*, that is, an approximation to a function that gets relatively better as the argument of the function increases.

Because N! is a product, it is convenient to deal with $\ln N$! because the latter is a sum. The asymptotic approximation to $\ln N$! is called *Stirling's approximation* and is given by

$$\ln N! = N \ln N - N \tag{21.20}$$

which is surely a lot easier to use than calculating N! and then taking its logarithm. Table 21.3 shows the value of $\ln N$! versus Stirling's approximation for a number of values of N. Note that the agreement, which we express in terms of relative error, decreases markedly with increasing N.

Table 21.3. A numerical comparison of $\ln N$! with Stirling's approximation.

N	$\ln N!$	$N \ln N - N$	relative error[a]
10	15.104	13.026	0.1376
50	148.48	145.60	0.0194
100	363.74	360.52	0.0089
500	2611.3	2607.3	0.0015
1000	5912.1	5907.7	0.0007

[a] relative error $= (\ln N! - N \ln N + N)/\ln N!$

EXAMPLE 21–6

A more refined version of Stirling's approximation says that

$$\ln N! = N \ln N - N + \ln (2\pi N)^{1/2}$$

Use this version of Stirling's approximation to calculate $\ln N$! for $N = 10$ and compare the relative error with that in Table 21.3.

SOLUTION: For $N = 10$,

$$\ln N! = N \ln N - N + \ln (2\pi N)^{1/2} = 15.092$$

and using the value of $\ln 10$! from Table 21.3, we see that

$$\text{relative error} = \frac{15.104 - 15.096}{15.104} = 0.0005$$

The relative error is significantly smaller than that in Table 21.3. The relative errors for the other entries in Table 21.3 are essentially zero for this extended version of Stirling's approximation.

The proof of Stirling's approximation is not difficult. Because $N! = N(N-1)(N-2)\ldots(2)(1)$, $\ln N!$ is given by

$$\ln N! = \sum_{n=1}^{N} \ln n \tag{21.21}$$

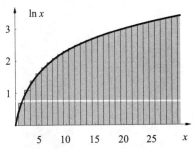

Figure 21.6. A plot of $\ln x$ versus x. The sum of the areas under the rectangles up to N is $\ln N!$.

Figure 21.6 shows $\ln x$ plotted versus x for integer values of x. According to Equation 21.21, the sum of the areas under the rectangles up to N in Figure 21.6 is $\ln N!$. Figure 21.6 also shows the continuous curve $\ln x$ plotted on the same graph. Thus, $\ln x$ is seen to form an envelope to the rectangles, and this envelope becomes a steadily smoother approximation to the rectangles as x increases. Therefore, we can approximate the area under these rectangles by the integral of $\ln x$. The area under $\ln x$ will poorly approximate the rectangles only in the beginning. If N is large enough (we are deriving an asymptotic approximation), this area will make a negligible contribution to the total area. We may write, then,

$$\ln N! = \sum_{n=1}^{N} \ln n \approx \int_{1}^{N} \ln x \, dx = N \ln N - N \quad (N \text{ large}) \tag{21.22}$$

which is Stirling's approximation to $\ln N!$. The lower limit could just as well have been taken as 0 in Equation 21.22, because N is large. (Remember that $x \ln x \to 0$ as $x \to 0$ (Problem 21–13).)

21.3 Continuous Distributions

So far we have considered only discrete distributions, but continuous distributions are perhaps even more important than discrete distributions in physical chemistry. It is convenient to use the unit mass analogy. Consider a unit mass to be distributed continuously along the x axis, or along some interval on the x axis. We define the linear mass density $\rho(x)$ by

$$dm(x) = \rho(x) \, dx$$

where dm is the fraction of the mass lying between x and $x + dx$. By analogy, then, we say that the probability that some quantity x, such as the position of a particle in a box, lies between x and $x + dx$ is

$$\text{Prob}\{x \le X \le x + dx\} = p(x) \, dx \tag{21.23}$$

and that

$$\text{Prob}\{a \le X \le b\} = \int_{a}^{b} p(x) \, dx \tag{21.24}$$

In the mass analogy, Prob$\{a \leq X \leq b\}$ is the fraction of mass that lies in the interval $a \leq x \leq b$. The normalization condition is

$$\int_{-\infty}^{\infty} p(x)\,dx = 1 \tag{21.25}$$

Following Equations 21.5 through 21.7, we have the definitions

$$\langle x \rangle = \int_{-\infty}^{\infty} x p(x)\,dx \tag{21.26}$$

$$\langle x^2 \rangle = \int_{-\infty}^{\infty} x^2 p(x)\,dx \tag{21.27}$$

and

$$\sigma_x^2 = \langle (x - \langle x \rangle)^2 \rangle = \int_{-\infty}^{\infty} (x - \langle x \rangle)^2 p(x)\,dx \tag{21.28}$$

EXAMPLE 21–7

Perhaps the simplest continuous distribution is the so-called uniform distribution, where

$$p(x) = \begin{cases} \text{constant} = A & a \leq x \leq b \\ 0 & \text{otherwise} \end{cases}$$

Show that A must equal $1/(b-a)$. Evaluate $\langle x \rangle$, $\langle x^2 \rangle$, σ_x^2, and σ_x for this distribution.

SOLUTION: Because $p(x)$ must be normalized,

$$\int_a^b p(x)\,dx = 1 = A \int_a^b dx = A(b-a)$$

Therefore, $A = 1/(b-a)$ and

$$p(x) = \begin{cases} \dfrac{1}{b-a} & a \leq x \leq b \\ 0 & \text{otherwise} \end{cases}$$

The mean of x is given by

$$\langle x \rangle = \int_a^b x p(x)\,dx = \frac{1}{b-a} \int_a^b x\,dx$$
$$= \frac{b^2 - a^2}{2(b-a)} = \frac{b+a}{2}$$

and the second moment of x by

$$\langle x^2 \rangle = \int_a^b x^2 p(x)\,dx = \frac{1}{b-a} \int_a^b x^2\,dx$$
$$= \frac{b^3 - a^3}{3(b-a)} = \frac{b^2 + ab + a^2}{3}$$

Last, the variance is given by Equation 21.28, and so

$$\sigma_x^2 = \langle x^2 \rangle - \langle x \rangle^2 = \frac{(b-a)^2}{12}$$

and the standard deviation is

$$\sigma_x = \frac{(b-a)}{\sqrt{12}}$$

Certainly the most important continuous probability distribution is the *Gaussian distribution*, both from a theoretical and a practical point of view. The Gaussian distribution is given by

$$p(x)\,dx = \frac{1}{\sqrt{2\pi\sigma^2}}e^{-x^2/2\sigma^2}\,dx \qquad -\infty < x < \infty \qquad (21.29)$$

The factor in front assures that $p(x)$ is normalized.

EXAMPLE 21–8
Determine the mean and variance of the Gaussian distribution given by Equation 21.29.

SOLUTION: The mean is given by

$$\langle x \rangle = \int_{-\infty}^{\infty} x p(x)\,dx = \frac{1}{\sqrt{2\pi\sigma^2}}\int_{-\infty}^{\infty} x e^{-x^2/2\sigma^2}\,dx$$

which equals zero because $x e^{-x^2/2\sigma^2}$ is an odd function. The variance is given by

$$\sigma_x^2 = \langle (x - \langle x \rangle)^2 \rangle = \frac{1}{\sqrt{2\pi\sigma^2}}\int_{-\infty}^{\infty} x^2 e^{-x^2/2\sigma^2}\,dx$$

$$= \sigma^2$$

Thus, we see that the parameter σ^2 in the Gaussian distribution is the variance.

We saw in Sections 4.4 and 10.1 that the value of σ controls the width of a Gaussian distribution: the smaller the value of σ, the narrower and more peaked the distribution. Figure 21.7 shows a Gaussian distribution plotted for several values of σ.

The probability that x lies within a certain interval $(-a, a)$ is given by

$$\text{Prob}\{-a \le X \le a\} = \frac{1}{\sqrt{2\pi\sigma^2}}\int_{-a}^{a} e^{-x^2/2\sigma^2}\,dx \qquad (21.30)$$

Using tables of the error function (Section 4.3), we can see that the probability that $-\sigma \le x \le \sigma$ is 0.6827; for $-2\sigma \le x \le 2\sigma$ is 0.9545, and for $-3\sigma \le x \le 3\sigma$

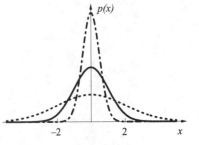

Figure 21.7. A plot of a Gaussian distribution (Equation 21.29) for several values of σ. The dotted curve corresponds to $\sigma = 2$, the solid curve to $\sigma_x = 1$, and the dash-dotted curve to $\sigma = 0.5$.

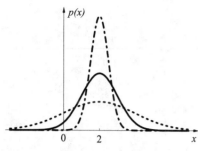

Figure 21.8. A plot of a Gaussian distribution (Equation 21.31) for $\langle x \rangle = 2$ and several values of σ. The dotted curve corresponds to $\sigma = 2$, the solid curve to $\sigma = 1$, and the dash-dotted curve to $\sigma = 0.5$.

is 0.9973; in other words, 99% of the area under a Gaussian curve lies within ± 3 multiples of σ (Problem 21–27).

A more general version of a Gaussian distribution is

$$p(x)dx = (2\pi\sigma^2)^{-1/2}e^{-(x-\langle x \rangle)^2/2\sigma^2}\, dx \tag{21.31}$$

The plots of Equation 21.31 look like those in Figure 21.7 except that the curves are centered at $x = \langle x \rangle$ rather than $x = 0$ (Figure 21.8).

We learned in Section 4.4 that as $\sigma \to 0$, the Gaussian distribution becomes progressively narrower, yet maintains the fixed unit area under the curve. This is one of the physical descriptions of the Dirac delta function, and one of the definitions of $\delta(x)$ is

$$\delta(x-a) = \lim_{\sigma\to 0} \frac{1}{\sqrt{2\pi\sigma^2}}e^{(x-a)^2/2\sigma^2} \tag{21.32}$$

A Gaussian distribution is one of the most important and commonly used probability distributions in all of science. It's impossible to overstate how important the Gaussian distribution is in the physical sciences, as well as in the biological sciences. Almost everyone has heard of a bell curve or a normal curve. The underlying reason that the Gaussian distribution is so important is a theorem in statistics called the central limit theorem. The *central limit theorem* says that if $X_1, X_2, \ldots, X_n$ are n independent random variables that have the same distribution, their mean $(X_1 + X_2 + \cdots + X_n)/n$ is approximately normally distributed, and becomes more so as n increases. Because many experiments are made on collections of molecules, the quantity that we observe is actually the mean value of many molecular quantities. Furthermore, the theory of errors in experimental measurements uses a Gaussian distribution because mean quantities are almost always normally distributed.

21.4 Joint Probability Distributions

We can extend the ideas in this chapter to more than one random variable. Let X and Y be two continuous random variables. The joint probability density is

$$p(x, y)dxdy = \text{Prob}\,\{x \le X \le x+dx \quad \text{and} \quad y \le Y \le y+dy\} \tag{21.33}$$

If we integrate $p(x, y)$ over the entire range of one of its variables, say y, we get

$$p(x) = \int_{-\infty}^{\infty} dy\, p(x, y) \tag{21.34}$$

which is called the *marginal density function* of X. The expected value of a function of both X and Y is

$$\langle f(x, y) \rangle = \int_{-\infty}^{\infty} dy \int_{-\infty}^{\infty} dx\, f(x, y)p(x, y) \tag{21.35}$$

If the value of X that we observe has no effect on the value of Y that we observe, and vice versa, then we say that X and Y are *independent*, in which case

$$p(x, y) = p(x)p(y) \quad \text{(independent)} \tag{21.36}$$

Problem 21–28 has you show that

$$\langle x + y \rangle = \langle x \rangle + \langle y \rangle \tag{21.37}$$

if X and Y are any two random variables and that

$$\text{Var}\,[\,X + Y\,] = \langle (x + y - \langle x \rangle - \langle y \rangle)^2 \rangle = \text{Var}\,[\,X\,] + \text{Var}\,[\,Y\,]$$
$$= \langle (x - \langle x \rangle)^2 \rangle + \langle (y - \langle y \rangle)^2 \rangle \tag{21.38}$$

if X and Y are independent. Furthermore, if X and Y are independent, then

$$\langle xy \rangle = \langle x \rangle \langle y \rangle \tag{21.39}$$

An important measure of the degree of lack of independence of two random variables is the *correlation coefficient*. Let

$$\langle (x - \langle x \rangle)^2 \rangle = \sigma_x^2 \tag{21.40}$$
$$\langle (y - \langle y \rangle)^2 \rangle = \sigma_y^2 \tag{21.41}$$

The quantity we wish to discuss is $\text{Cov}\,[X, Y] = \langle (x - \langle x \rangle)(y - \langle y \rangle) \rangle$, called the *covariance of X and Y*. We define the *correlation coefficient* ρ_{xy} by

$$\text{Cov}\,[X, Y] = \int_{-\infty}^{\infty} dy \int_{-\infty}^{\infty} dx \,(x - \langle x \rangle)(y - \langle y \rangle)p(x, y) = \rho_{xy}\sigma_x\sigma_y \tag{21.42}$$

If X and Y are independent, then $\rho_{xy} = 0$, and X and Y are said to be uncorrelated. It turns out that $-1 \le \rho_{xy} \le 1$, and the deviation of ρ_{xy} from zero is an indication of the degree of correlation between X and Y.

Problems

Probability problems can be somewhat tricky to solve, and sometimes the results are fairly counterintuitive. The first two problems are examples whose results many people find surprising (or even unconvincing).

21–1. If heads comes up 10 times in a row, many people feel strongly that the next toss is more likely to be tails than heads. How do you explain to them that they are dead wrong?

21–2. Consider a group of n persons. Calculate the probability that at least two of them have a birthday on the same day of the year for $n = 50$ (exclude leap year). What is the smallest value for which the probability is greater than 1/2? *Hint*: This is a good example where it is easier to calculate the opposite probability first and then subtract the result from 1; that is, calculate the probability that all n people have a different birthday, and then subtract that result from 1 to give the probability that at least two people have the same birthday.

21-3.   Enumerate the numbers of heads and tails obtained when a coin is tossed four times.

21-4.   Derive Equation 21.13.

21-5.   Show that the binomial distribution (Equation 21.9) is normalized and determine $\langle m \rangle$ and σ_m^2 in terms of n and p.

21-6.   An experimental paper contains a table of 14 measurements, and in 13 of them the final digit is even. What is the chance of this, assuming that even and odd final digits are equally likely?

21-7.   The coefficients of the expansion of $(1 + x)^n$ can be arranged in the following form:

$$\frac{n}{\quad}$$

n									
0					1				
1				1		1			
2			1		2		1		
3		1		3		3		1	
4	1		4		6		4		1

What is the relation between the numbers on a line and the numbers in the line above it? The triangular arrangement here is called *Pascal's triangle*. Use this result to write out the expansion of $(x + y)^5$.

21-8.   In how many ways can a committee of three be chosen from nine people?

21-9.   Show that the Poisson distribution (Equation 21.16) is normalized and determine $\langle m \rangle$ and σ_m^2 in terms of a.

21-10. Over a period of years, a professor has determined that a mean value of 4.3 students come to an office hour. What is the probability that no one comes to a given office hour? How about more than five? (Assume that there is no exam imminent.)

21-11. In a certain experiment, events are detected by bursts on a fluorescent screen. Over a period of time, it is observed that a mean value of 6.7 bursts per second occur. What is the probability of observing 6 or 7 bursts in any given second?

21-12. Calculate the relative error for $N = 50$ using the formula for Stirling's approximation given in Example 21-6, and compare your result with that given in Table 21.3 using Equation 21.20. Take $\ln N!$ to be 148.47777.

21-13. Prove that $x \ln x \to 0$ as $x \to 0$.

21-14. An important continuous distribution is the exponential distribution

$$p(x)\,dx = ce^{-\lambda x}\,dx \qquad 0 \le x < \infty$$

Evaluate c, $\langle x \rangle$, and σ_x^2, and the probability that $x \ge a$ in terms of λ.

21-15. Integrals of the type

$$I_n(\alpha) = 2 \int_0^\infty x^{2n} e^{-\alpha x^2}\,dx$$

occur frequently in a number of applications. We can simply either look them up in a table of integrals or continue this problem. We learned in Section 13.1 how

to evaluate $I_0(\alpha)$ by using the trick of expressing $I_0^2(\alpha)$ as a double integral in x
and y and then converting to plane polar coordinates. The result comes out to be

$$I_0(\alpha) = \left(\frac{\pi}{\alpha}\right)^{1/2}$$

Now prove that the $I_n(\alpha)$ may be obtained by repeated differentiation of $I_0(\alpha)$
with respect to α and, in particular, that

$$\frac{d^n I_0(\alpha)}{d\alpha^n} = (-1)^n I_n(\alpha)$$

Use this result and the fact that $I_0(\alpha) = (\pi/\alpha)^{1/2}$ to generate $I_1(\alpha)$, $I_2(\alpha)$, and so
forth.

21–16. Using the results derived in the previous problem, calculate $\langle x^4 \rangle$ for a Gaussian
distribution.

21–17. Consider a particle that is constrained to lie along a potential-free, one-
dimensional interval 0 to a. The probability that the particle is found to lie
between x and $x + dx$ is given by

$$p(x)\,dx = \frac{2}{a}\sin^2\frac{n\pi x}{a}dx$$

where $n = 1,\ 2,\ 3,\ \ldots$. First show that $p(x)$ is normalized. Now show that the
mean position of the particle along the interval 0 to a is $a/2$. Is this result
physically reasonable? Finally, show that $\langle x^2 \rangle = \left(\frac{a}{n\pi}\right)^2 \left(\frac{n^2\pi^2}{3} - \frac{1}{2}\right)$.

21–18. Use the result of the previous problem to show that $\sigma_x = (\langle x^2 \rangle - \langle x \rangle^2)^{1/2}$ for a
particle in a box is less than a, the width of the box, for any value of n. If σ_x is
the uncertainty in the position of the particle, could σ_x ever be larger than a?

21–19. Using the probability distribution given in Problem 21–17, calculate the
probability that the particle will be found between 0 and $a/2$. Does this result
make sense to you?

21–20. You learn in physical chemistry that the molecules in a gas travel at various
speeds and that the probability that a molecule has a speed between v and $v + dv$
is given by

$$p(v)\,dv = 4\pi \left(\frac{m}{2\pi k_B T}\right)^{3/2} v^2 e^{-mv^2/2k_B T}\,dv \qquad 0 \le v < \infty$$

where m is the mass of the particle, k_B is the Boltzmann constant (the molar
gas constant R divided by the Avogadro constant), and T is the kelvin
temperature. The probability distribution of molecular speeds is called the
Maxwell–Boltzmann distribution. First show that $p(v)$ is normalized, and then
determine the mean speed as a function of temperature. The necessary integrals
are (Problem 21–16)

$$\int_0^\infty x^{2n} e^{-\alpha x^2}\,dx = \frac{1 \cdot 3 \cdot 5 \cdots (2n-1)}{2^{n+1}\alpha^n}\left(\frac{\pi}{\alpha}\right)^{1/2} \qquad n \ge 1$$

and

$$\int_0^\infty x^{2n+1} e^{-\alpha x^2}\,dx = \frac{n!}{2\alpha^{n+1}}$$

21–21. Use the Maxwell–Boltzmann distribution in the previous problem to determine the mean kinetic energy of a gas-phase molecule as a function of temperature. The necessary integral is given in the previous problem.

21–22. The Maxwell–Boltzmann distribution for the *component*, v_x, of the velocity of a molecule in the gas phase is given by

$$p(v_x)\,dv_x = \left(\frac{m}{2\pi k_B T}\right)^{1/2} e^{-mv_x^2/2k_B T}\,dv_x \qquad -\infty < v_x < \infty$$

Notice that the component of a velocity can vary from $-\infty$ to ∞, whereas the *speed* of a molecule can vary only from 0 to ∞. Speed, being the magnitude of the velocity, is an intrinsically positive number, whereas the component of velocity can be positive or negative, reflecting the direction in which the molecule is moving. First show that $p(v_x)$ is normalized, and then calculate $\langle v_x \rangle$ and $\langle v_x^2 \rangle$. Compare your result for $\langle v_x \rangle$ with the result for $\langle v \rangle$ that you obtained in Problem 21–20. Interpret your result for $\langle v_x \rangle$ physically. The necessary integrals are given in Problem 21–20.

21–23. Using the distribution in the previous problem, derive an expression for Prob $\{-v_{x0} \le V_x \le v_{x0}\}$. Express your result in terms of the error function of $w_0 = (m/2k_B T)^{1/2} v_{x0}$. Calculate Prob $\{(-2k_B T/m)^{1/2} \le V_x \le (2k_B T/m)^{1/2}\}$.

21–24. Use the result of the previous problem to show that Prob $\{|\,V_x\,| \ge v_{x0}\} = 1 -$ erf(w_0). Calculate Prob $\{|\,V_x\,| \ge (k_B T/m)^{1/2}\}$ and Prob $\{|V_x| \ge (2k_B T/m)^{1/2}\}$.

21–25. Use the result of Problem 21–23 to plot Prob $\{-v_{x0} \le V_x \le v_{x0}\}$ against $v_{x0}/(2k_B T/m)^{1/2}$.

21–26. Another distribution that is frequently used in the kinetic theory of gases is the distribution of energy, or kinetic energy in particular. Let $\epsilon = mu^2/2$ and use the distribution in Problem 21–20 to show that Prob $\{\epsilon \le E \le \epsilon + d\epsilon\} = F(\epsilon)d\epsilon = \dfrac{2\pi}{(\pi k_B T)^{3/2}} \epsilon^{1/2} e^{-\epsilon/k_B T}\,d\epsilon$. First show that $F(\epsilon)$ is normalized and then calculate $\langle \epsilon \rangle$, and compare your result to $\left\langle \frac{1}{2}mv^2 \right\rangle$ that you obtained in Problem 21–21.

21–27. Use a table of the error function or numerical integration (Chapter 23) to show that

$$\frac{1}{\sqrt{2\pi\sigma^2}}\int_{-n\sigma}^{n\sigma} e^{-x^2/2\sigma^2}\,dx = \begin{cases} 0.6827 & n = 1 \\ 0.9545 & n = 2 \\ 0.9973 & n = 3 \end{cases}$$

21–28. Show that $\langle x + y \rangle = \langle x \rangle + \langle y \rangle$ for any two random variables X and Y and that Var $[X + Y] =$ Var $[X] +$ Var $[Y]$ if X and Y are independent.

21–29. Show that Equation 21.14 becomes Equation 21.15 when n becomes very large and Δt becomes very small such that $\Delta t/t = \lambda \Delta t/n$. This problem is a little challenging.

CHAPTER 22

Statistics: Regression and Correlation

A common and practical experiment in all sciences involves the repeated measurement of two different physical quantities for the purpose of determining a relationship between them. We often seek a linear relationship, even transforming variables to obtain a linear relationship. For example, Figure 22.1a shows the vapor pressure P of water plotted against the Celsius temperature, which is hardly linear. Thermodynamics tells us, however, that we should obtain a linear (or at least almost linear) plot if we plot $\ln P$ against the reciprocal of the kelvin temperature, as shown in Figure 22.1b.

This chapter involves regression and correlation. In regression analysis, we assume that one of the two variables, call it x, can be measured without appreciable error, and so we regard it as an ordinary variable. The other variable, y, which is a function of x, is subject to some imprecision or uncertainty and is regarded as a random variable. We shall restrict our discussion to linear regression analysis, where we assume that y and x are linearly related by $y = \alpha x + \beta$. We shall use the method of least squares to find the best estimates of α and β from the data at hand. Then we'll introduce a method that you can use to determine confidence intervals for α and β; in other words, we'll be able to say that there is a certain probability, say, 95%, that α lies in the interval $\alpha_1 \leq \alpha \leq \alpha_2$ and that β lies in the interval $\beta_1 \leq \beta \leq \beta_2$.

In the next section, we'll discuss correlation analysis, where both y and x are considered to be random variables, and we wish to determine if they are, in fact, related. The key quantity in this section is the correlation coefficient, which is a

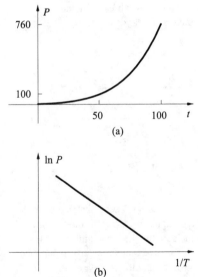

Figure 22.1. (a) The vapor pressure of water plotted against the Celsius temperature. (b) The logarithm of the vapor pressure of water plotted against the reciprocal of the kelvin temperature.

direct measure of how well y and x are correlated. We'll also present a method to determine confidence intervals for the correlation coefficients.

Table 22.1. Typical data for the resistivity ρ of a nichrome wire as a function of the Celsius temperature. The resistivity is expressed in units of meters·ohms $\times 10^{-7}$.

$t/°C$	20	25	30	35	40	45	50
ρ	9.137	8.913	8.665	8.528	8.242	8.203	7.972

22.1 Linear Regression Analysis

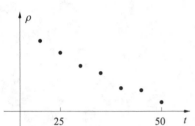

Figure 22.2. The resistivity ρ of a nichrome wire as a function of the Celsius temperature. The resistivity is expressed in units of meters·ohms $\times 10^{-7}$.

Table 22.1 lists some typical data for the resistivity ρ of a nichrome wire as a function of the Celsius temperature t. These data are plotted in Figure 22.2. Both Figure 22.2 and theory suggest that the relation between ρ and t is linear. We want to draw a straight line through the data in Figure 22.2 in the most objective way. Figure 22.3 shows a set of points with a straight line drawn through them. If the equation of the line is $y = a + bx$, then the vertical distance d_j of the point y_j from the straight line is

$$d_j = y_j - (a + bx_j) \tag{22.1}$$

We're going to choose a and b such that the sum of all the distances d_j is a minimum. In particular, we'll minimize the sum of the d_j^2, so that we have

$$S(a, b) = \sum_{j=1}^{n} d_j^2 = \sum_{j=1}^{n} (y_j - a - bx_j)^2 = \text{minimum} \tag{22.2}$$

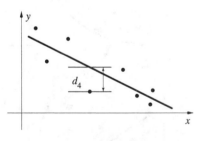

Figure 22.3. A set of data points with a straight line drawn through them. If the equation of the straight line is $y = a + bx$, then the vertical distance of y_j from the line is $d_j = y_j - a - bx_j$.

where n is the number of pairs of data points. This standard procedure is called the *method of least squares*. To minimize $S(a, b)$ in Equation 22.2, we simply set both $\partial S/\partial a$ and $\partial S/\partial b$ equal to zero and solve for a and b to obtain (Problem 22–1)

$$a = \overline{y} - b\overline{x} \quad \text{and} \quad b = \frac{\displaystyle\sum_{j=1}^{n} x_j y_j - n\overline{x}\,\overline{y}}{\displaystyle\sum_{j=1}^{n} x_j^2 - n\overline{x}^2} \tag{22.3}$$

where $\overline{y}$ and $\overline{x}$ are the means of y and x, respectively. Note that the first of Equations 22.3 says that $\overline{y} = a + b\overline{x}$.

Equations 22.3 are perfectly useful as they stand, but it is conventional to express b as

$$b = \frac{s_{xy}}{s_x^2} \tag{22.4}$$

where

$$s_x^2 = \frac{1}{n-1} \sum_{j=1}^{n} (x_j - \overline{x})^2 \tag{22.5}$$

and

$$s_{xy} = \frac{1}{n-1} \sum_{j=1}^{n} (x_j - \overline{x})(y_j - \overline{y}) \qquad (22.6)$$

The sum s_x^2 is called the *sample variance* and s_{xy} is called the *sample covariance*. The resultant straight line is called the *least-squares straight line of the sample*.

For the data in Table 22.2, $\overline{t} = 35$, $\overline{\rho} = 8.523$, $s_t^2 = 116.7$, and $s_{t\rho} = -4.449$, and so

$$b = \frac{-4.449}{116.7} = -0.03812$$

and

$$a = 8.523 + (0.03812)(35) = 9.858$$

Therefore, the least-squares straight line is $\rho = 9.858 - 0.03812\,t$ (Figure 22.4).

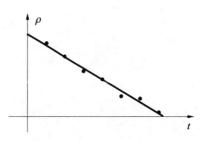

Figure 22.4. The data in Table 22.1 along with the least-squares straight line $\rho = 9.858 - 0.03812\,t$.

Table 22.2. The molar heat capacity ($J \cdot mol^{-1} \cdot K^{-1}$) of carbon monoxide as a function of the kelvin temperature.

T	600	650	700	750	800	850	900	950	1000
C_P	30.93	31.54	31.32	32.18	32.25	32.27	33.41	33.21	33.97

EXAMPLE 22–1

Table 22.2 lists some data for the molar heat capacity of carbon monoxide as a function of the kelvin temperature. Determine the least-squares straight line for these data.

SOLUTION: The necessary quantities from these data are

$$\overline{T} = 800 \qquad \overline{C}_P = 32.34$$

$$s_T^2 = 18750 \qquad s_{TC} = 134.0$$

and so $b = s_{TC}/s_T^2 = 0.007\,147$ and $a = 26.62$. The least-squares straight line is

$$C_P = 26.62 + 0.007\,147\,T$$

These data and results are plotted in Figure 22.5.

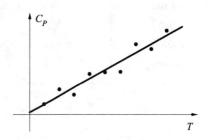

Figure 22.5. The data in Table 22.2 along with the least-squares straight line $C_P = 26.64 + 0.007\,147\,T$.

The values of a and b given by Equations 22.3 depend upon the sample that we choose and so will vary from sample to sample. We are really interested in the values of these parameters for the population from which these samples were drawn. Let's denote these population parameters by α and β; we are using the values of a and b that we determine from a least-squares analysis as our best estimates of α and β. Assuming that the y_j are normally distributed (which is a standard assumption in elementary statistics) with a mean $\mu(x) = \alpha + \beta x$ and have the same variance, we can determine what are called *confidence intervals* for α and β. In other words, we choose a probability η, close to unity, (say, 95% or 99%) and then determine pairs of quantities (α_1, α_2) and (β_1, β_2) such that

$$\text{Prob}\,\{\alpha_1 \leq \alpha \leq \alpha_2\} = \eta$$
$$\text{Prob}\,\{\beta_1 \leq \beta \leq \beta_2\} = \eta \tag{22.7}$$

The above intervals are the *confidence intervals* of α and β at a confidence level of η.

A step-by-step procedure for determining the confidence intervals for α and β is outlined below. This is a slight simplification of a standard procedure that can be found in most elementary statistics books.

Step 1: Calculate the two quantities

$$\sigma_a^2 = \frac{\left[(n-1)s_x^2 + n\overline{x}^2\right]\left(s_y^2 - b^2 s_x^2\right)}{n(n-2)s_x^2} \tag{22.8}$$

and

$$\sigma_b^2 = \frac{s_y^2 - b^2 s_x^2}{(n-2)s_x^2} \tag{22.9}$$

Step 2: Choose a confidence level, η (for example, 95%, 99%, ...).

Step 3: Use a table of the normal distribution, $p(z)$, with $\sigma = 1$ to determine a value of γ such that

$$\text{Prob}\,\{-\gamma \leq Z \leq \gamma\} = \int_{-\gamma}^{\gamma} p(z)\,dz = \eta \tag{22.10}$$

(See Table 22.3.)

Step 4: Calculate $\gamma\,\sigma_a$ and $\gamma\,\sigma_b$. The confidence intervals of α and β are given by

$$\text{Conf}\,\{a - \gamma\,\sigma_a < \alpha < a + \gamma\,\sigma_a\}$$

and

$$\text{Conf}\,\{b - \gamma\,\sigma_b < \beta < b + \gamma\,\sigma_b\}$$

Table 22.3. A few values of γ that satisfy Equation 22.10.

η	γ
0.9000	1.645
0.9500	1.960
0.9800	2.326
0.9900	2.576
0.9950	2.810

EXAMPLE 22–2
Use the data in Table 22.2 to determine 95% confidence intervals for α and β.

SOLUTION: Example 22–1 gives us $\overline{T} = 800$, $\overline{C}_P = 32.34$, $s_T^2 = 18750$, $s_{TC} = 134.0$, $n = 9$, $a = 26.62$, and $b = 0.007\,147$. Furthermore, the data in Table 22.2 yield $s_C^2 = 1.0332$. We now follow the step-by-step procedure presented above.

Step 1:

$$\sigma_a^2 = \frac{[(n-1)s_T^2 + n\overline{T}^2][s_C^2 - b^2 s_T^2]}{n(n-2)s_T^2}$$

$$= \frac{[(8)(18750) + (9)(800)^2][1.0332 - (0.007\,147)^2(18750)]}{(9)(7)(18750)}$$

$$= 0.3781$$

$$\sigma_b^2 = \frac{s_C^2 - b^2 s_T^2}{(n-2)s_T^2} = \frac{1.0332 - (0.007\,147)^2(18750)}{(7)(18750)}$$

$$= 5.73 \times 10^{-7}$$

or

$$\sigma_a = 0.6149 \qquad \text{and} \qquad \sigma_b = 0.000\,757$$

Step 2: $\eta = 0.95$

Step 3: Using Table 22.3, we find that $\gamma = 1.960$.

Step 4:

$$\text{Conf}\{25.41 < \alpha < 27.83\}$$

and

$$\text{Conf}\{0.00566 < \beta < 0.00863\}$$

Thus, there is a 0.95 probability that the values of α and β lie within these intervals.

22.2 Correlation Analysis

Up to this point, we have assumed that the independent variables (the x_j) are determined with little or no uncertainty, so that we may regard them as ordinary variables. From here on, we'll treat both X and Y as random variables. We define a *correlation coefficient of the sample* by

$$r = \frac{s_{xy}}{s_x s_y} \qquad s_x > 0, \ s_y > 0 \tag{22.11}$$

Since s_{xy} may be either positive or negative and $s_x > 0$ and $s_y > 0$, r may be either positive or negative. Problem 22–4 has you show that

$$s_y^2 \geq \frac{s_{xy}^2}{s_x^2} \tag{22.12}$$

and so we see that $r^2 \leq 1$, or that $-1 \leq r \leq 1$. Furthermore, if $r^2 = 1$, then σ_β given by Equation 22.9 equals zero, and all the sample pairs $(x_1, y_1), (x_2, y_2), \ldots, (x_n, y_n)$

will lie on a straight line. The converse is also true; that is, if all the sample pairs lie on a straight line, then $r^2 = 1$ (Problem 22–5). Suppose, on the other hand, that there is no relationship between the x_j and the y_j. Then the terms in the sum

$$\sum_{j=1}^{n}(x_j - \bar{x})(y_j - \bar{y})$$

will be equally likely to be positive and negative, and so s_{xy}, and consequently r, will be zero. In this case, the x_j and the y_j are said to be *uncorrelated*. This all suggests that the value of r is a measure of the *linear correlation* between the x_j's and the y_j's. (It is interesting to note that $r = 0$ does *not* imply that there is no relationship between x and y; it simply implies that there is no *linear* relationship between them.)

EXAMPLE 22–3
Determine the value of r^2 for the data given in Table 22.2.

SOLUTION:

$$r^2 = \frac{s_{TC}^2}{s_T^2 s_C^2} = \frac{(134.0)^2}{(18750)(1.0332)} = 0.9269$$

indicating that there is a fairly strong linear relationship between the molar heat capacity of carbon monoxide and the kelvin temperature over the given temperature range.

The sample correlation coefficient r is an estimate of the population correlation coefficient ρ, which is a true measure of the linear correlation between y and x. After all, we might have chosen n sample pairs that fortuitously laid on a straight line, and r would equal one even though y and x are not linearly related. We defined the covariance of two random variables

$$\sigma_{XY} = \langle (x - \langle x \rangle)(y - \langle y \rangle) \rangle \tag{22.13}$$

in Section 21.4. In Equation 22.13, $\langle x \rangle$ and $\langle y \rangle$ are the means of X and Y, respectively. In analogy with Equation 22.11 for r, we define the population correlation function ρ by

$$\rho = \frac{\sigma_{XY}}{\sigma_X \sigma_Y} \tag{22.14}$$

where $\sigma_X^2 = \langle (x - \langle x \rangle)^2 \rangle$ and $\sigma_Y^2 = \langle (y - \langle y \rangle)^2 \rangle$. The population correlation coefficient has properties similar to that of the sample correlation coefficient. For example, $-1 \le \rho \le 1$, and Y and X are linearly related if and only if $\rho^2 = 1$. If $\rho = 0$, then the random variables X and Y are said to be *uncorrelated*.

Equation 22–13 shows that $\sigma_{XY} = 0$ if X and Y are independent (Problem 22–6). Thus, if the random variables X and Y are independent, then they are uncorrelated.

We can calculate confidence intervals for ρ assuming that X and Y are jointly normally distributed. We now present the stepwise procedure without proof. This procedure is sometimes called the *large-sample procedure*.

Step 1: Calculate the quantity

$$z = \frac{1}{2} \ln \frac{1+r}{1-r}$$

Step 2: Choose a confidence level, η (for example, 95%, 99%, ...).

Step 3: Using tables of the normal distribution, $p(z)$ (Table 22.3), with $\sigma = 1$, determine a value of γ such that

$$\text{Prob}\{-\gamma \le Z \le \gamma\} = \int_{-\gamma}^{\gamma} p(z)\,dz = \eta$$

Step 4: Calculate

$$\rho_1 = \tanh\left(z - \frac{\gamma}{\sqrt{n-3}}\right) \quad \text{and} \quad \rho_2 = \tanh\left(z + \frac{\gamma}{\sqrt{n-3}}\right)$$

The confidence interval for ρ is given by

$$\text{Conf}\{\rho_1 \le \rho \le \rho_2\}$$

EXAMPLE 22–4

Determine 95% confidence intervals for the population correlation coefficient ρ using the data in Table 22.2.

SOLUTION: According to Example 22–3, $r = (0.9269)^{1/2} = 0.9628$.

Step 1: $z = \frac{1}{2} \ln \frac{1+r}{1-r} = \frac{1}{2} \ln \frac{1.9627}{0.0373} = 1.982$

Step 2: $\eta = 0.950$

Step 3: Using Table 22.3, we find that $\gamma = 1.960$

Step 4: $\rho_1 = \tanh\left(1.982 - \frac{1.960}{\sqrt{6}}\right) = \tanh(1.18) = 0.828$

$\rho_2 = \tanh\left(1.982 + \frac{1.960}{\sqrt{6}}\right) = \tanh(2.78) = 0.992$

Therefore,

$$\text{Conf}\{0.828 \le \rho \le 0.992\}$$

at a confidence level of 95%.

22.3 Error Propagation of Measurements

Before we finish this chapter, we should discuss the important topic of error prop-
agation in measurements. Let $f(x, y)$ be some quantity whose value we can determine by measuring x and y separately. For example, $f(x, y)$ might be an area
of a rectangle, which we determine by measuring its width, x, and length, y,
to give $A = xy$. Suppose now we measure x and y, so that we have the pairs
$(x_1, y_1), (x_2, y_2), \ldots, (x_n, y_n)$. We can calculate the means and sample variances
of x and y according to

$$\overline{x} = \frac{1}{n}\sum_{j=1}^{n} x_j \qquad \text{and} \qquad \overline{y} = \frac{1}{n}\sum_{j=1}^{n} y_j \qquad (22.15)$$

$$s_x^2 = \frac{1}{n-1}\sum_{j=1}^{n}(x_j - \overline{x})^2 \qquad \text{and} \qquad s_y^2 = \frac{1}{n-1}\sum_{j=1}^{n}(y_j - \overline{y})^2 \qquad (22.16)$$

We can also calculate n values of $f(x, y)$ from the n pairs (x_j, y_j) according to
$f_j = f(x_j, y_j)$, from which we can calculate

$$\overline{f} = \frac{1}{n}\sum_{j=1}^{n} f(x_j, y_j) \qquad \text{and} \qquad s_f^2 = \frac{1}{n-1}\sum_{j=1}^{n}(f_j - \overline{f})^2 \qquad (22.17)$$

Just as s_x and s_y are measures of the imprecision of the values of the x_j's and y_j's,
we consider s_f to be a measure of the imprecision of the $f(x_j, y_j)$.

Assuming, as usual, that the imprecisions are small, we can expand $f_j = f(x_j, y_j)$ in a Taylor series (Problem 3–26) about $x_j = \overline{x}$ and $y_j = \overline{y}$ to obtain

$$f_j = f(x_j, y_j) = f(\overline{x}, \overline{y}) + \left(\frac{\partial f}{\partial x}\right)_{\overline{x}, \overline{y}}(x_j - \overline{x}) + \left(\frac{\partial f}{\partial y}\right)_{\overline{x}, \overline{y}}(y_j - \overline{y}) + \cdots \qquad (22.18)$$

If we divide both sides of Equation 22.18 by n and then sum over j, we obtain
(Problem 21–18)

$$\overline{f} = f(\overline{x}, \overline{y}) \qquad (22.19)$$

Thus, if $f(x, y) = xy$ represents the area of the rectangle, then $\overline{A} = \overline{x}\,\overline{y}$.

We can also determine s_f^2 in terms of the imprecisions of the (x_j, y_j) data
pairs. Substitute Equations 22.18 and 22.19 into s_f^2 (Equation 22.17) to get (Problem 22–18)

$$s_f^2 = \left(\frac{\partial f}{\partial x}\right)_{\overline{x}, \overline{y}}^2 s_x^2 + \left(\frac{\partial f}{\partial y}\right)_{\overline{x}, \overline{y}}^2 s_y^2 + 2\left(\frac{\partial f}{\partial x}\right)_{\overline{x}, \overline{y}}\left(\frac{\partial f}{\partial y}\right)_{\overline{x}, \overline{y}} s_{xy}^2 \qquad (22.20)$$

where s_x^2, s_y^2, and s_{xy}^2 have been defined previously (Equations 22.5 and 22.6). If
the values of the x_j and the y_j are independent, then $s_{xy} = 0$ and Equation 22.20
reduces to

$$s_f^2 = \left(\frac{\partial f}{\partial x}\right)_{\overline{x}, \overline{y}}^2 s_x^2 + \left(\frac{\partial f}{\partial y}\right)_{\overline{x}, \overline{y}}^2 s_y^2 \qquad (22.21)$$

a standard formula in the theory of measurements.

EXAMPLE 22–5

Suppose we determine the density of a sphere by measuring its mass and its diameter. We find after a number of measurements that the mean mass is 106.4 grams with a standard deviation of 0.50 grams and that the mean diameter is 4.321 cm with a standard deviation of 0.0010 cm. Determine the mean density and its standard deviation.

SOLUTION: The density of the sphere is given by $\rho = m/(\pi d^3/6) = 6m/\pi d^3$. The mean density is

$$\overline{\rho} = \frac{6\overline{m}}{\pi\overline{d}^3} = \frac{6(106.4\text{ g})}{\pi(4.321\text{ cm})^3} = 2.519\text{ g·cm}^{-3}$$

The variance is given by

$$s_\rho^2 = \left(\frac{\partial\rho}{\partial m}\right)^2_{\overline{m},\overline{d}} s_m^2 + \left(\frac{\partial\rho}{\partial d}\right)^2_{\overline{m},\overline{d}} s_d^2$$

$$= \left(\frac{6}{\pi\overline{d}^3}\right)^2 s_m^2 + \left(-\frac{18\overline{m}}{\pi\overline{d}^4}\right)^2 s_d^2$$

$$= (5.604 \times 10^{-4})(0.50)^2 + (3.058)(0.0010)^2$$

$$= 1.436 \times 10^{-4}$$

or $s_\rho = \pm 0.01198$. This result is often expressed as $\rho = 2.519 \pm 0.012$ g·cm^{-3}.

Problems

22–1. Derive Equations 22.3.

22–2. Show that Equations 22.3 and Equations 22.4 through 22.5 are consistent.

22–3. Show that the least-squares line always passes through the point $(\overline{x}, \overline{y})$.

22–4. Show that $S(a, b)$ in Equation 22.2 can be expressed as

$$S = (n-1)(s_y^2 - b^2 s_x^2) = (n-1)\left(s_y^2 - \frac{s_{xy}^2}{s_x^2}\right),\text{ where}$$

$$s_y^2 = \frac{1}{n-1}\sum_{j=1}^{n}(y_j - \overline{y})^2.\text{ Now show that } s_y^2 \geq \frac{s_{xy}^2}{s_x^2}.$$

22–5. Show that the correlation coefficient $r = \pm 1$ if the sample pairs lie on a straight line. *Hint*: Use the result of the previous problem.

22–6. Show that σ_{XY} given by Equation 22.13 equals zero if X and Y are independent random variables.

22–7.   Use the following data to determine the least-squares fit to these data:

x	0.00	0.25	0.50	0.75	1.00	1.25	1.50	1.75	
y	−0.2765	0.0605	1.132	1.854	2.300	2.925	4.422	3.248	
x	2.00	2.25	2.50	2.75	3.00	3.25	3.50	3.75	4.00
y	4.120	4.453	5.631	5.125	5.412	6.684	7.307	7.726	8.400

22–8.   Use the data in the previous problem to determine a 95% confidence interval for α and β.

22–9.   Use the data in Problem 22–7 to determine the correlation coefficient r.

22–10. Let η' be the viscosity of a polymer solution and η be the viscosity of the solvent. Then $(\eta' - \eta)/\eta = \eta_{sp}$ is called the intrinsic viscosity of the solution. Theory says that η_{sp}/c should vary linearly with c, where c is the concentration of the polymer in the solution. Table 22.4 gives some typical data. Use these data to determine the least-squares fit.

Table 22.4. The intrinsic viscosity of cellulose nitrate in alcohol. The units of concentration are grams per liter.

c	0.00	0.10	0.20	0.30	0.40	0.50	0.60	0.70	0.80	0.90	1.00
η_{sp}/c	9.04	10.2	11.7	13.6	14.8	16.4	17.7	19.9	20.5	22.3	23.9

22–11. Use the data in the previous problem to determine a 99% confidence interval for α and β.

22–12. Use the data in Problems 22–7 and 22–10 to determine a 99% confidence interval for the population correlation coefficient ρ.

22–13. In the photoelectric effect, a metallic surface is irradiated with electromagnetic radiation and electrons are ejected from the surface. The energies of the ejected electrons are determined by measuring the potential at which the photoelectric current drops to zero. According to the theory of the photoelectric effect, the stopping potential ϕ_s is linearly related to the frequency of the radiation by $\phi_s = a\nu + b$. Use the data in Table 22.5 to determine the least-squares values of a and b and the value of the correlation coefficient r.

Table 22.5. The stopping potential for the photoelectric effect on sodium metal. The frequency ν is given as multiples of 10^{15} Hz and the stopping potential ϕ_s is given in volts.

ν	0.46	0.48	0.50	0.52	0.54	0.56	0.58	0.60
ϕ_s	0.0834	0.315	0.182	0.340	0.786	0.352	0.731	0.430

22–14. Use the data in the previous problem to determine a 90% confidence interval for α and β.

22–15. Use the data in Problem 22–13 to determine a 90% confidence interval for the population correlation coefficient ρ.

22–16. Table 22.6 gives the vapor pressure of water P (in torr) and the corresponding temperature t (in °C). Thermodynamics tells us that a plot of $\ln P$ against $1/T$ should approximate a straight line. Determine the least-squares fit to these data.

Table 22.6. The vapor pressure of water, P (in torr), and the corresponding Celsius temperature, t.

t	0	5	10	15	20	25	30
P	4.6	6.5	9.2	12.8	17.4	23.8	31.6
t	35	40	45	50	55	60	65
P	42.2	55.3	71.9	92.5	118.0	149.4	187.5
t	70	75	80	85	90	95	100
P	233.7	289.1	355.1	433.6	525.8	633.9	760

22–17. Use the data in the previous problem to determine a 99.5% confidence interval for α and β.

22–18. Substitute Equations 22.18 and 22.19 into Equation 22.17 to obtain Equation 22.20.

22–19. Suppose we determine the volume of a right circular cylinder by measuring its height and its radius. We find after a number of measurements that the mean height is 16.06 cm with a standard deviation of 0.015 cm and that the mean radius is 3.751 cm with a standard deviation of 0.018 cm. Determine the mean volume of the cylinder and its standard deviation.

22–20. The pressure of a gas at low pressures satisfies the ideal gas equation, $P = \dfrac{nRT}{V}$, where n is the number of moles, T is the kelvin temperature, V is the volume, and R is a constant whose value is 0.083145 L·bar·mol^{-1}·K^{-1}. Suppose that we determine the pressure by measuring n, T, and V. The data give us $\bar{n} = 0.1025$ mol, $s_n = 0.00652$ mol, $\bar{T} = 286.30$ K, $s_T = 0.196$ K, $\bar{V} = 30.444$ L, and $s_V = 0.0959$ L. Determine the mean pressure of the gas and its standard deviation.

CHAPTER 23

Numerical Methods

Most of the problems that you encounter in your physical chemistry courses are such that the equations can be solved analytically. Most of the algebraic equations are quadratic equations, which can be easily solved using the quadratic formula; most of the integrals are easy enough that they can be found in tables; simultaneous linear algebraic equations involve only two or three unknowns, at most; and so on. This is often not the case in research, however, and we have to resort to numerical methods to obtain numerical answers. There are numerous routines to solve large systems of algebraic equations, to evaluate integrals, to find eigenvalues and eigenvectors of large matrices, and to solve many other types of problems. There is a huge literature on numerical methods, and we can give only a glimpse of this field in this chapter. We shall discuss four types of problems: the determination of the roots of algebraic equations, the numerical evaluation of integrals, the numerical summation of series, and solving systems of linear algebraic equatons.

23.1 Roots of Equations

You learned in high school that a quadratic equation has two roots, given by the quadratic formula:

$$x = \frac{-b \pm \sqrt{b^2 - 4ac}}{2a}$$

Thus, the two values of x (called *roots*) that satisfy the equation $x^2 + 3x - 2 = 0$ are

$$x = \frac{-3 \pm \sqrt{17}}{2}$$

Although there are general formulas for the roots of cubic and quartic equations, they are very inconvenient to use; furthermore, there are no formulas for equations of the fifth degree or higher. Unfortunately, in practice we encounter such equations frequently and must learn to deal with them. Fortunately, with the advent of hand calculators and personal computers, the numerical solution of polynomial equations and other types of equations, such as $x - \cos x = 0$, is routine. Although these and other equations can be solved by "brute force" trial and error, much more organized procedures can arrive at an answer to almost any desired degree of accuracy. Perhaps the most widely known procedure is the *Newton–Raphson method*, which is best illustrated by a figure. Figure 23.1 shows a function $y = f(x)$ plotted against x. The solution to the equation $f(x) = 0$ is denoted by x_*. The idea behind the Newton–Raphson method is to guess an initial value of x (call it x_0) "sufficiently close" to x_* and draw the tangent to the curve $f(x)$ at x_0, as shown in Figure 23.1. Very often, the extension of the tangent line through the horizontal axis will lie closer to x_* than does x_0. We denote this value of x by x_1 and repeat the process using x_1 to get a new value of x_2, which will lie even closer to x_*. By repeating this process (called iteration) we can approach x_* to essentially any desired degree of accuracy.

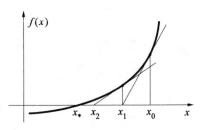

Figure 23.1. A graphical illustration of the Newton–Raphson method.

We can use Figure 23.1 to derive a convenient formula for the iterative values of x. The slope of $f(x)$ at x_n, $f'(x_n)$, is given by

$$f'(x_n) = \frac{f(x_n) - 0}{x_n - x_{n+1}}$$

Solving this equation for x_{n+1} gives

$$x_{n+1} = x_n - \frac{f(x_n)}{f'(x_n)} \tag{23.1}$$

which is the iterative formula for the Newton–Raphson method. As an application of this formula, consider the chemical equation

$$2\,NOCl(g) \rightleftharpoons 2\,NO(g) + Cl_2(g)$$

whose related equilibrium constant is 2.18 at a certain temperature. If 1.00 atm of $NOCl(g)$ is introduced into a reaction vessel, then at equilibrium $P_{NOCl} = 1.00 - 2x$, $P_{NO} = 2x$, and $P_{Cl_2} = x$; these pressures satisfy the equilibrium constant expression

$$\frac{P_{NO}^2 P_{Cl_2}}{P_{NOCl}^2} = \frac{(2x)^2 x}{(1.00 - 2x)^2} = 2.18$$

which we write as

$$f(x) = 4x^3 - 8.72x^2 + 8.72x - 2.18 = 0$$

Because of the stoichiometry of the reaction equation, the value of x we are seeking must be between 0 and 0.5, so let's choose 0.250 as our initial guess (x_0). Table 23.1 shows the results of using Equation 23.1. Notice that we have converged to three significant figures in just three steps.

Table 23.1. The results of the application of the Newton–Raphson method to the solution of the equation $f(x) = 4x^3 - 8.72x^2 + 8.72x - 2.18 = 0$.

n	x_n	$f(x_n)$	$f'(x_n)$
0	0.250	-4.825×10^{-1}	5.110
1	0.344	-4.772×10^{-2}	4.137
2	0.356	-7.491×10^{-4}	4.003
3	0.356		

EXAMPLE 23–1

The cubic equation

$$x^3 + 3x^2 + 3x - 1 = 0$$

occurs in a discussion of imperfect gases. Use the Newton–Raphson method to find the real root of this equation to five significant figures.

SOLUTION: We write the equation as

$$f(x) = x^3 + 3x^2 + 3x - 1 = 0$$

By inspection, a solution lies between 0 and 1. Using $x_0 = 0.5$ results in the following table:

n	x_n	$f(x_n)$	$f'(x_n)$
0	0.500 000	1.375 00	6.7500
1	0.296 300	0.178 294	5.04118
2	0.260 930	0.004 809	4.76983
3	0.259 920	−0.000 005	4.76220
4	0.259 920		

The answer to five significant figures is $x = 0.25992$. Note that $f(x_n)$ is significantly smaller at each step, as it should be as we approach the value of x that satisfies $f(x) = 0$, but that $f'(x_n)$ does not vary appreciably. The same behavior can be seen in Table 23.1.

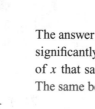

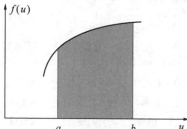

Figure 23.2. A plot of $y = x^{1/3}$, illustrating that the Newton–Raphson method fails in this case.

As powerful as it is, the Newton–Raphson method does not always work; when it does work, it is obvious the method is working, and when it doesn't work, it may be even more obvious that it's not working. A spectacular failure is provided by the equation $f(x) = x^{1/3} = 0$, for which $x_* = 0$. If we begin with $x_0 = 1$, we will obtain $x_1 = -2$, $x_2 = +4$, $x_3 = -8$, and so on. Figure 23.2 shows why the method is failing to converge. The message here is that you should always plot $f(x)$ first to get an idea of where the relevant roots lie and to see that the function does not have any peculiar properties. You should do Problems 23–1 to 23–8 to become proficient with the Newton–Raphson method.

23.2 Numerical Integration

There are also numerical methods to evaluate integrals. The net area between a curve and the horizontal axis is given by the integral

$$I = \int_a^b f(u)\,du \tag{23.2}$$

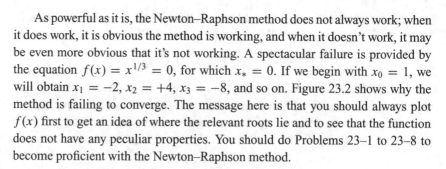

Figure 23.3. The integral of $f(u)$ from a to b is given by the shaded area.

and is shown as the shaded area in Figure 23.3. The fundamental theorem of calculus says that if $F(x) = \int_a^x f(u)\,du$, then $\dfrac{dF}{dx} = f(x)$. If there is no elementary function $F(x)$ whose derivative is $f(x)$, we say that the integral of $f(x)$ cannot be evaluated analytically. By elementary function, we mean a function that can be expressed as a finite combination of polynomial, trigonometric, exponential, and logarithmic functions.

It turns out that numerous integrals cannot be evaluated analytically. A particularly important example of an integral that cannot be evaluated in terms of elementary functions is

$$\phi(x) = \int_0^x e^{-u^2} du \tag{23.3}$$

Equation 23.3 serves to define the error function (Section 3.3). The value of $\phi(x)$ for any value of x is given by the area under the curve $f(u) = e^{-u^2}$ from $u = 0$ to $u = x$.

Let's consider the more general case given by Equation 23.2 or the shaded area in Figure 23.3. We can approximate this area in a number of ways. First divide the interval (a, b) into n equally spaced subintervals $u_1 - u_0$, $u_2 - u_1$, $\ldots$, $u_n - u_{n-1}$ with $u_0 = a$ and $u_n = b$. We let $h = u_{j+1} - u_j$ for $j = 0, 1, \ldots, n - 1$. Figure 23.4 shows a magnification of one of the subintervals, say, the u_j, u_{j+1} subinterval. One way to approximate the area under the curve is to connect the points $f(u_j)$ and $f(u_{j+1})$ by a straight line, as shown in Figure 23.4. The area under the straight-line approximation to $f(u)$ in the interval is the sum of the area of the rectangle $[hf(u_j)]$ and the area of the triangle $\left\{\frac{1}{2}h[f(u_{j+1}) - f(u_j)]\right\}$. The total area under the curve from $u = a$ to $u = b$ is given by the sum

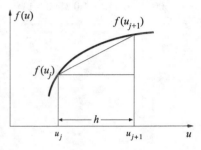

Figure 23.4. An illustration of the area of the $j + 1$st subinterval for the trapezoidal approximation.

$$
\begin{aligned}
I \approx I_n = \ & hf(u_0) + \frac{h}{2}[f(u_1) - f(u_0)] \\
& + hf(u_1) + \frac{h}{2}[f(u_2) - f(u_1)] \\
& \vdots \\
& + hf(u_{n-2}) + \frac{h}{2}[f(u_{n-1}) - f(u_{n-2})] \\
& + hf(u_{n-1}) + \frac{h}{2}[f(u_n) - f(u_{n-1})] \\
= \ & \frac{h}{2}[f(u_0) + 2f(u_1) + 2f(u_2) + \cdots + 2f(u_{n-1}) + f(u_n)] \tag{23.4}
\end{aligned}
$$

Note that the coefficients in Equation 23.4 go as 1, 2, 2, ..., 2, 1. Equation 23.4 is easy to implement on a hand calculator for $n = 10$ or so and on a personal computer for larger values of n. The approximation to the integral given by Equation 23.4 is called the *trapezoidal approximation*. The error goes as Ah^2, where A is a constant that depends upon the nature of the function $f(u)$. In fact, if M is the largest value of $|f''(u)|$ in the interval (a, b), then the error is *at most* $M(b - a)h^2/12$. Table 23.2 shows the values of

$$\phi(1) = \int_0^1 e^{-u^2} du \tag{23.5}$$

Table 23.2. The application of the trapezoidal approximation (Equation 23.4) to the evaluation of $\phi(1)$ given by Equation 23.5. The accepted value to eight significant figures is 0.746 82413.

n	h	I_{2n}
10	0.1	0.746 21800
100	0.01	0.746 81800
1000	0.001	0.746 82407

Table 23.3. The application of Simpson's approximation (Equation 23.6) to the evaluation of $\phi(1)$ given by Equation 23.5. The accepted value to eight significant figures is 0.746 82413.

n	h	I_{2n}
10	0.1	0.746 82494
100	0.01	0.746 82414
1000	0.001	0.746 82413

for $n = 10$ ($h = 0.1$), $n = 100$ ($h = 0.01$), and $n = 1000$ ($h = 0.001$). The "accepted" value (using more sophisticated numerical integration methods) is 0.746 82413, to eight significant figures.

We can develop a more accurate numerical integration routine by approximating $f(u)$ in Figure 23.4 by something other than a straight line. If we approximate $f(u)$ by a quadratic function, we have Simpson's approximation, whose formula is

$$I_{2n} = \frac{h}{3}[f(u_0) + 4f(u_1) + 2f(u_2) + 4f(u_3) + 2f(u_4) + \cdots$$
$$+ 2f(u_{2n-2}) + 4f(u_{2n-1}) + f(u_{2n})] \tag{23.6}$$

Note that the coefficients go as 1, 4, 2, 4, 2, 4, ..., 4, 2, 4, 1. We write I_{2n} in Equation 23.6 because Simpson's approximation requires that there be an even number of intervals. Table 23.3 shows the values of $\phi(1)$ in Equation 23.5 for $n = 10$, 100, and 1000.

Note that with $n = 100$, the result differs from the "accepted" value by only one unit in the eighth decimal place. The error for Simpson's approximation goes as h^4 compared with h^2 for the trapezoidal approximation. In fact, if M is the largest value of $|f^{(4)}(u)|$ in the interval (a, b), then the error is at most $M(b - a)h^4/180$. Problems 23–9 through 23–12 illustrate the use of the trapezoidal approximation and Simpson's approximation.

EXAMPLE 23–2

The Debye theory of the molar heat capacity of a monatomic crystal gives

$$\overline{C}_V = 9R \left(\frac{T}{\Theta_D}\right)^3 \int_0^{\Theta_D/T} dx \, \frac{x^4 e^x}{(e^x - 1)^2}$$

where R is the molar gas constant (8.314 J·K^{-1}·mol^{-1}) and Θ_D, the Debye temperature, is a parameter characteristic of the crystalline substance. Given that $\Theta_D = 309$ K for copper, calculate the molar heat capacity of copper at $T = 103$ K.

SOLUTION: At $T = 103$ K, the basic integral to evaluate numerically is

$$I = \int_0^3 dx \, \frac{x^4 e^x}{(e^x - 1)^2}$$

Using the trapezoidal approximation (Equation 23.4) and Simpson's approximation (Equation 23.6), we find the following values of I:

n	h	I_n (trapezoidal)	I_{2n} (Simpson's approximation)
10	0.3	5.9725	5.9648
100	0.03	5.9649	5.9648
1000	0.003	5.9648	5.9648

The molar heat capacity at 103 K is given by

$$\overline{C}_V = 9R \left(\frac{103 \text{ K}}{309 \text{ K}} \right)^3 I$$

or $\overline{C}_V = 16.5 \text{ J·mol}^{-1}\text{·K}^{-1}$, in agreement with the experimental value.

23.3 Summing Series

Suppose we have a convergent series

$$s = \sum_{n=1}^{\infty} u_n \tag{23.7}$$

of positive terms, $u_n \geq 0$, for all n. Let $S_N = \sum_{n=1}^{N} u_n$ be the Nth partial sum of this series and R_N be the remainder after N terms, so that

$$s = S_N + R_N \tag{23.8}$$

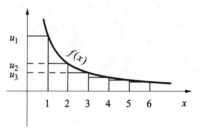

A practical question is how close is S_N to s, or, to put it another way, how small is R_N (relative to s). To answer this question, refer to Figure 23.5. The solid curve in this figure is that of a continuous function $f(x)$ such that $f(n) = u_n$ for $n = 1, 2, \ldots$. You can see in the figure that $f(1) = u_1$, $f(2) = u_2$, and so on. Choosing any value of N, Figure 23.5 shows that the area under the curve of $f(x)$ starting at N is greater than the area of the rectangles starting at $N + 1$. In an equation, we have

Figure 23.5. A pictorial aid to the validity of Equation 23.9.

$$\int_N^{\infty} f(x)\,dx \geq \sum_{n=N+1}^{\infty} u_n = R_N \tag{23.9}$$

Let's use this result to assess the approximation of S_N to s for the series $s = \sum_{n=1}^{\infty} n^{-2}$. In this case, $f(x) = 1/x^2$, and so

$$R_N \leq \int_N^{\infty} \frac{dx}{x^2} = \frac{1}{N} \tag{23.10}$$

Table 23.4 compares S_N, R_N, and the integral bound given by Equation 23.10. Notice that the series converges very slowly. It would take 100 000 terms to achieve an accuracy of ± 1 in the fifth decimal place.

EXAMPLE 23–3
Calculate $S = \sum_{n=1}^{\infty} n^{-4}$ to four decimal places with an error of not more than one unit in the fourth place.

SOLUTION: Equation 23.9 reads in this case

$$R_N = \sum_{n=N+1}^{\infty} \frac{1}{n^4} \leq \int_N^{\infty} \frac{dx}{x^4} = \frac{1}{3N^3}$$

This quantity will be less than 0.00005 if $N = 19$. The partial sum using 19 terms is 1.08228, compared to the exact value of $\pi^4/90 = 1.08232$. Notice that this series converges much faster than the series $\sum_{n=1}^{\infty} n^{-2}$.

Table 23.4. A comparison of the partial sums (S_N) and the remainder after N terms (R_N) of the series $\sum_{n=1}^{\infty} n^{-2}$, and its integral bound given by Equation 23.10. The exact value of the series is $\pi^2/6 = 1.64493\ldots$.

N	S_N	R_N	integral bound
500	1.64294	0.001 9980	0.002 000
1000	1.64393	0.000 9995	0.001 000
2000	1.64443	0.000 4999	0.000 500
3000	1.64460	0.000 3328	0.000 333
4000	1.64468	0.000 2500	0.000 250
5000	1.64473	0.000 2000	0.000 200
10000	1.64483	0.000 1000	0.000 100

We can also apply this method to a power series. Let's evaluate e^3 using its power series, which in this case reads

$$e^3 = S_N + R_N = \sum_{n=0}^{N} \frac{3^n}{n!} + \sum_{n=N+1}^{\infty} \frac{3^n}{n!} \tag{23.11}$$

The bounding function $f(x) = 3^x/x!$, and Equation 23.9 becomes

$$R_N \leq \int_N^{\infty} dx \, \frac{3^x}{x!} \tag{23.12}$$

This integral must be evaluated numerically, but that's no problem. Table 23.5, which can be easily generated with any computer algebra system, compares R_N and its integral bound for several values of N. Table 23.5 shows, for example, that 13 terms will give four-place accuracy.

The method that we have described is particularly suited for series consisting of all positive terms. Let's consider an *alternating series*, one in which the signs of successive terms alternate, such as

$$s = \sum_{n=0}^{\infty} (-1)^n u_n$$

where the $u_n \geq 0$. In the case of an alternating series, the error incurred when the series is truncated after the $n = N$ term is less in magnitude than the first term dropped; in other words,

$$\left| s - \sum_{n=0}^{N} (-1)^n u_n \right| < |u_{N+1}| \tag{23.13}$$

Table 23.5. A comparison of R_N given by Equation 23.11 and its integral bound given by Equation 23.12. The accepted value of the series is 20.08554 . . . .

N	S_N	R_N	integral bound
10	20.0797	0.005 872	0.012 344
11	20.0841	0.001 434	0.003 168
12	20.0852	0.000 324	0.000 751
13	20.0855	0.000 068	0.000 165
14	20.0855	0.000 013	0.000 034

Let's apply Equation 23.13 to

$$e^{-x} = \sum_{n=0}^{\infty} \frac{(-x)^n}{n!} = \sum_{n=0}^{\infty} \frac{(-1)^n x^n}{n!}$$

for positive values of x. If we truncate this series after the $n = N$ term, then the magnitude of the error $|R_N|$ will satisfy

$$|R_N| \le \frac{x^{N+1}}{(N+1)!} \tag{23.14}$$

Table 23.6 assesses Inequality 23.14 for $x = 3$, for which $e^{-3} = 0.049\ 787 \ldots$.

Table 23.6. An assessment of Inequality 23.14 for $x = 3$.

N	S_N	$|R_N|$	$|x|^{N+1}/(N+1)!$
9	0.037 054	0.012 7335	0.016 272
11	0.048 888	0.000 8991	0.001 109
13	0.049 741	0.000 0456	0.000 055
15	0.049 786	0.000 0012	0.000 002
17	0.049 787	$< 10^{-7}$	$< 10^{-7}$

23.4 Systems of Linear Algebraic Equations

Although Cramer's rule (Section 17.3) provides a systematic, compact approach to solving simultaneous linear algebraic equations, it is not a convenient computational procedure because of the necessity of evaluating numerous determinants. In this section, we shall present an alternative method of solving simultaneous equations, called *Gaussian elimination*, that is more computationally convenient than Cramer's method. As with all the numerical techniques that we have presented in this chapter, Gaussian elimination is just the first of a hierarchy of increasingly sophisticated algorithms. Before we present this method, however, we shall discuss some general ideas about systems of linear algebraic equations.

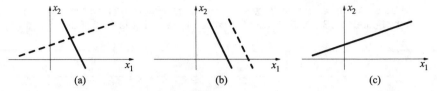

Figure 23.6. The three geometric possibilities of two linear equations in two unknowns, x_1 and x_2. (a) The solid line ($2x_1 + x_2 = 3$) and the dashed line ($x_1 - 3x_2 = -2$) have a unique point of intersection. (b) The solid line ($2x_1 + x_2 = 3$) and the dashed line ($2x_1 + x_2 = 5$) are parallel and have no point of intersection. (c) The two lines ($2x_1 - x_2 = 1$ and $4x_1 - 2x_2 = 2$) superimpose, and so there is an infinite number of solutions.

Let's start off with two equations in two unknowns.

$$a_{11}x_1 + a_{12}x_2 = h_1$$

$$a_{21}x_1 + a_{22}x_2 = h_2$$

Geometrically, we have three possibilities: (1) the graphs of the two straight lines intersect and we have a unique solution; (2) the lines are parallel and we have no solution; and (3) the lines coincide and we have an infinite number of solutions (Figure 23.6). An example of the first case is

$$2x_1 + x_2 = 3$$

$$x_1 - 3x_2 = -2$$

with $x_1 = 1$ and $x_2 = 1$ as its unique solution. An example of the second case is

$$2x_1 + x_2 = 3$$

$$2x_1 + x_2 = 5$$

These two lines are parallel and have no point in common. An example of the third case is

$$2x_1 + x_2 = 3$$

$$4x_1 + 2x_2 = 6$$

These two lines actually coincide and the solution can be written as $x_2 = 3 - 2x_1$, where x_1 can take on any value.

Let's now consider a general $n \times n$ system:

$$
\begin{aligned}
a_{11}x_1 + a_{12}x_2 + \cdots + a_{1n}x_n &= h_1 \\
a_{21}x_1 + a_{22}x_2 + \cdots + a_{2n}x_n &= h_2 \\
&\vdots \\
a_{n1}x_1 + a_{n2}x_2 + \cdots + a_{nn}x_n &= h_n
\end{aligned}
\tag{23.15}
$$

We can write Equations 23.15 as $AX = H$, where A is the *coefficient matrix*, X is the column vector of unknowns, and H is the constant vector of the system. If all the h_j in Equations 23.15 are equal to zero ($H = 0$), the system of equations is called *homogeneous*. If at least one $h_j \neq 0$, the system is called *nonhomogeneous*.

We can solve Equation 23.15 formally by multiplying from the left by A^{-1} (if it exists) to write

$$X = A^{-1}H \tag{23.16}$$

Recall that A^{-1} will exist if A is nonsingular (in other words, if $|A| \neq 0$). Careful analysis of Equation 23.16 says that the nonhomogeneous $n \times n$ system $AX = H$ has a unique solution if and only if A is nonsingular.

If $H = 0$, that is, if the system is homogeneous, then $x_1 = x_2 = \cdots = x_n = 0$ (the trivial solution) is always a solution. But we said that a solution is unique if and only if A is nonsingular, so if A is nonsingular, there is *only* a trivial solution to a homogeneous system. To have a nontrivial solution to an $n \times n$ set of homogeneous equations, the coefficient matrix must be singular. Recall that we deduced the same thing in Section 17.3.

We shall now spend the rest of this section actually finding solutions to systems of linear equations. Let's consider the equations

$$\begin{aligned} 2x_1 + x_2 + 3x_3 &= 4 \\ 2x_1 - 2x_2 - x_3 &= 1 \\ -2x_1 + 4x_2 + x_3 &= 1 \end{aligned} \tag{23.17}$$

The coefficient matrix and the constant vector are

$$A = \begin{pmatrix} 2 & 1 & 3 \\ 2 & -2 & -1 \\ -2 & 4 & 1 \end{pmatrix} \quad \text{and} \quad H = \begin{pmatrix} 4 \\ 1 \\ 1 \end{pmatrix}$$

We now form a new matrix, called the *augmented matrix*, by adjoining H to A so that it is the last column.

$$A|H = \begin{pmatrix} 2 & 1 & 3 & \bigm| & 4 \\ 2 & -2 & -1 & \bigm| & 1 \\ -2 & 4 & 1 & \bigm| & 1 \end{pmatrix} \tag{23.18}$$

Clearly, this matrix contains *all* the information in Equations 23.17 and is just a succinct expression of them. Just as we may multiply any of the equations in Equations 23.17 by a nonzero constant without jeopardizing the solutions, we may multiply any row of $A|H$ without altering its content. Similarly, we may interchange any two rows of either Equations 20.17 or $A|H$ and replace any row by the sum of that row and a constant times another row. These three operations are called *elementary (row) operations*:

1. We may multiply any row by a nonzero constant.

2. We may interchange any pair of rows.

3. We may replace any row by the sum of that row and a constant times another row.

The key point is that these elementary operations produce an *equivalent system*, that is, a system with the same solution as the original system. Matrices that differ by a set of elementary operations are said to be *equivalent*.

We are now going to manipulate A|H in Equation 23.18 by elementary operations so that there are zeros in the lower left positions of A|H. Add -1 times row 1 to row 2 and add row 1 to row 3 in Equation 23.18 to get

$$\begin{pmatrix} 2 & 1 & 3 & | & 4 \\ 0 & -3 & -4 & | & -3 \\ 0 & 5 & 4 & | & 5 \end{pmatrix}$$

Now add 5/3 times row 2 to row 3 to get

$$\begin{pmatrix} 2 & 1 & 3 & | & 4 \\ 0 & -3 & -4 & | & -3 \\ 0 & 0 & -8/3 & | & 0 \end{pmatrix}$$

Write out the corresponding system of equations:

$$2x_1 + x_2 + 3x_3 = 4$$

$$-3x_2 - 4x_3 = -3$$

$$-8/3\, x_3 = 0$$

and work your way from bottom to top to find $x_3 = 0$, $x_2 = 1$, and $x_1 = 3/2$.

The final form of A|H is said to be in *echelon form*. The following Examples provide two other applications of Gaussian elimination.

EXAMPLE 23–4
Solve the equations

$$x_1 + x_2 - x_3 = 2$$

$$2x_1 - x_2 + 3x_3 = 5$$

$$3x_1 + 2x_2 - 2x_3 = 5$$

SOLUTION: The augmented matrix is

$$\begin{pmatrix} 1 & 1 & -1 & | & 2 \\ 2 & -1 & 3 & | & 5 \\ 3 & 2 & -2 & | & 5 \end{pmatrix}$$

Add -2 times row 1 to row 2 and -3 times row 1 to row 3 to obtain

$$\begin{pmatrix} 1 & 1 & -1 & | & 2 \\ 0 & -3 & 5 & | & 1 \\ 0 & -1 & 1 & | & -1 \end{pmatrix}$$

To avoid introducing fractions, interchange rows 2 and 3 and then add -3 times the new row 2 to the new row 3 to get

$$\begin{pmatrix} 1 & 1 & -1 & | & 2 \\ 0 & -1 & 1 & | & -1 \\ 0 & 0 & 2 & | & 4 \end{pmatrix}$$

The corresponding set of equations is

$$x_1 + x_2 - x_3 = 2$$
$$-x_2 + x_3 = -1$$
$$2x_3 = 4$$

Solving these equations from bottom to top gives $x_3 = 2$, $x_2 = 3$, and $x_1 = 1$.

EXAMPLE 23–5

Solve the equations

$$x_1 + x_2 + x_3 = -2$$
$$x_1 - x_2 + x_3 = 2$$
$$-x_1 + x_2 - x_3 = -2$$

SOLUTION: The augmented matrix is

$$A|H = \begin{pmatrix} 1 & 1 & 1 & -2 \\ 1 & -1 & 1 & 2 \\ -1 & 1 & -1 & -2 \end{pmatrix}$$

Add -1 times row 1 to row 2 and add row 1 to row 3 to obtain

$$\begin{pmatrix} 1 & 1 & 1 & -2 \\ 0 & -2 & 0 & 4 \\ 0 & 2 & 0 & -4 \end{pmatrix}$$

Now add row 2 to row 3 to get

$$\begin{pmatrix} 1 & 1 & 1 & -2 \\ 0 & -2 & 0 & 4 \\ 0 & 0 & 0 & 0 \end{pmatrix}$$

In this case, the corresponding set of equations is

$$x_1 + x_2 + x_3 = -2$$
$$-2x_2 = 4$$
$$0x_3 = 0$$

The solutions are $x_3 = $ arbitrary, $x_2 = -2$, and $x_1 = -x_3$, so the solution is not unique. Note that $|A| = 0$ in this case, so we should not expect a unique solution.

EXAMPLE 23–6
Solve the equations

$$2x_1 - x_3 = -1$$
$$3x_1 + 2x_2 = 4$$
$$4x_2 + 3x_3 = 6$$

SOLUTION: The augmented matrix is

$$\mathsf{A} \,|\, \mathsf{H} = \begin{pmatrix} 2 & 0 & -1 & \bigm| & -1 \\ 3 & 2 & 0 & \bigm| & 4 \\ 0 & 4 & 3 & \bigm| & 6 \end{pmatrix}$$

Add $-3/2$ times row 1 to row 2 to obtain

$$\begin{pmatrix} 2 & 0 & -1 & \bigm| & -1 \\ 0 & 2 & 3/2 & \bigm| & 11/2 \\ 0 & 4 & 3 & \bigm| & 6 \end{pmatrix}$$

Now add -2 times row 2 to row 3 to get

$$\begin{pmatrix} 2 & 0 & -1 & \bigm| & -1 \\ 0 & 2 & 3/2 & \bigm| & 11/2 \\ 0 & 0 & 0 & \bigm| & -5 \end{pmatrix}$$

This last line says that $-5 = 0$, meaning that there is no solution to the above equations. They are inconsistent.

We have considered only $n \times n$ systems of equations, but Gaussian elimination is applicable to nonsquare systems of equations as well.

The numerical methods, such as Newton–Raphson, Simpson's approximation, and Gaussian elimination, that we have discussed in this chapter are sometimes called pedagogical methods, in the sense that they are useful to illustrate how you go about finding numerical solutions to certain problems, but are not often used in practice. As we said in the introduction, there is a huge literature on numerical methods, where you will find much more sophisticated methods than we have used here. These methods are much faster for a desired accuracy and can also handle situations in which the methods that we have discussed fail. The computer algebra systems, such as *MathCad*, *Mathematica*, and *Maple*, that we have encouraged you to use throughout the book use these more sophisticated methods to find roots of equations, to integrate numerically, to evaluate series expansions, and to solve systems of algebraic equations, and much, much more.

Problems

23–1.   Solve the equation $x^5 + 2x^4 + 4x = 5$ to four significant figures for the root that lies between 0 and 1.

23–2. Use the Newton–Raphson method to derive the iterative formula

$$x_{n+1} = \frac{1}{2}\left(x_n + \frac{A}{x_n}\right)$$

for the value of $\sqrt{A}$. This formula was discovered by a Babylonian mathematician more than 2000 years ago. Use this formula to evaluate $\sqrt{2}$ to five significant figures.

23–3. Use the Newton–Raphson method to solve the equation $e^{-x} + (x/5) = 1$ to four significant figures. This equation occurs in the theory of blackbody radiation.

23–4. Consider the chemical reaction described by the equation

$$CH_4(g) + H_2O(g) \rightleftharpoons CO(g) + 3\,H_2(g)$$

at 300 K. If 1.00 atm of $CH_4(g)$ and $H_2O(g)$ are introduced into a reaction vessel, the pressures at equilibrium obey the equation

$$\frac{P_{CO}\,P_{H_2}^3}{P_{CH_4}\,P_{H_2O}} = \frac{(x)(3x)^3}{(1-x)(1-x)} = 26$$

Solve this equation for x.

23–5. In a discussion of imperfect gases, the following cubic equation arises

$$64x^3 + 6x^2 + 12x - 1 = 0$$

Use the Newton–Raphson method to find the only real root of this equation to five significant figures.

23–6. In a discussion of imperfect gases, the following cubic equation arises

$$V^3 - 0.1231\,V^2 + 0.02056\,V - 0.001271 = 0$$

Use the Newton–Raphson method to find the root to this equation that is near $V = 0.120$.

23–7. In a discussion of imperfect gases, the following cubic equation arises

$$V^3 - 0.36636\,V^2 + 0.038020\,V - 0.0012102 = 0$$

Use the Newton–Raphson method to show that the three roots to this equation are 0.07073, 0.07897, and 0.2167.

23–8. The Newton–Raphson method is not limited to polynomial equations. For example, the equation

$$\varepsilon^{1/2} \tan \varepsilon^{1/2} = (12 - \varepsilon)^{1/2}$$

occurs in a quantum-mechanical study of a particle in a box with a finite depth. One way to solve this equation for ε is to plot $\varepsilon^{1/2} \tan \varepsilon^{1/2}$ and $(12 - \varepsilon)^{1/2}$ versus ε on the same graph and noting the intersections of the two curves. When you do this, you find that $\varepsilon = 1.47$ and 11.37. Solve the above equation using the Newton–Raphson method and obtain the same values of ε.

You should use a CAS such as MathCad, Maple or Mathematica for the next four problems.

23–9.   Use the trapezoidal approximation and Simpson's approximation to evaluate

$$I = \int_0^1 \frac{dx}{1 + x^2}$$

to an accuracy of ± 0.0001. What must n be to assure this accuracy in each case? This integral can be evaluated analytically; it is given by $\tan^{-1}(1)$, which is equal to $\pi/4$, so $I = 0.785\,39816$ to eight significant figures.

23–10. Use the trapezoidal approximation and Simpson's approximation to evaluate $\ln 2$ to five significant figures by evaluating

$$\ln 2 = \int_1^2 \frac{dx}{x}$$

What must n be to assure five-digit accuracy in each case?

23–11. Use Simpson's approximation to evaluate

$$I = \int_0^5 e^{-x^2}\, dx$$

to five-place accuracy. What must the value of n be to assure this accuracy?

23–12. Evaluate the integral

$$S = 4\pi^{1/2} \left(\frac{2\alpha}{\pi} \right)^{3/4} \int_0^\infty r^2 e^{-r} e^{-\alpha r^2}\, dr$$

for values of α between 0.200 and 0.300 and show that S has a maximum value at $\alpha = 0.271$. This problem occurs in the theory of chemical bonding.

23–13. How many terms of the series $\sum_{n=1}^\infty n^{-5/2}$ are needed to calculate the sum of the series to an accuracy of ± 0.001?

23–14. How many terms of the series $\sum_{n=1}^\infty n^{-3}$ are needed to calculate the sum of the series to an accuracy of ± 0.0001?

23–15. How many terms of the series $\sum_{n=1}^\infty n^{-3/2}$ are needed to calculate the sum of the series to an accuracy of ± 0.001?

23–16. How many terms are needed to calculate the value of e^4 to an accuracy of ± 0.0001 from its power series?

23–17. How many terms are needed to calculate the value of e^{-2} to an accuracy of ± 0.001 from its power series?

23–18. Solve the equations

$$x_1 + 2x_2 - 3x_3 = 4$$
$$2x_1 - x_2 + x_3 = 1$$
$$3x_1 + 2x_2 - x_3 = 5$$

23–19. Solve the equations

$$2x + 5y + z = 5$$
$$x + 4y + 2z = 1$$
$$4x + 10y - z = 1$$

23–20. Solve the equations

$$x + y \quad\;\; = 1$$
$$x \quad\;\; + z = 1$$
$$2x + y + z = 0$$

23–21. Solve the equations

$$2x_1 + x_2 - x_3 + x_4 = -2$$
$$x_1 - x_2 - x_3 + x_4 = 1$$
$$x_1 - 4x_2 - 2x_3 + 2x_4 = 6$$
$$4x_1 + x_2 - 3x_3 + 3x_4 = -1$$

23–22. Solve the equations

$$x + 2y - 6z = 2$$
$$x + 4y + 4z = 1$$
$$3x + 10y + 2z = -1$$

23–23. Solve the equations

$$x + 2y - z = 3$$
$$x + 3y + z = 5$$
$$3x + 8y + 4z = 17$$

23–24. Use any CAS to solve the equations in Problems 23–18 through 23–23.

REFERENCES

The following books are textbooks.

Courant, R., *Differential and Integral Calculus*, Wiley-Interscience: New York, 1992. This two-volume set was first published in 1934 and has never been equaled.

Ayres, F., Jr., and Mendleson, E., *Calculus*, 4th ed., Schaum's Outline Series, McGraw-Hill: New York, 1999. A good, inexpensive introduction to elementary calculus.

Edwards, C. H., and Penney, D. E., *Elementary Differential Equations*, 5th ed., Prentice Hall: Englewood Cliffs, NJ, 2003. A solid, readable book with many applications.

Murphy, G., *Ordinary Differential Equations and Their Solutions*, Van Nostrand: New York, 1960. This book is to ordinary differential equations as a table of integrals is to integrals. It is out of print, but your library may have a copy.

Spiegel, M., *Fourier Analysis*, Schaum's Outline Series, McGraw-Hill: New York, 1974. An inexpensive introduction to the gamma function, beta function, and orthogonal polynomials as well as Fourier series and Fourier integrals.

Tolstov, G., *Fourier Series*, Dover: New York, 1976. A very nice treatment of Fourier series and orthogonal functions.

Lipschutz, S. and Lipson, M., *Linear Algebra*, Schaum's Easy Outline Series, McGraw-Hill: New York, 2002. A fairly short, pleasant introduction to linear algebra.

Farlow, S., *Partial Differential Equations for Scientists and Engineers*, Dover: New York, 1993. An inexpensive, very readable book with many applications.

Spiegel, M., Schiller, J., and Srinivasan, R. A., *Probability and Statistics*, 2nd ed., Schaum's Outline Series, McGraw-Hill: New York, 2000. A good introduction to probability and statistics at the level of physical chemistry.

Taylor, J., *Introduction to Error Analysis*, 2nd ed., University Science Books: Sausalito, CA, 1997. An excellent discussion of the application of statistics to experimental data. (This book has the greatest cover of any textbook.)

Cheung, C-K., Keough, G. E., Landraitis, C., and Gross, R., *Getting Started with Math-ematica*, 2nd ed., Wiley: New York, 2005. A fairly short and practical introduction to *Mathematica*.

McQuarrie, D., and Simon, J., *Physical Chemistry: A Molecular Approach*, University Science Books: Sausalito, CA, 1997.

McQuarrie, D., *Quantum Chemistry*, 2nd ed., University Science Books: Sausalito, CA, 2007.

McQuarrie, D., *Mathematical Methods for Scientists and Engineers*, University Science Books: Sausalito, CA, 2003. What can I say about this one and the two above?

Abramowitz, M., and Stegun, I., *Handbook of Mathematical Functions with Formulas, Graphs, and Mathematical Tables*, Dover: New York, 1965. A Dover reprint of a standard mathematical handbook. This book is also available online (see the introduction to Chapter 4).

Pauling, L., and Wilson, E. B., Jr., *Introduction to Quantum Mechanics with Applications to Chemistry*, Dover: New York, 1985. A Dover reprint of a classic early text on quantum chemistry.

The following books are not textbooks, but are books for a general (albeit mathematically pretty literate) reader.

Maor, E., *e: The Story of a Number*, Princeton University Press: Princeton, NJ, 1994. A nicely crafted history of the development of calculus.

Dunham, W., *The Calculus Gallery*, Princeton University Press: Princeton, NJ, 2005. A slightly more demanding history of calculus than Maor's by one of the best general mathematics authors.

Dunham, W., *Journey through Genius*, Wiley: New York, 1990. Any book by Dunham is worth reading; this one focuses on the history of mathematics through some selected theorems.

Nahin, P., *An Imaginary Tale: The Story of $\sqrt{-1}$*, Princeton University Press: Princeton, NJ, 1998. An engaging accounting of the infiltration of $\sqrt{-1}$ into mathematics.

Nahin, P., *Dr. Euler's Fabulous Formula*, Princeton University Press: Princeton, NJ, 2006. Not as easy a read as some of the other books here, but an excellent general discussion of Fourier series and Fourier integrals.

Sawyer, W. W., *Mathematician's Delight*, Dover: New York, 2007. A Dover reprint of a delightful book originally published over 50 years ago and still a great read.

Sawyer, W. W., *Prelude to Mathematics*, Dover: New York, 1982. Like Sawyer's *Mathematician's Delight*, a Dover reprint of a book published over 50 years ago and still an engaging book.

Weaver, W., *Lady Luck*, Dover: New York, 1982. A Dover reprint of a classic popular tour of probability theory first published in 1963. A little dated, but still a fine read.

Salsburg, D., *The Lady Tasting Teas*, Owl Books: New York, 2002. An entertaining discussion of the influence of statistics on science.

Websites

There are many good websites out there, but these are probably still going to be around by the time you read this.

http://en.wikipedia.org/wiki/Computer_algebra_system

http://en.wikipedia.org/wiki/Trigonometric_identity

http://www.sosmath.com/trig/Trig5/trig5/trig5.html

http://en.wikipedia.org/wiki/Table_of_derivatives

http://en.wikipedia.org/wiki/Table_of_integrals

http://en.wikibooks.org/wiki/Calculus/Taylor_series

http://en.wikipedia.org/wiki/Fourier_series

http://en.wikipedia.org/wiki/Vector_space

http://www.sosmath.com/tables/tables.html

http://www.convertit.com/Go/Convertit/Reference/AMS55.ASP

http://www-history.mcs.st-and.ac.uk/

http:///www.wolfram.com

http://www.maplesoft.com

http://www.mathsoft.com

ANSWERS TO SELECTED PROBLEMS

Chapter 1

1–1.   (a) looks like a vee; (c) looks like a curved vee

1–2.   looks like a step

1–3.   looks like $y = 2x$ for $x < 0$ and it equals zero for $x > 0$

1–4.   (a) odd; (b) neither; (c) even; (d) neither

1–5.   It equals 1 for $0 < x < 1$ and zero for $x > 1$

1–6.   looks like a triangular wave

1–7.   (a) periodic, 2π; (b) periodic, $\pi/2$; (c) periodic, π; (d) not periodic

1–8.   (a) $1 < x < 2$; (b) $-3/5 < x < 1$; (c) all values of x

1–9.   Both functions approach $e^x/2$ for large x

1–10. yes; no

1–11. (a) 1; (b) 0

1–12. (a) $\alpha = \beta = -1/2$

1–13. (a) $(1 - 4x - 2x^2)e^{-x^2}$; (b) $(x \cos x - \sin x)/x^2$; (c) $2x \tan 2x + 2x^2 \sec^2 2x$; (d) $-e^{-\sin x} \cos x$

1–14. no

1–15. (a) $2 \cos x - x^2 \cos x - 4x \sin x$; (b) $-2e^{-x} \cos x$; (c) $3 + 2 \ln x$

1–16. (a) $y'(x) = -2xe^{-x^2}$, $y''(x) = (4x^2 - 2)e^{-x^2}$;
       (b) $y'(x) = -e^{-x} \cos e^{-x}$, $y''(x) = e^{-x} \cos e^{-x} - e^{-2x} \sin e^{-x}$;
       (c) $y'(x) = -e^{-\tan x} \sec^2 x$, $y''(x) = e^{-\tan x}(\sec^4 x - 2 \sec^2 x \tan x)$

1–17. $y'(x) = -3/2$, $y = -\frac{3}{2}x + \frac{5}{2}$

1–18. a maximum at $x = -1$, a minimum at $x = 1$, and an inflection point at $x = 0$

1–19. There is an inflection point at $x = 0$

1–20. a minimum at $x = 2$, a maximum at $x = -1$, and an inflection point at $x = 1/2$

1–21. a maximum at $x = 1$, a minima at $x = 2$ and $x = -2$, and inflection points at $x = (1 \pm \sqrt{13})/3$

1–22. $x^x(1 + \ln x)$

1–23. a maximum height of 5120 at $t = 2$

1–24. Minimize $A = ab$, with $2a + 2b = p$ and get $a = b = p/4$

1–25. Minimize $D = (x^2 + 1/x)^{1/2}$ to get $x = \pm(1/2)^{1/3}$ and $y = \pm 2^{1/6}$; $D = 1.3747$

1–26. When you set $d\rho_\lambda/d\lambda$ equal to zero, you get $xe^x = 5(e^x - 1)$ where $x = hc/\lambda_{max}k_B T$, or $x = 4.965$

1–30. The starting point is $\frac{1}{2} \sin x \cos x \le \frac{1}{2}x \le \frac{1}{2} \tan x$

Chapter 2

2–1. (a) $x^{n+1}/n + 1 + c$; (b) $\ln x + c$; (c) $-e^{-x} + c$; (d) $-\cos x + c$; (e) $\ln(1 + x) + c$

2–2. (a) $\ln T_2/T - 1$; (b) 2; (c) $(e - 1)/e$; (d) $1/T_1 - 1/T_2$

2–3. (a) $\frac{1}{2} \ln 2$; (b) $\frac{1}{2} \ln 2$; (c) $\frac{1}{2} \ln 5$

2–5. (a) $-(1 + x)e^{-x} + c$; (b) $\sin x - x \cos x + c$; (c) $x \ln x - x + c$; (d) $x^3(3 \ln x - 1)/9 + c$

2–6. (a) -2π; (b) $\ln 4 - 3/4$; (c) 2; (d) $\pi - 2$

2–7. (a) 0; (b) 0; (c) 0

2–8. (a) 0; (b) 1/2; (c) $\sin^2 \pi^2/8$; (d) 0

2–9. 4/3

2–10. $\pi a^2/2$

2–13. Let $\beta^{1/2}x = u$

2–14. Let $\alpha x = u$

2–16. $F(1) = 1$ and $F(2) = 3$

2–17. 2

2–18. $F(x) = \begin{cases} 0 & x < 0 \\ x^2/2 & 0 < x < 1 \\ x - \dfrac{1}{2} & 1 < x < 2 \\ 3x - \dfrac{x^2}{2} - \dfrac{5}{2} & 2 < x < 3 \\ 2 & x > 3 \end{cases}$

2–19. $\lim\limits_{x \to \infty} x(x^2 + 1)/(x^6 + 1)^{1/2} = K = 1$ with $p = 1$

2–20. $\lim\limits_{x \to \infty} x^{3/2}x^2/(x^4 + 1) = K = 0$ with $p = 3/2$

2–21. Let $x - a = u$ and investigate the behavior of $\displaystyle\int_0^{b-a} \frac{du}{u^p}$

2–22. The $-x^2$ dominates $6x$ as $x \to \infty$

Chapter 3

3–1. 3/2

3–2. 1/3

3–3. 27/99

3–4. 142/999

3–6. 1/24

3–7. (a) converges; (b) converges; (c) diverges; (d) converges

3–8.   (a) converges; (b) converges; (c) diverges; (d) converges

3–9.   (a) $|x| < 1/2$; (b) $|x - 1| < 1$, or $0 < x < 2$; (c) $|2x - 1| < 3$, or $-1 < x < 2$; (d) $x < 0$

3–10. no

3–13. (a) $|x| < 1$; (b) all values of x; (c) all values of x; (d) $|x| < 1$

3–14. Differentiate the geometric series two times

3–15. Subtract $\ln(1 - x) = -x - x^2/2 - x^3/3 + x^4/4 + \cdots$ from $\ln(1 + x) = x - x^2/2 + x^3/3 - x^4/4 + \cdots$

3–16. $(x + x^2)/(1 - x)^3$

3–17. Use the geometric series with $x = e^{-h\nu/k_B T}$

3–22. (a) 1; (b) 0; (c) $-1/2$

3–23. $-1/2$

3–25. $a^3/3 - a^4/4 + O(a^5)$

3–27. One series converges for $x > 1$ and the other for $x < 1$; their sum converges for no values of x

3–28. (a) converges; (b) converges; (c) converges; (d) diverges

3–30. Another way is to use the fact that $\sinh bx = (e^{bx} - e^{-bx})/2$

3–31. The integrand is well behaved at $x = 0$, and the integral converges. In fact, it is equal to 0.42872.

3–32. The integrand goes as $1/x$ as $x \to 0$, and so the integral diverges

3–33. You get $e^{-1/x^2} = 0 + 0 + 0 + \cdots$

Chapter 4

4–1. $3\sqrt{\pi}/4a^{5/2}$

4–2.   (a) 9/8; (b) 3/8

4–3. $\Gamma(1)/2a$

4–4. $(-1)^n \Gamma(n + 1)$

4–6.   (a) 3840; (b) 105

4–7. $\Gamma[(m + 1)/n]/n$

4–8.   Let $ax^2 = u$

4–10. Write $\ln ab = \displaystyle\int_1^{ab} \frac{du}{u} = \int_1^a \frac{du}{u} + \int_a^{ab} \frac{du}{u}$ and let $u = az$ in the second integral

4–14. $5\pi/8$

4–15. 4

4–19. Integrate by parts

4–20. Complete the square of $ax^2 + 2bx + c$ and write it as $a\,[\,(x + b/a)^2 + (ac - b^2)/a^2\,]$

4–21. Let $(t + x^2)^{1/2} = u$

4–23. $\displaystyle\int_{-\infty}^{\infty} x f(x)\delta(x)\, dx = 0$ and $\displaystyle\int_{-\infty}^{\infty} x f(x)\delta'(x)\, dx = \left[x f(x)\delta(x) \right]_{-\infty}^{\infty} - \int_{-\infty}^{\infty} \delta(x)[\,x f'(x) + f(x)\,]\, dx$

$\displaystyle = -\int_{-\infty}^{\infty} \delta(x) f(x)\, dx$ for a continuous function $f(x)$

4–24. Let $ax = u$

4–25. $I(\sigma) = e^{-\sigma/2} \sin x_0$

4–26. $I = \cos b$

4-30. $\text{erf}(x) = \dfrac{2}{\sqrt{\pi}} \sum_{n=0}^{\infty} \dfrac{(-1)^n x^{2n+1}}{n!(2n+1)} = \dfrac{2}{\sqrt{\pi}} \left(x - \dfrac{x^3}{3} + \dfrac{x^5}{10} + \cdots \right)$

Chapter 5

5-1. (a) $2, -11$; (b) 1; (c) $1/e^2$; (d) $2, -\sqrt{2}$

5-2. (a) x; (b) $x^2 - 4y^2$; (c) $4xy$; (d) $x^2 + 4y^2$; (e) 0

5-3. (a) $225° = 5\pi/4$; (b) $135° = 3\pi/4$; (c) $315° = 7\pi/4$; (d) $270° = 3\pi/2$

5-4. (a) $6e^{i\pi/2}$; (b) $(18)^{1/2}e^{-0.340\,i}$; (c) $(5)^{1/2}e^{(1.107+\pi)i}$, note quadrant; (d) $\sqrt{2}e^{i\pi/4}$

5-5. (a) $(1 - i)/\sqrt{2}$; (b) $-3 + i3\sqrt{3}$; (c) $\sqrt{2}(1 - i)$; (d) 2

5-6. Multiplying by i is equivalent to multiplying by $e^{i\pi/2}$

5-7. $e^{i\pi} = \cos\pi + i\sin\pi = -1$

5-9. The region between circles of radius 1 and 3 centered at the point $(0, -1)$

5-11. Equate the real and imaginary parts of each side of $e^{in\theta} = \cos n\theta + i\sin n\theta = (e^{i\theta})^n = (\cos\theta + i\sin\theta)^n$ for $n = 2, 3, \ldots$

5-12. (a) $1 + i = \sqrt{2}e^{i\pi/4}$, and so $(1 + i)^{10} = 2^5 e^{10i\pi/4} = 32\,i$; (b) $(1 - i)^{12} = 2^6 e^{-12i\pi/4} = 64(-1) = -64$

5-15. Use the fact that $e^{2i\theta} = (e^{i\theta})^2$ and that $\cos^2\theta + \sin^2\theta = 1$; then use $e^{4i\theta} = (e^{i\theta})^4$

5-18. Start with $\cos\alpha\cos\beta = (e^{i\alpha} + e^{-i\alpha})(e^{i\beta} - e^{i\beta})/4$

5-19. $i^i = (e^{i\pi/2})^i = e^{-\pi/2}$; better yet, $i^i = (e^{i\pi/2 + 2\pi ni})^i = e^{-\pi/2 - 2\pi n}$ for $n = 0, \pm1, \pm2, \ldots$

5-20. All the roots lie on a unit circle with one root at the point $(1, 0)$. The remaining roots are uniformly distributed over the unit circle.

5-21. $2, -1 \pm i\sqrt{3}$

Chapter 6

6-1. (a) $y(x) = 1/3 + ce^{-x^3}$; (b) $y(x) = x^3/5 + 2x/3 + c/x^2$; (c) $s(t) = t^3 e^{3t}/2 + cte^{3t}$; (d) $x(y) = ce^y - y^2 - 2y - 2$

6-2. $y(x) = x$

6-3. $m(t) = \dfrac{2}{5}(20 + t) + \dfrac{3(20)^5}{5(20 + t)^4}$

6-4. $A(t) = k_2(A_0 + B_0)/(k_1 + k_2) + (k_1 A_0 - k_2 B_0)e^{-(k_1+k_2)t}/(k_1 + k_2)$

6-5. $m(t) = 10(4000 + 40t + t^2)/(20 + t)$. This function has a minimum at $t = 40$ min.

6-6. 348 s

6-7. (a) $y(x) = c_1 e^{2x} + c_2 e^{-x}$; (b) $y(x) = (c_1 + c_2 x)e^{3x}$; (c) $y(x) = e^{-2x}(c_1 e^{\sqrt{3}x} + c_2 e^{-\sqrt{3}x})$

6-8. (a) $y(x) = 3e^{2x}/4 + e^{-2x}/4$; (b) $y(x) = e^{-x}(\cos\sqrt{3}x + 2\sin\sqrt{3}x/\sqrt{3})$; (c) $y(x) = \cos 3x + \sin 3x/3$

6-9. (a) $y(x) = (1 - e^{-6x})/6$; (b) $y(x) = (e^{3x} - e^x)/2$; (c) $y(x) = \sin 2x/2 = \cos x \sin x$

6-10. (a) $y(x) = 2e^{2x}$; (b) $y(x) = 2e^{3x} - 3e^{2x}$; (c) $y(x) = 2e^{2x}$

6-11. $y(x) = c/x^2$

6-12. $y(x) = x \ln x$

6-13. A frequency of $\omega/2\pi$ translates into a period of $2\pi/\omega$. If you replace t by $t + 2\pi/\omega$ in $\cos\omega t$, you get $\cos(\omega t + 2\pi) = \cos\omega t$. The same is true for $A\cos\omega t + B\sin\omega t$.

6-14. (a) $x(t) = v_0 \sin\omega t/\omega$; (b) $x(t) = x_0 \cos\omega t + v_0 \sin\omega t/\omega$

6-16. $\theta(0) = \theta_0$ gives $c_1 = \theta_0$ and $\theta'(0) = 0$ gives $c_2 = \gamma\theta_0/2\omega$

6-20. $v(t) = mg(1 - e^{-\gamma t/m})/\gamma$; $\lim_{t\to\infty} v(t) = mg/\gamma$

Chapter 7

7-1.   (a) $\displaystyle\sum_{n=0}^{\infty}(n+1)a_{n+1}x^n$; (b) $\displaystyle\sum_{n=0}^{\infty}(n+1)(n+2)a_{n+2}x^n$; (c) $\displaystyle x^2\sum_{n=0}^{\infty}nc_nx^n$

7-2.   (a) $a_n = (-1)^n 2^n a_0/n!$; (b) $a_n = (n+1)a_0/2^n$; (c) $a_n = (-1)^n/(n!)^2$

7-3. $a_{2n} = (-1)^n a_0/(2n)!$; $a_{2n+1} = (-1)^n a_1/(2n+1)!$

7-4. $\displaystyle y(x) = \sum_{n=0}^{\infty}\frac{(-1)^n x^n}{n!} = e^{-x}$

7-5. $\displaystyle y(x) = a_1x + a_0\left(1 - x\sum_{n=0}^{\infty}\frac{x^{2n+1}}{2n+1}\right) = a_1x + a_0\left(1 - \frac{x}{2}\ln\frac{1+x}{1-x}\right)$

7-6. $\displaystyle y(x) = a_0\sum_{n=0}^{\infty}(n+1)x^{2n} + a_1\sum_{n=0}^{\infty}\frac{2n+3}{3}x^{2n+1}$

7-7.   Use the result of Example 3–5

7-8.   Use the ratio test

7-9.   (a) $|x| < 1$; (b) all values of x; (c) $|x| < 2$

7-12. $f_4(x) = a_0(1 - 10x^2 + 70x^4/6)$ and $f_5(x) = a_1(x - 14x^3/3 + 21x^5/5)$

7-14. $P_4(1) = 1$ gives $a_0 = 3/8$ and $P_5(1) = 1$ gives $a_1 = 15/8$

7-19. Legendre polynomials are either even or odd functions of x

7-20. $\displaystyle \Theta''(\theta) + \frac{\cos\theta}{\sin\theta}\Theta'(\theta) + \alpha(\alpha+1)\Theta(\theta) = 0$. This result is often written as

$$\sin\theta\frac{d}{d\theta}\left(\sin\theta\frac{d\Theta}{d\theta}\right) + \alpha(\alpha+1)\sin^2\theta\,\Theta(\theta) = 0$$

7-23. $\displaystyle p(x) = 0$ and $\displaystyle q(x) = -8/(1+4x^2) = -8\sum_{n=0}^{\infty}(-1)^n(4x^2)^n$ and so we predict that the radius of convergence is at

least as large as $1/2$ $(4x^2 < 1)$

7-24. $\displaystyle u(x) = \int e^{3x^2/2}\,dx$

Chapter 8

8-1.   Use $-[(1-x^2)P_n'(x)]' = -(1-x^2)P_n''(x) + 2xP_n'(x)$

8-2. $\displaystyle \int_{-1}^{1}P_1(x)P_2(x)\,dx = \int_{-1}^{1}P_1(x)P_4(x)\,dx = 0$ because the integrands are odd functions of x

8-3. $4P_4(x) = 7xP_3(x) - 3P_2(x) = (35x^4 - 30x^2 + 3)/8$

8-5. $\displaystyle \int_{-1}^{1}P_n(x)P_m(x)\,dx = h_n\delta_{nm}$ means that only terms with $n = m$ in the double summation survive, or that

$G(x,t)G(x,u)$ is a function of tu only

8-7.   Let $x = 1$ in Equation 8.9 to get $\displaystyle G(1,t) = \frac{1}{1-t} = \sum_{n=0}^{\infty}t^n$; let $x = 1$ to get $\displaystyle G(-1,t) = \frac{1}{1+t} = \sum_{n=0}^{\infty}(-1)^n t^n$

8-9.   Use Equation 8.8 for $xP_n(x)$ and then use orthogonality of the $\{P_n(x)\}$

8-10. Use Equations 8.9 and 8.10

8-11. Differentiate D_N^2 with respect to α_j and set the result equal to zero

8-13. Use the fact that $D_N^2 \geq 0$

8-14. $\phi_0(x) = 1$, $\phi_1(x) = 2x$, $\phi_2(x) = 4x^2 - 2$; these are Hermite polynomials

8–15. $H_4(x) = 2x H_3(x) - 6H_2(x) = 16x^4 - 48x^2 + 12$

8–17. Use the recursion formula in Table 8.2 for $x H_m(x)$ and then use orthogonality of the $\{H_n(x)\}$

8–18. Use the technique of the preceding problem sequentially

8–19. $L_3(x) = -(x - 1 - 4)L_2(x) - 4L_1(x) = -x^3 + 9x^2 - 18x + 6$

8–21. $L_1^1(x) = -1;\ L_2^1(x) = 2x - 4;\ L_3^1(x) = -18 + 18x - 3x^2;\ L_2^2(x) = 2;\ L_3^2(x) = 18 - 6x;\ L_3^3(x) = -6$

8–24. $H_0(x) = 1;\ H_1(x) = 2x;\ H_2(x) = 4x^2 - 2;\ H_3(x) = 8x^3 - 12x$

8–25. $L_0(x) = 1;\ L_1(x) = 1 - x;\ L_2(x) = x^2 - 4x + 2;\ L_3(x) = 6 - 18x + 9x^2 - x^3$

8–26. The function $f(x) - a_0 = f(x) - 1/2$ is an odd function of x

Chapter 9

9–1. Use $\sin ax \sin bx = (e^{iax} - e^{-iax})(e^{ibx} - e^{-ibx})/(2i)^2$ and $\cos ax \cos bx = (e^{iax} + e^{-iax})(e^{ibx} + e^{-ibx})/4$

9–2. Use $\sin^2 x = (e^{ix} - e^{-ix})^2/(2i)^2$ and $\cos^2 x = (e^{ix} + e^{-ix})^2/4$

9–5. Use $\cos ax \sin bx = (e^{iax} + e^{-iax})(e^{ibx} - e^{-ibx})/(4i)$

9–7. Let $x = \pi z/l$

9–10. $a_0 = 1,\ a_n = 0\ (n \ge 1)$ and $b_n = 2/n\pi$ for n odd and 0 for n even

9–11. $a_0 = 2\pi^2/3,\ a_n = (-1)^n 4/n^2$, and $b_n = 0$

9–12. $a_0 = 1,\ a_n = -4/n^2\pi^2$ for n odd, 0 for n even, and $b_n = 0$

9–13. $a_0 = 1,\ a_1 = 2/\pi,\ a_2 = 0,\ a_3 = -2/3\pi,\ a_4 = 0,\ a_5 = 2/5\pi,\ \dots$ and $b_1 = 2/2\pi,\ b_2 = 2/\pi,\ b_3 = 2/3\pi,$
$b_4 = 0, \dots$

9–20. $f(0) = \dfrac{2l^2}{3} + \dfrac{4l^2}{\pi^2} \displaystyle\sum_{n=0}^{\infty} \dfrac{(-1)^{n+1}}{n^2} = \dfrac{2l^2}{3} + \dfrac{4l^2}{\pi^2}\dfrac{\pi^2}{12} = l^2\ f(l) = \dfrac{2l^2}{3} - \dfrac{4l^2}{\pi^2} \displaystyle\sum_{n=0}^{\infty} \dfrac{1}{n^2} = \dfrac{2l^2}{3} - \dfrac{4l^2}{\pi^2}\dfrac{\pi^2}{6} = 0$

9–21. $f(2\pi) = \dfrac{4\pi^2}{3} + 4 \displaystyle\sum_{n=1}^{\infty} \dfrac{1}{n^2} = 2\pi^2 = f(0)$. This result makes sense because the periodic function $f(x)$ has
discontinuities at $x = n\pi$ for $n = 0,\ \pm2,\ \pm4, \dots$ and $2\pi^2 = (2\pi)^2/2$ is the average value of $f(x)$ at these points.

9–23. (a) $1/n^2$ (actually $1/(n^2 - 1)$, which goes as $1/n^2$ as n gets large); (b) $1/n$; (c) $1/n^3$; (d) $1/n$ (the b_n go as $1/n$)

9–24. $f(x) = a(x^4 - 2x^3 + x^2 + \alpha)$

Chapter 10

10–1. $\hat{F}(\omega) = (\pi/2a^2)^{1/2} e^{-|a\omega|}$

10–2. $\hat{F}(\omega) = (2/\pi)^{1/2} \sin a\omega/\omega$

10–3. $\hat{F}(\omega)\dfrac{1}{(2\pi)^{1/2}} \left[\dfrac{\tau}{\tau^2 + (\omega + \omega_0)^2} + \dfrac{\tau}{\tau^2 + (\omega - \omega_0)^2} \right]$

10–8. Use the inverse of $\hat{F}_C(\omega)$ to do the second part of the problem

10–9. Differentiate the integral in Problem 10–8 with respect to a

10–11. Set $d\hat{F}(\omega)/d\omega$ equal to zero to get $\omega_{max} = \omega_0$. Now use $\hat{F}(\omega_0) = (2/\pi\alpha^2)^{1/2}$ and $\hat{F}(\omega_0 \pm \alpha) = \hat{F}(\omega_0)/2$. The
width at half maximum is $\omega_0 + \alpha - (\omega_0 - \alpha) = 2\alpha$.

10–12. Use the inverse of $\hat{F}(\omega)$ in Problem 10–2

10–13. Let $a = 1$ and $x = 0$ and use the fact that $\sin az/z$ is an odd function of z

10–15. The Cosine transform displays the frequencies well

10–18. Use Equation 10.20 and the fact that $\hat{F}(\omega) = (2/\pi)^{1/2}/(1 + \omega^2)$

10–19. Let $x = u/\alpha$ to get $\pi/4\alpha^3$

Chapter 11

11-1.   (a) $\pm x^2$; (b) $(x^3 - a^3)e^{-ax}$; (c) 9/4; (d) $6xy^2z^4 + 2x^3z^4 + 12x^3y^2z^2$

11-2.   (a) nonlinear; (b) nonlinear (unless the coefficients are real); (c) nonlinear; (d) nonlinear

11-3.   (a) $-\omega^2$; (b) $i\omega$; (c) $\alpha^2 + 2\alpha + 3$; (d) 6

11-4. $-(a^2 + b^2 + c^2)$

11-5.   (a) d^4/dx^4; (b) $d^2/dx^2 + 2x d/dx + (1 + x^2)$; (c) $d^4/dx^4 - 4x d^3/dx^3 + (4x^2 - 2)d^2/dx^2 + 1$

11-6.   (a) commute; (b) do not commute; (c) do not commute (because of the $\pm$ associated with SQRT); (d) commute

11-7.   Only if $\hat{P}$ and $\hat{Q}$ commute

11-8.   The integral consists of integrals over complete cycles of the cosine and sine

11-10. (a) $[\hat{A}, \hat{B}] = 2d/dx$; (b) $[\hat{A}, \hat{B}] = -2x^2$; (c) $-f(0)$; (d) $4x d/dx + 3$

11-11. id/dx, d^2/dx^2, and x are Hermitian

11-14. $\int \psi_m^* \hat{A}\hat{B}\psi_n \, dx = \int (\hat{A}^*\psi_m^*)\hat{B}\psi_n \, dx = \int (\hat{B}^*\hat{A}^*\psi_m^*)\psi_n \, dx = \int \psi_n \hat{B}^*\hat{A}^*\psi_m^* \, dx$

11-15. From the previous problem, $\int \psi_m^*(\hat{A}\hat{B})\psi_n \, dx = \int \psi_n(\hat{A}\hat{B})^*\psi_m^* \, dx$ only if $\hat{A}$ and $\hat{B}$ commute

11-16. Use the fact that $\hat{A}^n\psi = \beta^n\psi$

11-17. Use the fact that $\hat{A}$ commutes with itself

11-18. Only if $\hat{A}$ and $\hat{B}$ commute

11-19. $[\hat{A}, \hat{B}]\hat{C} + \hat{B}[\hat{A}, \hat{C}] = \hat{A}\hat{B}\hat{C} - \hat{B}\hat{A}\hat{C} + \hat{B}\hat{A}\hat{C} - \hat{B}\hat{C}\hat{A} = \hat{A}\hat{B}\hat{C} - \hat{B}\hat{C}\hat{A} = [\hat{A}, \hat{B}\hat{C}]$

Chapter 12

12-1.   (a) $\dfrac{\partial f}{\partial x} = e^y$, $\dfrac{\partial f}{\partial y} = xe^y + 1$, $\dfrac{\partial^2 f}{\partial x^2} = 0$, $\dfrac{\partial^2 f}{\partial y^2} = xe^y$, $\dfrac{\partial^2 f}{\partial x \partial y} = e^y$

   (b) $\dfrac{\partial f}{\partial x} = y\cos x + 2x$, $\dfrac{\partial f}{\partial y} = \sin x$, $\dfrac{\partial^2 f}{\partial x^2} = -y\sin x + 2$, $\dfrac{\partial^2 f}{\partial y^2} = 0$, $\dfrac{\partial^2 f}{\partial x \partial y} = \cos x$

   (c) $\dfrac{\partial f}{\partial x} = -2xe^{-(x^2+y^2)}$, $\dfrac{\partial f}{\partial y} = -2ye^{-(x^2+y^2)}$, $\dfrac{\partial^2 f}{\partial x^2} = (4x^2 - 2)e^{-(x^2+y^2)}$, $\dfrac{\partial^2 f}{\partial y^2} = (4y^2 - 2)e^{-(x^2+y^2)}$,

   $\dfrac{\partial^2 f}{\partial x \partial y} = 4xye^{-(x^2+y^2)}$

12-2.   (a) $f_{xy} = f_{yx} = -4xye^{-y^2}$; (b) $f_{xy} = f_{yx} = ye^{-y}\sin xy - e^{-y}\sin xy - xye^{-y}\cos xy$;
   (c) $f_{xy} = f_{yx} = \cos xy - xy\sin xy$

12-3. $\partial^2 P/\partial T \partial V = \partial^2 P/\partial V \partial T = -R/(V - b)^2$

12-6.   (a) $\left(\dfrac{\partial V}{\partial T}\right)_{n,P} = \dfrac{nR}{P}$, $\left(\dfrac{\partial T}{\partial V}\right)_{n,P} = \dfrac{P}{nR}$ (b) $\left(\dfrac{\partial V}{\partial T}\right)_{n,P} = \dfrac{nR}{P}$, $\left(\dfrac{\partial T}{\partial V}\right)_{n,P} = \dfrac{P}{nR}$

12-7.   0 and a/V^2

12-10. 0 in both cases

12-11. Yes. It is the total differential of $\pi r^2 h$.

12-12. No; dx/T is an exact differential. (It is dS for an ideal gas.)

12-15. $2t + t^2(2t^2 + 3)e^{t^2}$

12-16. $\partial u/\partial s = (te^s + \cos s)e^{ts^2 + \sin s}$ and $\partial u/\partial t = e^{s+te^s+\sin s}$

12-17. $Y = n_1\left(\dfrac{\partial Y}{\partial n_1}\right) + n_2\left(\dfrac{\partial Y}{\partial n_2}\right) + \cdots = n_1\overline{Y}_1 + n_2\overline{Y}_2 + \cdots$. The $\overline{Y}_j$ are called partial molar quantities.

12-18. $A = -PV + \mu n$, or $\mu n = G = A + PV$

12-19. (a) a minimum at each critical point; (b) a maximum at each critical point; (c) a maximum; (d) a maximum

12-20. (a) a minimum at $(-1, 2)$; (b) a saddle at $(-1, -2)$; (c) a minimum at $(1, 1)$

12-21. (a) a saddle at $(0, 0)$; (b) a minimum at $(3, -1)$; (c) a saddle at $(-2, 0)$ and a minimum at $(-2, 1)$

12-22. $4a^{1/2}(a - 1)/3$

12-23. $4/3$

12-24. (a) $1/2$; (b) 1

12-26. $(e - 2)/2$

12-27. $4[\sqrt{2} + \sinh^{-1}(1)]/3 = 3.0608\ldots$

12-31. $(\partial H/\partial P)_T = 0$ for an ideal gas.

12-33. $z(x, y) = x^3/3 - y^2 + x \sin y + \text{constant}$

12-34. $z(x, y) = x^2 \sin y + y^2 + e^x(1 + y) + \text{constant}$

Chapter 13

13-1. $(14)^{1/2}$ and $(x^2 + y^2 + z^2)^{1/2}$

13-2. $\mathbf{u} \cdot \mathbf{v} = 6 - 16 + 10 = 0$

13-3. $\mathbf{v} \cdot \mathbf{j} = 0$

13-4. $\cos \theta = -3/(6)^{1/2}(14)^{1/2} = -0.327$, or $\theta = 109° = 1.904$ rad

13-6. $\mathbf{u} \times \mathbf{v} = 5\mathbf{i} + 5\mathbf{j} - 5\mathbf{k}$; $\mathbf{v} \times \mathbf{u} = -5\mathbf{i} - 5\mathbf{j} + 5\mathbf{k}$

13-9. $\theta = 90°$ and $\sin \theta = 1$ for circular motion

13-11. Note that $\dot{\mathbf{u}} \times \dot{\mathbf{u}} = \mathbf{0}$

13-12. $\mathbf{r} \times \mathbf{F}$ is the torque

13-14. $3\mathbf{i} - \mathbf{j} + \mathbf{k}$

13-16. (a) $y^2 + 2xz - x^2$; (b) 3

13-21. $\mathbf{E} = 3\mu x \left(\dfrac{x\mathbf{i} + y\mathbf{j} + z\mathbf{k}}{r^5} \right) - \dfrac{\mu \mathbf{i}}{r^3}$

13-22. (b) $y^3 + x^2 y$

Chapter 14

14-1. (a) $120°$; (b) $225°$; (c) $45°$; (d) $330°$

14-2. It is a circle of radius $a/2$ centered at the point $r = a/2$ and $\theta = \pi/2$. Area $= \pi a^2/4$

14-3. The curve consists of lobes directed along the $\pm x$ axes. The area of each lobe is $\pi/2$. (The limits of the integration over the right lobe are $-\pi/4$ to $\pi/4$ and those for the left lobe are $3\pi/4$ to $5\pi/2$.)

14-4. $(1, \frac{\pi}{2}, 0)$; $(1, \frac{\pi}{2}, \frac{\pi}{2})$; $(1, 0, \phi)$; $(1, \pi, \phi)$

14-5. (a) a circle of radius 5 centered at the origin; (b) a cone of angle $\pi/4$ about the z axis with apex at the origin; (c) the plane containing the y and z axes

14-6. $2\pi a^3/3$

14-7. $2\pi a^2$

14-8. $4/15$

14-12. 0 and $4\pi/3$

14-13. $8\pi/3$

14-16. $\nabla^2 f = \dfrac{\partial^2 f}{\partial r^2} + \dfrac{1}{r} \dfrac{\partial f}{\partial r} + \dfrac{1}{r^2} \dfrac{\partial^2 f}{\partial \theta^2}$

14-20. $\hat{F}(k) = e^{-\alpha k^2}/(2\pi)^{3/2}$

Chapter 15

15-4. $u(x, t) = \cos \dfrac{3\pi vt}{l} \sin \dfrac{3\pi x}{l}$

15-5. Using the relation $\sin^3 \theta = \frac{3}{4} \sin \theta - \frac{1}{4} \sin 3\theta$, we predict that the $n = 1$ and $n = 3$ modes will be excited

15-7. $E_{n_x,n_y,n_z} = \dfrac{n_x^2 h^2}{8ma^2} + \dfrac{n_y^2 h^2}{8mb^2} + \dfrac{n_z^2 h^2}{8mc^2}$

15-8. If $t \to t + t_1$, then the shape of the wave will be unaltered if $x \to vt_1$. Furthermore, at any instant of time, if $x \to x + \lambda$, then the wave is unaltered. In addition, if $t \to t + \lambda/v$, then the wave is unaltered.

15-9. $\cos \omega_n t \sin \dfrac{n\pi x}{l} = \dfrac{1}{2} \sin \left(\dfrac{n\pi vt}{l} + \dfrac{n\pi x}{l} \right) - \dfrac{1}{2} \sin \left(\dfrac{n\pi vt}{l} - \dfrac{n\pi x}{l} \right)$

$= \dfrac{1}{2} \sin \left[\dfrac{n\pi}{l}(x + vt) \right] + \dfrac{1}{2} \sin \left[\dfrac{n\pi}{l}(x - vt) \right]$

15-10. Use the fact that $\lambda_n = 2l/n$, or that $n/l = 2/\lambda_n$

Chapter 16

16-4. The introduction of the reduced mass reduces a two-body problem to a one-body problem

16-6. $\mu = m_1 m_2/(m_1 + m_2) = m/2$ when $m_1 = m_2 = m$

16-7. Use $m_1 r_1 = m_2 r_2$ and $r = r_1 + r_2$ to write $r_1 = m_2 r/(m_1 + m_2)$ and $r_2 = m_1 r/(m_1 + m_2)$

16-11. $m = |m| = 2m$ when m is positive and so $i^{m+|m|} = i^{2m} = (-1)^m = -1$ if m is odd; $m + |m| = 0$ when m is negative, and so $i^{m+|m|} = i^0 = 1$

16-13. $2P_3^1(x) = -3P_1^1(x) + 5x P_2^1(x) = 3(5x^2 - 1)(1 - x^2)^{1/2}$

16-17. The constant $= 5/4\pi$ for $l = 2$

16-19. $L_1(x) = e^x \dfrac{d}{dx}(xe^{-x}) = 1 - x$; $\quad L_2(x) = e^x \dfrac{d^2}{dx^2}(x^2 e^{-x}) = x^2 - 4x + 2$;

$L_3(x) = e^x \dfrac{d^3}{dx^3}(x^3 e^{-x}) = -x^3 + 9x^2 - 18x + 6$;

$L_4(x) = x^4 - 16x^3 + 72x^2 - 96x + 24$;

$L_5(x) = -x^5 + 25x^4 - 200x^3 + 600x^2 - 600x + 120$

$L_1^1(x) = \dfrac{dL_3(x)}{dx} = -1$; $\quad L_2^1(x) = \dfrac{dL_2(x)}{dx} = 2x - 4$;

$L_3^1(x) = \dfrac{dL_3(x)}{dx} = -3x^2 + 18x - 18$; $\quad L_3^3(x) = \dfrac{d^3 L_3(x)}{dx^3} = -6$;

$L_4^3(x) = \dfrac{d^3 L_4(x)}{dx^3} = 24(x - 4)$; $\quad L_5^5(x) = \dfrac{d^5 L_5(x)}{dx^5} = -120$

16-20. $\psi_{100}(r, \theta, \phi) = e^{-r/a_0}/(\pi a_0^3)^{1/2}$

16-21. $\psi_{310}(r, \theta, \phi) = \dfrac{1}{81} \left(\dfrac{2}{\pi a_0^3} \right)^{1/2} \rho (6 - \rho)e^{-\rho/3} \cos \theta$, where $\rho = r/a_0$

Chapter 17

17-1. $D = 5$

17-2. -5

17-3. -2

17-4. 0 (the first and third columns are equal); 0 (the first column is twice the third column)

17-5. 5

17–6. -1

17–7. $\pm\sqrt{3}, 0, 0$

17–8. $\pm2, 0, 0$

17–9. $2, 1, 1, -1, -1, -2$

17–10. 1

17–11. 1 and $1 \pm \sqrt{2}$

17–12. $x = 9/5$ and $y = 1/5$

17–13. $x = 1, y = 3$, and $z = -4$

17–14. $2, 0, 0, -2$ (see Problem 17–8)

17–15. $D = 0$; the equations are inconsistent and have no solution

17–16. $1 \pm i$

Chapter 18

18–1. Show that $R(c_1\mathbf{r}_1 + c_2\mathbf{r}_2) = c_1R\,\mathbf{r}_1 + c_2R\,\mathbf{r}_2$

18–2. (a) $\mathbf{u} \to -\mathbf{u}$, inversion through the origin; (b) $u_y \to -u_y$, reflection through the x axis; (c) $u_x \to -u_x$, reflection through the y axis

18–3. $x_2^2 + y_2^2 = (x_1 \cos\theta - y_1 \sin\theta)^2 + (x_1 \sin\theta + y_1 \cos\theta)^2 = x_1^2 + y_1^2$

18–4. $C = \begin{pmatrix} 5 & -3 & -2 \\ -11 & 4 & -6 \\ -3 & -1 & -1 \end{pmatrix}; D = \begin{pmatrix} -7 & 6 & 1 \\ 19 & -2 & 12 \\ 6 & 5 & 5 \end{pmatrix}$

$|A| = 3, |B| = 1, |C| = -117$, and $|D| = 456$

18–8. Show that $R^\mathsf{T}R = RR^\mathsf{T} = I$

18–9. Show that $S^\mathsf{T}S = SS^\mathsf{T} = I; D = S^{-1}AS = S^\mathsf{T}AS = \begin{pmatrix} 2 & 0 & 0 \\ 0 & 1 & 0 \\ 0 & 0 & 0 \end{pmatrix}$

D is diagonal.

18–10. $(AB)_{ij} = \sum_k a_{ik}b_{kj}; (AB)_{ij}^\mathsf{T} = \sum_k a_{jk}b_{ki} = \sum_k b_{ik}^\mathsf{T}a_{kj}^\mathsf{T} = (B^\mathsf{T}A^\mathsf{T})_{ij}$

18–11. $(AB)^{-1}(AB) = I, (AB)^{-1}A = B^{-1}, (AB)^{-1} = B^{-1}A^{-1}$

18–12. Start with $\sum_k a_{ik}a_{kj}^\mathsf{T} = \delta_{ij} = \sum_k a_{ik}a_{jk}$. This is a summation over columns and so it means that the i and j rows are orthogonal.

18–13. Does $(AB)^{-1} = (AB)^\mathsf{T}$ if $A^{-1} = A^\mathsf{T}$ and $B^{-1} = B^\mathsf{T}$? But $(AB)^{-1} = B^{-1}A^{-1}$ and $(AB)^\mathsf{T} = B^\mathsf{T}A^\mathsf{T}$ (Problems 18–10 and 18–11), and so does $B^{-1}A^{-1} = B^\mathsf{T}A^\mathsf{T}$? The equality is valid because $A^{-1} = A^\mathsf{T}$ and $B^{-1} = B^\mathsf{T}$.

18–15. $|A| = -2, |B| = 18, |AB| = -36$, and $|BA| = -36$

18–16. Start with $A^\mathsf{T}A = I$. Then $|A^\mathsf{T}| \cdot |A| = 1$. Now use $|A^\mathsf{T}| = |A|$ to show that $|A| = \pm1$.

18–17. Start with $A^{-1}A = I$. Then $|A^{-1}| \cdot |A| = 1$, or $|A^{-1}| = 1/|A|$.

18–18. Show that $A^\mathsf{T}A = AA^\mathsf{T} = I$

18–19. $(14)^{1/2}$

18–20. $(i)(i) + (1)(2) + (i)(i) = 0$

18–21. Show that $A^T = A^*$

18–22. Start with $A^\dagger A = I$. Then $|A^\dagger||A| = 1$. Now $|A^\dagger| = |A^*| = |A|^*$, and so $|A|$ is of absolute value unity.

18–23. Show that $A^\dagger A = AA^\dagger = I$, or that $A^\dagger = A^{-1}$

18–24. rows: $(2 + 4i)(-4i) + (-4i)(-2 - 4i) = 0$; columns: $(2 + 4i)(4i) + (4i)(-2 - 4i) = 0$

18–25. Show that $A^\dagger A = AA^\dagger = I$, or that $A^\dagger = A^{-1}$

18–27. $x_1 = 24/13$, $x_2 = -19/13$, $x_3 = -8/13$

Chapter 19

19–1. $\lambda = 2, 0$; $\mathbf{v}_1 = (1, 1)$, $\mathbf{v}_2 = (1, -1)$

19–2. $\lambda = 3, -1$; $\mathbf{v}_1 = (1, -1)$, $\mathbf{v}_2 = (1, 1)$

19–3. $\lambda = (1 + \sqrt{5})/2, 1, (1 - \sqrt{5})/2$; $\mathbf{v}_1 = ((1 + \sqrt{5})/2, 0, 1)$, $\mathbf{v}_2 = (0, 1, 0)$, $\mathbf{v}_3 = ((1 - \sqrt{5})/2, 0, 1)$

19–4. $\lambda = 2, 1, 0$; $\mathbf{v}_1 = (1, 0, -1)$, $\mathbf{v}_2 = (0, 1, 0)$, $\mathbf{v}_3 = (1, 0, 1)$

19–11. $x_1(t) = c_1 e^{-t} + c_2 e^{-3t}$; $x_2(t) = -2c_2 e^{-3t}$

19–12. $x_1(t) = c_1 e^{-t} + c_2 e^{3t}$; $x_2(t) = -2c_1 e^{-t} + 2c_2 e^{3t}$

19–13. $x_1(t) = 25e^t/6 - 7e^{7t}/6$; $x_2(t) = -5e^t/6 - 7e^{7t}/6$

19–14. $x_1(t) = e^{-2t}/3 + 2e^t/3$; $x_2(t) = 2e^{-2t}/3 + e^t/3$; $x_3(t) = e^{-2t}/3 - e^t/3$

19–15. If $x_1(0) = x_2(0)$, then Equations 19.30 give $b_2 = 0$ and $b_1 = [x_1(0) + x_2(0)]/2$. If $x_1(0) = -x_2(0)$, then Equations 19.30 give $b_1 = 0$ and $b_2 = [x(0) - x_2(0)]/2$.

19–16. Substitute $x_1 = (y_1 + y_2)/\sqrt{2}$ and $x_2 = (y_1 - y_2)/\sqrt{2}$ into Equation 19.22 to obtain $V(y_1, y_2) = k(y_1^2 + 3y_2^2)/2$

19–18. $\omega^2 = 2, 1$, and $1/2$

19–20. $S = \dfrac{1}{\sqrt{2}} \begin{pmatrix} 1 & 0 & 1 \\ 0 & \sqrt{2} & 0 \\ -1 & 0 & 1 \end{pmatrix}$; $S^{-1} = S^T = \dfrac{1}{\sqrt{2}} \begin{pmatrix} 1 & 0 & -1 \\ 0 & \sqrt{2} & 0 \\ 1 & 0 & 1 \end{pmatrix}$

19–22. $D = \begin{pmatrix} 2 & 0 \\ 0 & 0 \end{pmatrix}$

19–23. $D = \begin{pmatrix} -1 & 0 \\ 0 & 3 \end{pmatrix}$

19–24. $D = \begin{pmatrix} 2 & 0 & 0 \\ 0 & 1 & 0 \\ 0 & 0 & 0 \end{pmatrix}$

19–25. $\text{Tr } AB = \sum_{i=1}^{n} \sum_{j=1}^{n} a_{ij} b_{ji} = \sum_{j=1}^{n} \sum_{i=1}^{n} b_{ji} a_{ij} = \sum_{j=1}^{n} \sum_{i=1}^{n} a_{ij} b_{ji}$

$\text{Tr } S^{-1}AS = \text{Tr } S S^{-1}A = \text{Tr } A$

19–26. $\lambda = a + 2, a + 1, a + 1, a - 1, a - 1, a - 2$

Chapter 20

20–8. $\begin{vmatrix} 0 & 1 & 0 & 0 \\ 1 & 1 & 0 & 0 \\ 0 & 1 & 1 & 0 \\ 0 & 0 & 0 & 1 \end{vmatrix} = -1 \neq 0$. The only solution to $c_1\mathbf{v}_1 + c_2\mathbf{v}_2 + c_3\mathbf{v}_3 + c_4\mathbf{v}_4 = \mathbf{0}$ is a trivial solution, so the vectors are linearly independent.

20–9. $\begin{vmatrix} 1 & 1 & 1 \\ 1 & -1 & 1 \\ -1 & 1 & -1 \end{vmatrix} = 0$ so there is a nontrivial solution to $c_1\mathbf{v}_1 + c_2\mathbf{v}_2 + c_3\mathbf{v}_3 = \mathbf{0}$. The vectors are not linearly independent. ($\mathbf{v}_2 = -\mathbf{v}_3$)

20–10. There is no solution to $\begin{pmatrix} 1 \\ 0 \\ 2 \end{pmatrix} = c_1 \begin{pmatrix} 1 \\ 1 \\ 1 \end{pmatrix} + c_2 \begin{pmatrix} 1 \\ -1 \\ -1 \end{pmatrix} + c_3 \begin{pmatrix} 3 \\ 1 \\ 1 \end{pmatrix}$. Therefore, $(1, 0, 2)$ is not in the set.

20–11. $\begin{vmatrix} 1 & 1 & 1 & 1 \\ 1 & -1 & 1 & -1 \\ 1 & 2 & 3 & 4 \\ 1 & 0 & 2 & 0 \end{vmatrix} = 4 \neq 0$ and so the vectors are linearly independent

20–12. $(-2, -1, 4)$

20–14. $|\mathbf{u} \cdot \mathbf{v}| = 1 < 6$

20–15. $\langle f_1, f_2 \rangle = 5/6 < 7^{1/3}/3$

20–18. $\mathbf{u}_1 = (1, -1, 0)/\sqrt{2}$, $\mathbf{u}_2 = (1, 1, 0)/\sqrt{2}$, $\mathbf{u}_3 = (0, 0, 1)$

20–20. $i\,\mathbf{v}_1 = \mathbf{v}_3$

20–21. $\begin{vmatrix} 1 & 0 & 0 \\ 1 & i & 1 \\ -i & i & -1 \end{vmatrix} = -2i \neq 0.$ The vectors are linearly independent.

20–22. (a) $(1 + i)(2 + i) = 1 + 3i$; (b) $2i(2 + i) = -2 + 4i$

20–23. $\begin{vmatrix} 1 & 0 & 0 & 1 \\ 0 & 1 & -i & 0 \\ 0 & 1 & i & 0 \\ i & 0 & 0 & -1 \end{vmatrix} = 2 - 2i \neq 0.$ They are linearly independent.

20–24. (a) $\langle \mathbf{u} \,|\, \mathbf{v} \rangle = -i - 2$, $\langle \mathbf{v} \,|\, \mathbf{u} \rangle = i - 2$; (b) $\langle \mathbf{u} \,|\, \mathbf{v} \rangle = 2i$, $\langle \mathbf{v} \,|\, \mathbf{u} \rangle = -2i$

20–25. $\langle \mathbf{u} \,|\, i\,\mathbf{v} \rangle = i - 2i = -i = i\langle \mathbf{u} \,|\, \mathbf{v} \rangle$; $\langle i\,\mathbf{u} \,|\, \mathbf{v} \rangle = -i + 2i = i = i^*\langle \mathbf{u} \,|\, \mathbf{v} \rangle$

Chapter 21

21–1. Each toss is an independent event. Or, the probability of 10 heads followed by a head is the very same as the probability of 10 heads followed by a tail.

21–2. The probability that no two people have the same birthday is given by $\text{Prob} = \dfrac{365}{365} \cdot \dfrac{364}{365} \cdots \dfrac{365 - n + 1}{365} = \dfrac{365!}{(365)^n(365 - n)!}$, and the probability that two or more people have the same birthday is $1 - \text{Prob}$. For $n = 50$, the probability is 0.9704. The smallest value of n for which the probability is greater than $1/2$ is $n = 23$.

21–5. $\langle m \rangle = np$ and $\sigma_m^2 = np(1 - p)$

21–6. Use the binomial distribution: $p_1 = 14!/13!1!\,(1/2)^{14} = 8.54 \times 10^{-4}$

21–7. The number on each line is the sum of the two numbers above it; $x^5 + 5x^4y + 10x^3y^2 + 10x^2y^3 + 5xy^4 + y^5$

21–8. $9!/3!6! = 84$

21–9. $\langle m \rangle = a$ and $\sigma_m^2 = a$

21–10. Use the Poisson distribution with $a = 4.3$: $p_0 = e^{-4.3} = 0.0136$;
Prob$(m > 5) = 1 - p_0 - p_1 - p_2 - p_3 - p_4 - p_5 = 0.263$

21–11. Use the Poisson distribution with $a = 6.7$: $p_6 + p_7 = 0.303$

21–12. 1.12×10^{-5}

21-14. $c = \lambda$, $\langle x \rangle = 1/\lambda$, $\sigma_x^2 = 1/\lambda^2$, $\text{Prob}\{X > a\} = e^{-\lambda a}$

21-15. $I_1(\alpha) = \pi^{1/2}/2\alpha^{3/2}$, $I_2(\alpha) = 3\pi^{1/2}/4\alpha^{5/2}$

21-16. $3\sigma^4$

21-18. $\sigma_x = a\left(\dfrac{1}{12} - \dfrac{1}{2n^2\pi^2}\right)^{1/2} < a$ for $n = 1,\ 2,\ \ldots$

21-19. $1/2$

21-20. $\langle v \rangle = (8k_B T/\pi m)^{1/2}$

21-21. $\left\langle \dfrac{1}{2}mv^2 \right\rangle = \dfrac{m}{2}\langle v^2 \rangle = \dfrac{3}{2}k_B T$

21-22. $\langle v_x \rangle = 0$ and $\langle v_x^2 \rangle = k_B T/m$

21-23. $\text{Prob}\{-v_{x0} \leq V_x \leq v_{0x}\} = \text{erf}(w_0)$
$\text{Prob}\{-(2k_B T/m)^{1/2} \leq V_x \leq (2k_B T/m)^{1/2}\} = \text{erf}(1) = 0.8427$

21-24. $\text{Prob}\{V_x \geq +(k_B T/m)^{1/2}\} = 1 - \text{erf}(1/\sqrt{2}) = 0.3173$ and $\text{Prob}\{V_x \geq +(2k_B T/m)^{1/2}\} = 1 - \text{erf}(1) = 0.1573$

21-26. $\langle \varepsilon \rangle = 3k_B T/2$

Chapter 22

22-3. See the first of Equations 22.3

22-4. $s_y^2 \geq s_{xy}^2/s_x^2$ because $S \geq 0$

22-5. $S = 0$ if the sample pairs lie on a straight line. Therefore, $s_{xy}^2 = s_x^2 s_y^2$ (see previous problem), or $s_{xy} = \pm s_x s_y$, or $r = \pm 1$ (see Equation 22.11).

22-6. If X and Y are independent, then $\sigma_{XY} = E[X - \langle x \rangle] E[Y - \langle y \rangle] = 0 \cdot 0$

22-7. $y = 0.09074 + 2.0288x$

22-8. $0.09074 - (1.960)(0.2279) < \alpha < 0.09074 + (1.960)(0.2279)$
$2.0288 - (1.960)(0.09719) < \beta < 2.0288 + (1.960)(0.09719)$

22-9. $r = 0.9832$

22-10. $\eta_{sp}/c = 8.885 + 14.964c$

22-11. $8.885 - (2.576)(0.1452) < \alpha < 8.885 + (2.576)(0.1452)$
$14.964 - (2.576)(0.2454) < \beta < 14.964 + (2.576)(0.2454)$

22-12. For the data in Problem 22-7, $0.9351 < \rho < 0.9957$. For the data in Problem 22-10, $0.9926 < \rho < 0.9998$.

22-13. $\phi_s = -1.3208 + 3.2513v$

22-14. $-1.3208 - (1.645)(0.8255) < \alpha < -1.3208 + (1.6450(0.8255)$
$3.2513 - (1.645)(1.5517) < \beta < 3.2513 + (1.645)(1.5517)$

22-15. $0.0396 < \rho < 0.9071$

22-16. $\ln P = 20.61 - \dfrac{5201.9}{T}$

22-17. $20.612 - (2.810)(0.04345) < \alpha < 20.612 + (2.810)(0.04345)$
$-5201.9 - (2.810)(13.848) < \beta < -5201.9 + (2.810)(13.848)$

22-19. $V = (709.9 \pm 6.8)\ \text{cm}^3$

22-20. $P = (0.0801 \pm 0.0051)\ \text{bar}$

Chapter 23

23-1. 0.8596

23-2. 1.4142

23–3.   4.965

23–4.   0.6148

23–5.   0.077796

23–6.   0.07498

23–9.   For the trapezoidal approximation, $n > 40$; for Simpson's approximation, $m \approx 6$

23–10. For the trapezoidal approximation, $n > 130$; for Simpson's approximation, $m > 10$

23–11. $n > 14$ and $I = 0.88623$

23–13. $> 76, 1.341$

23–14. $> 70, 1.2021$

23–15. $\approx 4\,000\,000, 2.611$

23–16. $> 16, 54.5981$

23–17. $> 9.\ 0.135$

23–18. $x_1 = 13/12, x_2 = 7/12, x_3 = -7/12$

23–19. $x = 11, y = -4, z = 3$

23–20. There is no solution. Note that the determinant of the coefficients is equal to zero.

23–21. There is no solution. Note that the determinant of the coefficients is equal to zero.

23–22. There is no solution. Note that the determinant of the coefficients is equal to zero.

23–23. $x = 17/13, y = -2/3, z = 4/3$

INDEX

A

absolute value of a complex number, 56
algebraic function, 2
alternating series, 312
amplitude, 72, 191
angular frequency, 73
angular momentum, 172
antiderivative, 17
argument of a complex number, 56
associated Laguerre polynomials,
   100, 215
   differential equation, 103
   generating function, 101
   integral condition, 102
   recursion formula, 101
associated Legendre functions, 210*ff*
   normalization constant, 211
augmented matrix, 315
auxiliary equation, 70

B

basis, 268
basis function, 247
basis set, 247, 268
Bessel's inequality, 105
beta function, 44*ff*
binomial coefficient, 281

binomial distribution, 282
binomial expansion, 34
binomial series, 34
birthday problem, 291
blackbody radiation, 13, 24, 26, 35*ff*
Bohr radius, 215
boundary condition, 74, 192, 206
boundary value problem, 74

C

cardioid, 181
CAS, 19, 20, 77
Cayley–Hamilton theorem, 262
center-of-mass coordinates, 217
central limit theorem, 290
chain rule, single variable, 8
chain rules for partial differentiation,
   150*ff*
characteristic equation, 248
characteristic frequencies, 257
Chebyshev polynomials, 100
   differential equation, 103
   generating function, 101
   integral condition, 102
   recursion formula, 101
circular function, 2
classical harmonic oscillator, 72*ff*

classical wave equation, 191*ff*
coefficient matrix, 314
cofactor, 221, 238
column matrix, 239
commutative operation, 168
commutator, 134*ff*
commuting matrices, 236
commuting operators, 135
compatible matrices, 237
complementary error function, 47
complex Fourier series, 113
complex inner product space, 273*ff*
complex number, 55*ff*
compound interest, 13
computer algebra systems (CAS), 19,
   20, 77
concave downward, 10
concave upward, 10
confidence intervals, 298
continuous function, 5*ff*
continuous probability distribution, 287*ff*
convergence in the mean, 103
convergence of Fourier series, 114*ff*
convergence of infinite series, 27*ff*
convergent integral, 22
correlation analysis, 299*ff*
correlation coefficient, 291, 299

$$= uv \cos\theta \qquad \mathbf{u} \times \mathbf{v} = uv\,\mathbf{c}\,\sin\theta$$

$$\mathbf{u} \times \mathbf{v} = \begin{vmatrix} \mathbf{i} & \mathbf{j} & \mathbf{k} \\ u_x & u_y & u_z \\ v_x & v_y & v_z \end{vmatrix}$$

$$= \mathbf{i}\,(u_y v_z - v_y u_z) + \mathbf{j}\,(u_z v_x - v_z u_x + \mathbf{k}\,(u_x v_y - v_x u_y)$$

$$\mathbf{u} \cdot (\mathbf{v} \times \mathbf{w}) = \begin{vmatrix} u_x & u_y & u_z \\ v_x & v_y & v_z \\ w_x & w_y & w_z \end{vmatrix}$$

$$\mathbf{u} \times (\mathbf{v} \times \mathbf{w}) = \mathbf{v}\,(\mathbf{u} \cdot \mathbf{w}) - \mathbf{w}\,(\mathbf{u} \cdot \mathbf{v})$$

$$\operatorname{grad} f = \nabla f = \frac{\partial f}{\partial x}\mathbf{i} + \frac{\partial f}{\partial y}\mathbf{j} + \frac{\partial f}{\partial z}\mathbf{k}$$

$$\operatorname{div} \mathbf{v} = \nabla \cdot \mathbf{v} = \frac{\partial v_x}{\partial x} + \frac{\partial v_y}{\partial y} + \frac{\partial v_z}{\partial z}$$

$$\operatorname{div} \operatorname{grad} f = \nabla^2 f$$

Fourier Series

$$f(x) = \frac{a_0}{2} + \sum_{n=1}^{\infty} a_n \cos\frac{n\pi x}{l} + \sum_{n=1}^{\infty} b_n \sin\frac{n\pi x}{l}$$

$$a_n = \frac{1}{2l}\int_{-l}^{l} f(x)\cos\frac{n\pi x}{l}dx \qquad \text{and} \qquad b_n = \frac{1}{2l}\int_{-l}^{l} f(x)\sin\frac{n\pi x}{l}dx$$

Fourier Transforms

$$f(t) = e^{-a|t|} \qquad \hat{F}(\omega) = \left(\frac{2}{\pi}\right)^{1/2}\frac{a}{\omega^2 + a^2}$$

$$f(t) = e^{-a^2 t^2} \qquad \hat{F}(\omega) = \frac{1}{(2a^2)^{1/2}}e^{-\omega^2/4a^2}$$

$$f(t) = e^{-at}\cos\omega_0 t \quad (t > 0) \qquad \hat{F}(\omega) = \left(\frac{2}{\pi}\right)^{1/2}\frac{a}{a^2 + (\omega - \omega_0)^2} + \left(\frac{2}{\pi}\right)^{1/2}\frac{a}{a^2 + (\omega + \omega_0)^2}$$

Linear First-Order Differential Equations

$$y'(x) + p(x)y(x) = q(x)$$

$$y(x) = e^{-\int p(x)\,dx}\left[\int q(x)e^{\int p(x)\,dx}dx + c\right]$$